Time Domain Electron Spin Resonance

Time Domain Electron Spin Resonance

Edited by

Larry Kevan
Wayne State University

Robert N. Schwartz
University of Illinois at Chicago Circle

A Wiley-Interscience Publication

JOHN WILEY & SONS
New York · Chichester · Brisbane · Toronto

Library of Congress Cataloging in Publication Data:

Time domain electron spin resonance.

"A Wiley-Interscience publication."
Includes index.
1. Electron paramagnetic resonance spectroscopy.
I. Kevan, Larry.   II. Schwartz, Robert Nelson,
1940-

QD96.E4T55   543′.08   78-31128
ISBN 0-471-03814-8

Printed in the United States of America

10 9 8 7 6 5 4 3 2 1

PREFACE

Time domain (pulsed) electron spin resonance (ESR) spectroscopy has recently become an important tool for studying electron spin relaxation processes in liquids and solids. Electron spin relaxation data are important because they can provide detailed experimental information about the molecular motion of the paramagnetic probe as well as indirect information about the structural and dynamical properties of the environment surrounding the probe. This type of information is also important to theoreticians interested in testing various models of structural organization and molecular motion in liquids and solids.

The purpose of this volume is to present a survey of time domain ESR techniques and their application to the study of the structural and dynamical properties of both paramagnetic centers and the phases that they probe in liquid, crystalline, glassy, and polymeric systems.

To date most time domain ESR experiments have been confined to the laboratories of the physicist. The systems most commonly studied have been crystalline transition metal complexes and the measurements have been made at low temperatures. In recent years important advances have been made in microwave technology and digital electronics that have made it possible to build rapid-response pulsed ESR spectrometers. This has opened a whole new area for researchers who have been using conventional continuous wave (CW) ESR techniques to investigate paramagnetic systems of chemical, biological, and physical interest.

In this book several chapters focus on saturation recovery and electron spin-echo methods for studying electron spin relaxation because these methods have been little exploited by chemists and biologists. These two techniques are both competitive and complementary and current opinion is divided as to the relative advantages of each. In addition to dynamical information from electron spin echo measurements one chapter deals with structural information obtainable when the spin echo is modulated by weak hyperfine interaction. The formation mechanisms and chemical dynamics of radical and triplet state species are also discussed in two other chapters in which the focus is on the use of time domain ESR to study transient chemical intermediates.

In addition to a comprehensive description of the various time domain ESR techniques, specific applications of these techniques to a variety of chemical, biological, and physical problems are discussed in most chapters. Thus we hope that this book will prove useful to chemists, biochemists, biophysicists, and chemical physicists with some knowledge of ESR and give them new insights into how time domain ESR may be utilized for solving some of their scientific problems. Researchers in the related fields of biology, medicine, and materials science are also finding increasing applications for ESR and should also find much of interest in the present volume.

We would like to express our sincere thanks to our contributors for their enthusiasm and timely cooperation. We are also grateful to Dorothy Giannola, Agnes Wirtz, and the secretarial staff at the University of Illinois at Chicago Circle for their help in maintaining some consistency of format in the written and illustrated material of the various chapters.

LARRY KEVAN
ROBERT N. SCHWARTZ

Detroit, Michigan
Chicago, Illinois
August 1979

CONTENTS

Time Domain Electron Spin Resonance

1 SATURATION RECOVERY METHODOLOGY

James S. Hyde

National Biomedical ESR Center
Department of Radiology
Medical College of Wisconsin
Milwaukee, Wisconsin

1 INTRODUCTION

The basic idea of saturation recovery was mentioned in a nuclear magnetic resonance (NMR) context as early as 1948 by Bloembergen, Purcell, and Pound.[1] One perturbs the steady-state population difference of a pair of levels by the application of a pumping radio-frequency (rf) field (i.e., alters the z-component of magnetization) and then observes the time-dependent return to equilibrium of the population difference. The observation is not, however, a direct detection of that population difference or z-component, but an indirect detection. A weak observing rf field is used

1

to probe the spin system. The quantity actually detected is the response along y of the spin system (i.e., m_y) to the observing rf field (assumed to be along x in the rotating frame) after that spin system has initially been prepared by application of the pumping field.

A proper experimental methodology for any technique involves two aspects: (1) the apparatus itself and (2) procedures for the use of the apparatus based on theoretical modeling of the experiment. The first electron paramagnetic resonance (EPR) papers[2-6] employing saturation recovery methods are not entirely satisfactory when judged against these two criteria. None of them was primarily instrument oriented, since the spectrometers were very briefly described. Although many of the key practical procedures for obtaining reliable measurements were employed, this usage was highly intuitive, with no mathematical analysis of the experiments. Failure to be sufficiently critical of the experimental methods could have led to errors in some of the results presented in these early papers.

The author has written six papers on saturation recovery methods.[7-12] The thrust of these papers is the application of the method to systems of chemical and biological interest, including free radicals in solutions, frozen solutions of free radicals, and nitroxide radical spin-labeled proteins. This is in contrast to the early work, which was concerned with relaxation processes of transition metal and lanthanide elements in single crystals at cryogenic temperatures. Two of the papers, references 8 and 9, appeared in the *Review of Scientific Instruments*. They give considerable instrumental detail and also include theoretical analysis of the experimental method. The content of this chapter is essentially derived from these six papers.

The technique of saturation recovery is on the verge of very rapid increase in utilization. This is primarily due to the recent availability of fast digital integrated circuits, which permit detection of faster transient signals with methods that approach theoretical efficiency, from a signal-to-noise point of view, in the collection of data. It is also due in part to the historical development of EPR spectroscopy as applied to systems of chemical and biological interest. The rapid advances in the last 20 years have led us, in the opinion of the author, to a point where many important questions need to be answered concerning the interactions of these spin systems with the environment or lattice.

2 METHODOLOGY DERIVED FROM THE EARLY PAPERS

Two basic spectrometer types are employed: the pumping and observing microwave frequencies (1) coherent and (2) incoherent. Coherent spectrometers derive the pumping and observing power from either the same microwave source using switched microwave diodes to control the microwave field intensity or a traveling wave tube (TWT) amplifier that produces the pumping field with the input to the TWT derived from the observing source. Incoherent spectrometers generally used a pulsed magnetron for the pumping source and a klystron for the observing source. This classification into coherent and incoherent designs was not recognized in the early papers, but rather in references 8 and 9, where it was realized that circumstances exist where these two classes of instruments can yield different results.

The early papers were properly concerned with what remain as two of the most difficult questions concerning the method: (1) The conceptual basis of the method implies a single homogeneously broadened line, whereas many EPR spin systems are inhomogeneously broadened. Rather than say much about the saturation recovery response of inhomogeneously broadened spin systems, considerable care was taken in the five early papers to assure that the systems under investigation were homogeneous. (2) Spectral diffusion and pseudosecular relaxation processes are a potentially serious complication.

The main procedures for obtaining reliable measurements defined in the early papers are as follows:

1. Recoveries should be simple single exponentials. If they are not, a complication is occurring.
2. Measurements of the relaxation times using data from the tails of the recovery traces are more reliable than measurements taken in the early stages of the recovery signals.
3. The measured times should be independent of observing power.
4. The measured times should be independent of pumping power *duration*.
5. The measured times should be independent of pumping power *level*.

Measured times are observed to be dependent on observing levels that are high enough to cause partial saturation of the spin systems. These times, however, approach a constant value asymptotically as the observing level is decreased.

Recovery signals that depend on pumping power duration suggest the presence of spectral diffusion. (As an aside, such measurements could in principle be used to study spectral diffusion.)

To this list of five can be added the following, which are the result of more recent studies. They are given here for the convenience of the reader and are discussed in considerable detail later in the chapter.

6. Measurements should be made at exact resonance of a homogeneous line, and not, for example, at the peak of the first derivative of an absorption line. Thus when using field modulation it is preferable that the dispersion be detected.

7. In coherent spectrometers, the recovery signal should be independent of the *phase* of the pumping field with respect to the observing field.

8. In incoherent spectrometers the frequencies of the observing and pumping fields should be fairly close, but the results should be independent of small variations in the difference of the two frequencies.

All experiments should be performed using these eight procedural rules.

3 FREE-INDUCTION DECAY AND SATURATION RECOVERY

A central problem in pulsed EPR experiments is that it is very difficult to deliver a long pulse of pumping microwave power that is sufficiently intense to result in complete saturation (i.e., in negligible magnetization along x, y, and z in the rotating frame). This is because the available rf field intensity B_2 is much less than the spectral widths and often less than the width of inhomogeneously broadened lines. If m_x and m_y are finite after a pumping microwave pulse, there will be a free-induction decay signal that may compete with the saturation recovery signal. This observation seems to have been missed by the early workers, who were using population difference models rather than Bloch equation models in their discussion. A concern about complications of free-induction decay is the principal new insight in references 8 and 9. Free-induction decay will occur in a time T_2, the transverse ralaxation time. If $T_2 \ll T_1$, as may well have been the case in the early work, then very fast initial T_2 decays would be of little consequence. However, in spin systems of chemical and biological interest, T_2 and T_1 are often more nearly comparable.

It is important, therefore, that the reader thoroughly grasp the distinction between free-induction decay and saturation recovery. Free-induction decay is the time evolution toward zero intensity of the observable components of magnetization, m_x and m_y, in the absence of any microwave power incident on the sample. It is a transverse relaxation process or a dephasing of the spins. Free induction is a decay or disappearance of a signal; saturation recovery is a growth or appearance of a signal. Saturation recovery is independent of the phase of the pumping microwave power.

Free-induction decay is the signal from the *relaxing* magnetization; the saturation recovery signal comes from the *relaxed* magnetization.

In the apparatus described in references 8 and 9, three independent methods for experimentally distinguishing between free-induction decay and saturation recovery were described. They are immediately plausible on physical grounds and follow directly from the theoretical analysis as described later in this chapter. These methods are based on the following observations.

1. The free-induction decay signal depends on the phase of the pumping microwave signal relative to a reference microwave level. Its disappearance arises from loss of phase coherence. The saturation recovery signal is independent of the phase of the pumping microwave signal.
2. The free-induction decay signal is independent of the presence of a weak observing field incident on the sample, whereas saturation recovery requires such a field.
3. Depending on which harmonic of the field modulation is observed, including the zeroth harmonic, either the dispersion or the absorption is a maximum at the line center while the other component (i.e., the absorption or the dispersion, respectively) of the rf magnetization is zero. This is true both for steady-state and transient solutions of the Bloch equations. It is possible to arrange microwave phases such that the free-induction decay signal corresponds to such a zero signal for all time. To observe a saturation recovery signal, the phase of the observing microwave field is then set such that a maximum signal is detected.

These three methods become clearer by consideration in the next section of the actual microwave bridge used to exploit them.

Free-induction decay is itself an interesting spectroscopic observable of the response of the spin system to a perturbing microwave pulse. It is completely straightforward to observe both the free-induction decay signal and the saturation recovery signal for each sample to be investigated. Indeed this is a useful control procedure for the experiment.

4 INSTRUMENTATION FOR SATURATION RECOVERY

In this section the principles of instrument design are emphasized. The spectroscopist must construct his own spectrometer, since at the time this chapter is being written no commercial EPR saturation recovery spectrometer systems are available.

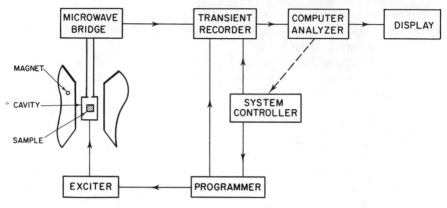

Figure 1 Time domain EPR spectrometer system.

The most general time domain EPR spectrometer system is illustrated in Fig. 1. The configuration applies equally well for saturation recovery, flash photolysis, spin echo, or any time-dependent perturbation of the spin system.

Referring to this figure, the exciter in saturation recovery is preferably a microwave source that is coherent with the source in the microwave bridge. For other time domain experiments it might be a laser, a heat source for temperature-jump experiments, or whatever.

Optimum efficiency in collection of data involves repeating the excitation cycle at a frequency of the order of $1/T_1$, where T_1 is the electron spin-lattice relaxation time. The system controller may provide the time base and control signals for both the transient recorder and the programmer. For saturation recovery, the programmer creates a pulse of selectable duration and a repetition rate corresponding to the overall repetition rate of the experiment. The transient recorder begins to collect data at the end of the excitation pulse.

There is another mode of operation in the block diagram of Fig. 1 called the handshake mode. The arrow from the programmer to the transient recorder we call a *handshake*. After the programmer has "told" the exciter what to do, the programmer shakes hands with the transient recorder, saying, "The exciter did it. Start collecting data." The programmer need not be on the same time base as the system controller in this mode. It is sufficient for the system controller to send only a start signal to the programmer.

The computer analyzer is, of course, not absolutely necessary, but it is the belief of the writer that no new instrument should be constructed at the present stage of EPR spectrometer development except with a design where a general purpose minicomputer is an integral part of the instrumental configuration. The computer analyzer in the diagram of Fig. 1 reads the memory of the transient recorder every second or so—that is, at a frequency very much slower than any transient processes in the spin system. The dotted arrow from the computer analyzer to the system controller is to indicate that the entire experiment can be controlled from the computer if desired.

This brings us to the transient recorder. It can be a single-channel boxcar integrator as in references 8 and 9. Ideally, however, it would be a very fast analog to digital (A–D) converter coupled to a very fast adder in order to signal-average, say, 500 channels at a rate of the order of 10^5 Hz for perhaps 10 s. The rate is governed by T_1 and the signal-to-noise ratio depends on the number of additions. The transient recorder is a data storage buffer between the microwave detector and the computer analyzer. At the time this chapter is being written it appears that no design meeting these specifications has been described in the literature, nor are commercial instruments available. The difficulty is not in fast A–D converters but in fast adders. Thus various compromises on either the overall repetition rate or the number of channels are necessary.

Specifications for the transient recorder are governed in part by the relaxation times that one expects to encounter. For several reasons a T_1 of 10^{-7} s is the shortest time that one can reasonably hope to measure with a device such as that illustrated in Fig. 1. These reasons are as follows: First, the ringing time of the cavity is $Q_0/2\pi\nu \simeq 10^{-7}$ s for a typical unloaded Q_0 of 6000 at X-band. Thus the microwave field cannot be cut off faster than this. One can, of course, degrade the cavity Q or use a slow-wave structure, but this results, in general, in degradation of the signal-to-noise ratio.

Second, there is an instrumental trade-off in present technology for the acquisition of short-duration repetitive signals between aperture width, number of channels, and repetition rate. One would like, as an ideal example, to have an aperture time $\leq \frac{1}{3}T_1$, at least 10 channels or "looks" for each transient signal, and an overall repetition rate of $(10T_1)^{-1}$. Decreasing the number of channels and/or decreasing the overall repetition rate lowers the signal-to-noise ratio. Single-channel boxcars with very narrow apertures are available commercially and permit in principle the measurements of relaxation times shorter than 10^{-7} s; however, the overall repetition rate of such devices is many orders of magnitude less than $(10T_1)^{-1} = 10^6$ Hz, information is collected in only one channel, and the experiment fails in many practical situations for lack of sufficient signal-to-noise ratio.

Third, if we assume that $T_1 = T_2 = 10^{-7}$ s, then the term $\gamma_e^2 B_2^2 T_1 T_2 = 1$

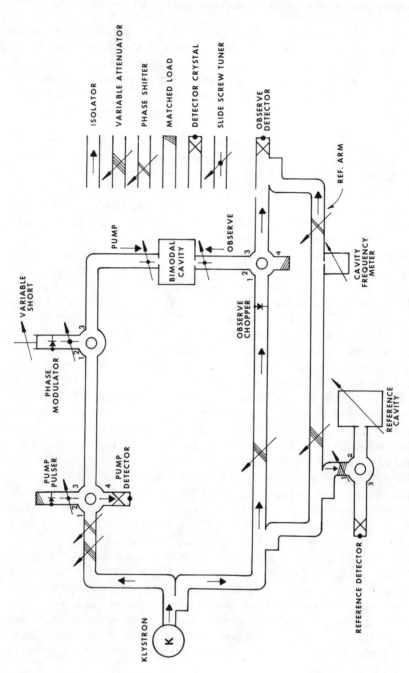

ISOLATOR

VARIABLE ATTENUATOR

PHASE SHIFTER

MATCHED LOAD

DETECTOR CRYSTAL

SLIDE SCREW TUNER

OBSERVE DETECTOR

REF. ARM

CAVITY FREQUENCY METER

VARIABLE SHORT

PHASE MODULATOR

PUMP

BIMODAL CAVITY

OBSERVE

OBSERVE CHOPPER

PUMP PULSER

PUMP DETECTOR

REFERENCE CAVITY

REFERENCE DETECTOR

KLYSTRON

K

Figure 2 Block diagram of the microwave bridge of the pulse spectrometer.[9]

(i.e., partial saturation) when the pumping microwave field $B_2 = 0.6$ G. This is about the value of the microwave field intensity available in the conventional EPR cavity and microwave bridge designs. To saturate systems with shorter relaxation times than 10^{-7} s requires microwave power in excess of 0.5 W.

Thus it is concluded that 10^{-7} s is a reasonable shortest time objective in specifying the transient recorder. Many instrumental problems arise if one strives for measurement of still shorter times.

The microwave bridge used in reference 9 is shown in Fig. 2. It is not necessarily suggested that this bridge be copied. Many variations are possible. A careful analysis of it does, however, bring out many of the principles of design.

Every saturation recovery bridge, including the early designs, is a *three-arm* type. One arm delivers pumping microwave power to the sample. The second arm is the observing arm through which resonance information passes. And the third is the reference arm. The reference arm provides a coherent microwave reference signal that serves to bias the detector into a favorable range for efficient detection of microwave power and also to provide a phase reference.

The bridge of Fig. 2 contains a bimodal cavity. Three-arm bridges can also be constructed using reflection and transmission cavities. Sometimes the arms are combined in different ways. But the functions provided by the three arms are always present: pumping, observing, and phase reference.

It is felt that the coherent pumping configuration has so many advantages over incoherent pumping that the latter will not be considered further. Since one has three arms with coherent microwave signals in each, adjustment of the bridge must always involve the setting of two phase shifters. One is present in the pumping arm and one in the reference arm of Fig. 2.

A primary problem in saturation recovery apparatus is in the ringing of the cavity. When the pumping power is turned off, the microwave energy of the cavity dies out with a time constant of the order of 10^{-7} s. For a matched cavity, half of the energy is lost in Joulean heating of the cavity walls and half is emitted from the cavity through the iris. It may find its way to the microwave detector resulting in damage to the detector crystal or saturation of an amplifier. Measurements of relaxation times cannot be made until this transient signal has died away. A central idea in using a bimodal cavity is that cavity ringing in one mode does not cause any ringing, assuming perfect isolation, in the orthogonal crossed mode. Thus detection of the saturation recovery signal can begin immediately after the end of the pulse and one is inherently able to measure shorter times than with, for example, a reflection cavity. When using transmission or reflection cavities, special "protect" diodes or other provisions must be introduced to avoid difficulties with the ringing pulses.

The bimodal cavity described in reference 8 was, like all bimodal cavities developed thus far, rather difficult to use. Balancing in order to achieve adequate isolation between modes was always possible but somewhat tedious.

If one is satisfied with the measurements of relaxation times of 10^{-6} s or longer, reflection or transmission cavities are probably easier to use. If one sets as a goal 10^{-7} s, then the bimodal geometry seems clearly preferable and one simply recognizes that a more difficult experiment is being performed that requires more care and more elaborate instrumentation.

Each of the three schemes for avoiding free-induction decay mentioned previously can be used in the spectrometer of Fig. 2.

1. The effect of the phase modulation in the pumping arm is to modulate the phase of the pumping microwaves with respect to the phase of the reference arm microwaves by 180° at a low audio frequency. Thus the free-induction decay signal is modulated at the audio rate and the saturation recovery signal is unmodulated. The receiver can be organized to detect either signal as shown in Fig. 3. Curves a and b show the combination of induction and recovery signals obtained for fixed pumping phases of 0° and of 180°. Pumping phase modulation and differential amplification of the signal obtained in the two modulation half-cycles gives the free-induction signal alone (c). For comparison, the free-induction signal was obtained in a different way in (d). Here the observing power was zero and square-wave field modulation on and off resonance replaced the phase modulation. Gains and pump power in all traces in Fig. 3 were the same. Finally, curve e shows the results of summing the signals detected with phase modulation: Only the recovery signal remains.

2. Since the saturation recovery signal amplitude depends on the observing level, modulation at the "observe chopper" results in transfer of modulation to the saturation recovery signal but not to the free-induction decay signal. This can be the basis for separation of these signals. However, diode chopping in the observing arm often resulted in spurious pulses reaching the detector crystals. This method has been abandoned in favor of phase modulation.

3. By setting the phase of the pumping arm with respect to the reference arm such that the free-induction decay *dispersion* signal is detected, setting the magnetic field at the line center where both the steady state

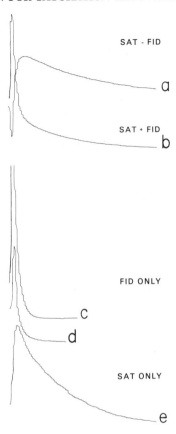

Figure 3 Separation of free-induction decay and saturation recovery signals using the pumping-phase modulation technique (see text).[9] The sample was TCNE radical anion at +31°C. $T_2 = 0.67 \ \mu$s and $T_1 = 7.8 \ \mu$s.

and transient dispersion are zero (no field modulation is employed) and setting the phase of the observing arm with respect to the reference arm such that the absorption mode is observed, free-induction and saturation recovery signals can be separated. This method is often used in conjunction with method 1 .

Provision to lock the klystron frequency to the resonant frequency of either the sample cavity or a reference cavity gives a certain element of flexibility in use of the equipment. It is extremely undesirable for any microwave pulses to reach the automatic frequency control (AFC) circuitry.

The many microwave isolators in Fig. 2 are to prevent pumping-arm pulses from reaching the detector crystal by any spurious path. Every such instrument employs many isolators.

5 THEORY

In this section various time-dependent solutions of the Bloch equations are given without detailed derivation. The mathematical methods for solution of the coupled differential equations are given in reference 9 and are based on the work of Atkins et al.[13]

At time $t = 0$, the magnetization has arbitrary magnitude and direction in the rotating frame:

$$\mathbf{m}(0) = \begin{pmatrix} x_0 \\ y_0 \\ z_0 \end{pmatrix} m_{z^\infty}. \tag{1}$$

Here m_{z^∞} is the equilibrium magnetization along the z-direction. The components $x_0 m_{z^\infty}$, $y_0 m_{z^\infty}$, and $z_0 m_{z^\infty}$ give the projections along x, y, and z in the rotating frame following an arbitrary preparation of the spin system with the pumping microwave field. The observing microwave field $\gamma_e B_1$ is along x, and the detected signal is along y. By changing the phase of the pumping microwave field $\gamma_e B_2$ with respect to the reference microwave field one can orient the pumping microwave field at any angle in the x-y plane.

CASE 1   Homogeneous line, exact resonance, $T_1 \gg T_2$.

$$\frac{m_y(t)}{m_{z\infty}} = y_0 \exp\left(-\frac{t}{T_2}\right) + \gamma_e B_1 T_2 \left\{ \left[\frac{(z_0 - 1)}{(1 + \gamma_e^2 B_1^2 T_1 T_2)} \right] \exp\left(-\frac{t}{T_1} - \gamma_e^2 B_1^2 T_2 t\right) \right.$$
$$\left. + \frac{1}{1 + \gamma_e^2 B_1^2 T_1 T_2} - z_0 \exp\left(-\frac{t}{T_2}\right) \right\}. \tag{2}$$

Although this is perhaps the simplest expression that can be derived, it obviously contains many complexities. Equations such as this are the basis for our criticism of the rather superficial theory given in the early papers.

The first term, $y_0 \exp(-t/T_2)$, is the free-induction decay. One can readily see how each of the three methods for suppression of this term outlined in the previous section apply. Pumping-phase inversion by 180° changes the sign of y_0. Addition of signals detected at 0 and 180° pumping phases eliminates this signal. Also, one can always adjust the pumping-arm phase relative to the reference-arm phase so that the projection of the magnetization on the x-y plane at the end of the pumping pulse is perpendicular to the direction of detection. This equation says that if y_0 is initially zero, free-induction decay terms at all later times are also zero. The saturation recovery terms are all proportional to $\gamma_e B_1$. Modulating $\gamma_e B_1$ and phase detecting at the modulation frequency suppresses the free-induction decay term.

Even having suppressed the first term, one still has difficulties. The last term $z_0 \exp(-t/T_2)$ can be suppressed if sufficient pumping power is available to make $z_0 = 0$. This term is the response of the residual magnetization along the z axis following the pumping pulse to the observing field $\gamma_e B_1$. One can separate out the last term by performing experiments at several pumping powers.

And finally the saturation recovery term $\exp[-t/(T_1^{-1} + \gamma_e^2 B_1^2 T_2)]$ only gives the correct T_1 recovery in the limit of zero observing microwave power. Thus $\gamma_e^2 B_1^2 T_2$ must be much less than T_1.

A complete solution for exact resonance with no assumption made concerning the relative magnitudes of T_1 and T_2 is given in reference 9. The methods for suppression of the free-induction decay signal still work. The term in z_0 still causes trouble but can be avoided by using sufficiently large pumping microwave fields.

The saturation recovery term always is a single exponential with a time constant T_1 if $\gamma_e B_1$ is sufficiently small.

Let us summarize then these important conclusions: If the line is homogeneous, if the observing power is small, if the pumping power is sufficiently large that $z_0 \simeq 0$ or analysis of results as a function of pumping power is performed, if one is at exact resonance, and if techniques are used to suppress free-induction decay, then in all Case 1 situations a true exponential recovery signal is observed with a time constant T_1.

CASE 2   Homogeneous line, off resonance, small observing field, $T_1 \gg T_2$.

$$\frac{m_y(t)}{m_{z^\infty}} = (y_0 \cos \delta\omega t - x_0 \sin \delta\omega t) \exp\left(-\frac{t}{T_2}\right)$$

$$- \gamma_e B_1 T_2 \left\{ \frac{(z_0 - 1)}{1 + \delta\omega^2 T_2^2} \exp\left(-\frac{t}{T_1}\right) \right.$$

$$+ \frac{1}{1 + \delta\omega^2 T_2^2}$$

$$\left. - \frac{z_0}{1 + \delta\omega^2 T_2^2} (\cos \delta\omega t - \delta\omega T_2 \sin \delta\omega t) \exp\left(-\frac{t}{T_2}\right) \right\}. \quad (3)$$

Here $\delta\omega$ is the difference between the resonance condition for exact resonance and the actual resonance condition. This expression can be compared term by term with Eq. (2).

Equation (3) shows that the pumping-phase modulation method will suppress the free-induction decay term (the first term) since change of phase of the pumping microwave field by 180° changes the sign of both x_0, y_0. The suppression technique using modulation of the observing power also works as before. However, the third suppression technique (adjustment of the pumping phase such that the resulting magnetization after the pumping pulse is perpendicular to the direction of detection) does not work. Off resonance, terms in both y_0 and x_0 are found.

Even though no term in $\gamma_e B_1$ appears in the saturation recovery exponentials in Eq. (3), extending the calculation to second order shows that, just as for exact resonance, reliable exponential recoveries can only be obtained in the limit of small observing power.

Thus using the same arguments as for Case 1, one can work one's way around the free-induction decay term and the term in z_0 and end up, even when off resonance, with a simple exponential recovery in T_1.

The oscillating terms in Eq. (3) are interesting since they are damped with a time constant T_2. This offers the potential of extracting both T_1 and T_2 from the experiment. This of course is a little forced, since a

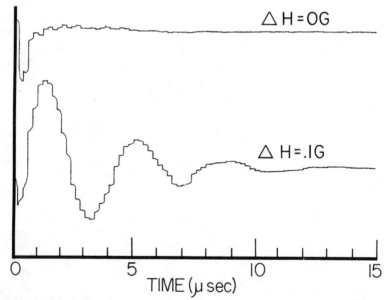

Figure 4   Free-induction decay off-resonance oscillations in time.[9] The sample was $10^{-4}M$ TCNE radical anion.

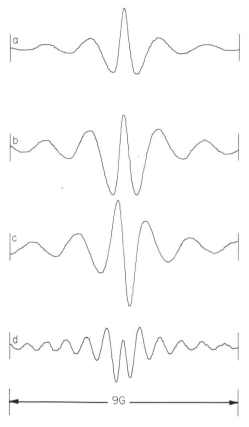

Figure 5 Pulsed induced EPR displays of free-induction decay oscillations in $\delta\omega$ observed at a fixed delay following a long (30 μs) pumping pulse.[9] The sample was $10^{-4}M$ TCNE radical anion. The aperture was 200 ns. (a) Absorption, 20 μs delay, 25 dB pumping power; (b) absorption, 20 μs delay, 19 dB pumping power; (c) dispersion, 20 μs delay, 19 dB pumping power; (d) absorption, 40 μs delay, 19 dB pumping power, gain two times higher than (a), (b), and (c).

homogeneous line was assumed and T_2 can therefore immediately be obtained from the ordinary EPR spectrum.

Note that the oscillations have the product $\delta\omega t$ as an argument. At fixed $\delta\omega$ there are oscillations in time. An example is shown in Fig. 4, which is the free-induction decay. Here $\gamma_e B_1 = 0$, so that the saturation recovery terms can be dropped. It is also possible to display the oscillations using the so-called pulse-induced EPR method. One displays the signal at a fixed delay after the pumping pulse as a function of the resonant condition $\delta\omega$. An example is shown in Fig. 5. Note that changing

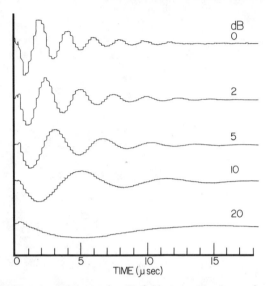

Figure 6 Oscillations about the pumping B_2 field at exact resonance for various attenuations of the pumping power.[9] The sample was $10^{-4}M$ TCNE radical anion. $B_1 = 0.185$ G at 0 dB.

the pumping power changes the relative magnitudes of x_0 and y_0, thus altering the *shape* of the display. Doubling of the delay doubles the oscillation frequency, traces c, d. These oscillations disappear at exact resonance, which can be used for precise determination of the resonance condition.

Oscillations are also found at exact resonance as can be seen from the complete solution to the Bloch equations given in reference 9. The condition is that $[(T_1^{-1} - T_2^{-1})^2 - 4\gamma_e^2 B_2^2]^{1/2}$ be imaginary. Oscillations occur about B_2 with a frequency $\gamma_e B_2$. An example is shown in Fig. 6. The frequency of oscillations permits the measurement of the absolute magnitude of the pumping microwave field, which can be a useful technique. One can also use these exact resonance oscillations to determine the combination of $\gamma_e B_2$ and pumping-pulse duration required to deliver a 90° or 180° pulse.

CASE 3   Inhomogeneous line, $T_1 \gg T_2$, small observing field, pumping field >inhomogeneity.

A distribution of resonance conditions given by $h(\omega - \omega_0)$ is assumed and Eq. (3) is integrated over this shape function. As before, we consider the free-induction decay term, the saturation recovery term, and the troublesome term in z_0.

All *three* techniques for suppression of the free-induction decay signal work. In particular, the terms involving $\sin \delta\omega t$ in Eq. (3) are odd functions that drop out when observing the center of a symmetrical inhomogeneously broadened line.

For an inhomogeneous Gaussian line shape $[T_2^*/(2\pi)^{1/2}]\exp[-\frac{1}{2}(\delta\omega T_2^*)]$, the free-induction decay is given by

$$y_0 \exp\left\{-\left[\frac{t}{T_2}+\frac{1}{2}\left(\frac{t}{T_2^*}\right)^2\right]\right\}.$$

In general, inhomogeneities shorten the free-induction decay and may cause departure from pure exponential decay.

As before, the z_0 term may significantly distort the true saturation recovery term at short times after the pumping pulse.

The time constant of the true recovery part of the signal remains unaffected by the inhomogeneity.

CASE 4 Inhomogeneous line, $T_1 \gg T_2$, small observing field, pumping field $\ll$ inhomogeneity.

The free-induction decay term has a time constant under reasonable approximations of $\gamma B_2 (T_1/T_2)^{1/2}$. By making B_2 sufficiently large, this decay can be made to occur in arbitrarily short times.

The term in z_0 can never be driven to zero, since we have assumed an inhomogeneous line of width much greater than the amplitude of the pumping microwave field. However, the time constant for this term is $\gamma_e B_2 (T_1/T_2)^{1/2}$, just as for the free-induction decay term and can be manipulated in the same way (i.e., by varying B_2).

Although the true recovery signal continues to have a T_1 time constant, it should clearly be recognized that the z_0 term is a potentially serious complication when working with inhomogeneously broadened lines. *It will always give rise to a fast initial transient signal.*

It is suggested that the reader refer back to the list of experimental controls given in Section 2. The theoretical basis for these rules can now be appreciated.

6 APPLICATIONS TO FREE RADICALS

In this section selected applications of the saturation recovery technique, again leaning heavily on our own publications, are reviewed. Our point of view is to use these applications to illustrate aspects of the methodology.

6.1   Free Radicals in Liquids

The experimental literature consists of references 7–11 by the author and his colleagues and references 14–17 by the group of Venkataraman.

Heisenberg exchange presents a potential complication in saturation recovery experiments on free radicals in liquids. If concentrations are sufficiently high that all hyperfine lines are effectively coupled, then recovery proceeds with a true T_1. However, it seems likely that at these concentrations *inter*molecular dipole-dipole contributions to T_1 will be dominant and reliable information on *intra*molecular T_1's cannot be obtained.

At sufficiently low concentrations, both dipole-dipole and Heisenberg exchange interactions are negligible, but the concentrations may be so low that there is a serious problem in the signal-to-noise ratio.

Huisjen and Hyde[7] give data on T_1 of the free radical galvinoxyl in *sec*-butyl benzene and toluene. At the lowest concentrations used in that work $(5 \times 10^{-5}M)$ there was still evidence of a concentration dependence of the T_1's, at least at the lowest viscosities. The instrument used in this work was the most sensitive one that has been described in the literature. It still deviates from theoretical sensitivity in that a single-channel rather than a multichannel boxcar was employed.[8] A conclusion then is that the sensitivity in the measurement of spin-lattice relaxation times of free radicals in solution is a serious but not insurmountable experimental problem.

The shortest spin-lattice relaxation time measured by the saturation recovery technique is 1.7×10^{-7} s. This was on the alcohol nitroxide tanol at $+31°C$ at rather high concentration $(7.8 \times 10^{-4}M)$. An actual recovery signal measured on this nitroxide at a concentration of $3 \times 10^{-3}M$ is shown in Fig. 7. It is clear that the time response of the equipment described in references 8 and 9 is adequate for the investigation of relaxation times of free radicals in solutions.

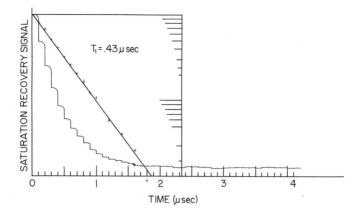

Figure 7 Typical saturation recovery signal. Experimental conditions: $3 \times 10^{-3}M$ tanol in *sec*-butyl benzene; room temperature; pumping pulse 300 mW and 1 μs duration; observing power 4 mW; 100 ns aperture.[8]

Whether or not intermolecular interactions need to be of concern depends on the magnitude of the interactions in the appropriate units of radians per second compared with T_1^{-1}. The measured value of T_1 for the nitroxide tanol was about 10 times shorter than that for the free radical galvinoxyl at the same concentrations, the same temperature, and in the same solvent.[7,10]

Thus *inter*molecular interactions for the spin label are in general an order of magnitude less serious. This interplay between concentrations, signal-to-noise ratio, and observed T_1's must be understood by the experimentalist.

It is reasonable to inquire about how continuous wave (cw) or progressive saturation techniques compare with saturation recovery techniques in the measurement of T_1 of free radicals in solution. Figure 8 shows an example of such a comparison using the nitroxide tanone in N-N dimethyl formamide (DMF). The solution contained $0.3M$ La^{3+} for other purposes and is not of concern here. The data are plotted as a function of rotational correlation time, which is in turn proportional to viscosity. These curves show completely different dependencies on viscosity. In discussing this apparent inconsistency, one must make a distinction between the *effective* T_1 and the *true* spin-lattice relaxation time. The effective T_1 arises from the sum of all relaxation paths including nuclear and cross relaxation, whereas the true relaxation time refers to direct relaxation between observed levels. The cw saturation method measures the effective T_1. Now it is not completely obvious that the saturation

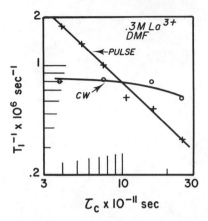

Figure 8 Saturation recovery and cw measurements of $T_{1_c}^{-1}$ for tanone in DMF with $0.3M$ La^{3+} added as a function of rotational correlation time.[11]

recovery method measures the true T_1 unaffected by these other relaxation paths. But in a wide range of circumstances it does. We return to this point in the next section.

With this background the rationalization for the difference of the results using the two methods at high viscosities is that nitrogen nuclear relaxation is expected to make T_1^{-1}(effective) greater than T_1^{-1} (true). This mechanism can give rise to a factor of 3 difference for nitroxides, since rapid nuclear relaxation couples the three lines and results in three parallel relaxation paths. One might attempt to introduce a correction by theoretically estimating the nuclear relaxation time as was done by Hwang et al.[18] Or one might measure the ratio of nuclear to electron spin-lattice relaxation times independently using electron-electron double resonance (ELDOR) and make a correction. There is no doubt but that the two techniques must, when the data are properly analyzed, yield the same results. But the pseudosecular relaxation processes are a serious complication for cw saturation.

At rotational correlation times of the order of 10^{-9} to 10^{-10} s, the nitroxide line is homogeneously broadened, it has a Lorentzian line shape, and T_2 can be immediately extracted from the EPR spectrum. At shorter correlation times, the nitroxide line becomes Gaussian in shape and inhomogeneously broadened as T_2^{-1} becomes less than the inhomogeneous $(T_2^*)^{-1}$ arising from the 12 methyl protons and the ring protons. One can attempt to analyze the resulting progressive saturation curves using the method of Castner, which is appropriate for a line shape that is intermediate between Lorentzian and Gaussian.[19] This was not done in Fig. 8 and is presumably the reason for the discrepancy.

We remark that for pure inhomogeneous lines where $(T_2^*)^{-1} \gg \gamma_e B_1$, it is not possible in principle to obtain T_1 using cw saturation methods. Only the product $T_1 T_2$ can be measured.

A conclusion then is that the saturation recovery technique yields directly the values for T_1 in systems exhibiting inhomogeneous broadening whereas sophisticated data processing with uncertain input parameters is necessary to obtain true values of T_1 from cw saturation measurements.

Except for the experiment described in Fig. 8, all the work discussed here was on free radicals in solvents of low dielectic loss, using conventional EPR sample tubes of 3 mm inner diameter. If relaxation studies on free radicals in solution are to be fully exploited, it is clear that the sensitivity of the equipment must be adequate to permit experiments in lossy solvents using 1 and 2 mm sample tubes.

It is the argument of the present section that the serious experimental problems in doing saturation recovery experiments on free radicals in solution are nearly overcome. The papers that have been published bring out many interesting physical or chemical problems that are now for the first time soluble. Thus this seems likely to be an important area of research in the immediate future. The following are some of the questions that have arisen in the course of the research that resulted in the first papers employing saturation recovery techniques on free radicals in solution.

1. In the study of galvinoxyl it was observed that the low-field hyperfine lines have a longer T_1, although the high-field side of the spectrum is better resolved. Although a dependence of T_1 and T_2 on nuclear quantum number is no surprise, the coefficients of the terms linear in M in the expression given by Kivelson[20] for T_1 and T_2 that result in the high-field–low-field asymmetry in both T_1 and T_2 cannot be of different sign.

2. In this same galvinoxyl study it was observed that the spin-lattice relaxation time became independent of the viscosity at a value of η/T of about 3×10^{-1}. Huisjen and Hyde (unpublished, but see Fig. 9) observed that T_1 of tanol in supercooled sec-butyl benzene was independent of viscosity whereas companion ELDOR experiments[21,22] showed that the rotational correlation time of the nitroxide was nevertheless governed by

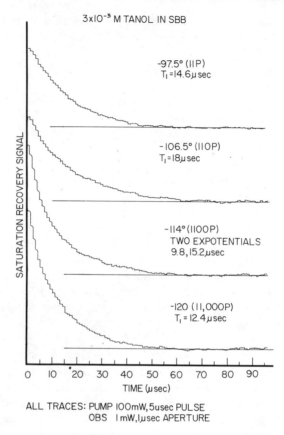

Figure 9 Saturation recovery signals for $3 \times 10^{-3}M$ tanol in supercooled *sec*-butyl benzene. The viscosities were determined from data given by A. J. Barlow, J. Lamb, and A. J. Matheson, *Proc. Roy. Soc.*, **292A**, 322 (1966). Experimental conditions: pumping power 100 mW, pumping duration 5 μs, observing power 1 mW, observing aperture 1 μs.

the macroscopic viscosity. One wonders how the liquid state and solid-state models of spin-lattice relaxation come together at high viscosities.

3. A conclusion from the work of Percival and Hyde[10] is that the dominant mechanism for spin-lattice relaxation of nitroxides remains unknown. Since nitroxides exhibit shorter relaxation times than other free radicals, it is natural to look at particular properties of these molecules, including intramolecular motional modes. Hwang et al.[18] reached a similar conclusion. In the light of the vast amount of research being done using nitroxide radical spin labels, this uncertainty in spin-lattice relaxation mechanisms is particularly vexing.

4. There is evidence of subtle differences in relaxation behavior depending on solvent. Even the data on galvinoxyl obtained with toluene and with sec-butyl benzene[7] showed differences that appear outside of the experimental error.

5. Viscosity dependencies of the intermolecular dipole-dipole contribution to T_1 vary approximately as the half-power. This dependence is consistent with the Torrey random flight model in the diffusive limit.[23,24] With the increased precision in this measurement offered by the saturation recovery technique, one can hope to construct more detailed models of the intermolecular collison dynamics.

6. Nonclassical jump diffusion motions, as occur for small molecules both in viscous and nonviscous liquids, result in considerable spectral densities in the microwave range but may be relatively ineffective in altering spectral line shapes in distinctive ways. The technique offers a good potential for the investigation of such motions.

6.2 Spin Labels and the Very-Slow-Tumbling Domain

If we introduce three critical frequencies, $\Delta\omega$, τ_R^{-1}, and T_1^{-1}, where $\Delta\omega$ is the order of magnitude of the anisotropy of the magnetic interactions of a spin label, τ_R is the rotational correlation time of the label, and T_1 is the spin-lattice relaxation time of the label, then the very-slow-tumbling domain is characterized by the relationships

$$\Delta\omega\tau_R \gg 1; \qquad 100T_1 > \tau_R > 0.01T_1.$$

The first inequality guarantees that the ordinary EPR spectrum is the same as that from a rigid powder. The second inequality leads to spectral diffusion of saturation if an intense microwave field is incident on the sample, since rotational diffusion is comparable to spin-lattice relaxation. Nitroxide radical spin labels fulfill this inequality in the range of correlation times of 10^{-3}–10^{-7} s. It is in this range of conditions that the new spin-label technique of saturation transfer spectroscopy functions.[25,26] This technique permits the measurements of correlation times from 10^{-3} to 10^{-7} s and has been widely applied to macromolecular complexes.

The extent of spectral diffusion that occurs in a time T_1 determines the nature of the observed spectra. The early progressive saturation experiments described by Hyde and Dalton[27] showed that the spin-lattice relaxation time is nearly independent of the motion. It was this discovery that made saturation transfer spectroscopy a practical technique. If T_1 were to show a strong dependence on rotational diffusion, the entire saturation transfer approach could well fail.

However, there is some variation of T_1 among different spin-labeled systems and precise interpretation of saturation transfer spectra requires knowledge of T_1. In a model study Huisjen and Hyde (unpublished) measured the spin-lattice relaxation time of tanol in supercooled sec-butyl benzene using the saturation recovery method. For rotational correlation times between 10^{-5} and 10^{-7} s, T_1 was unchanged and about equal to 1.5×10^{-5} s. (See Fig. 9.) When $\tau_R \sim T_1$, analysis of the recovery signal shows a complex decay that presumably arises from spectral diffusion competing with spin-lattice relaxation. A reliable value for the relaxation time is obtained from the tail of the recovery signal. (See the next section for further discussion of spectral diffusion effects.) In reference 8, Huisjen and Hyde report a value of 6.6×10^{-6} s for maleimide spin-labeled hemoglobin in water at room temperature. (See Fig. 10.) This experiment was very difficult from the point of view of signal-to-noise ratio. It represents the only time-dependent measurement of relaxation in a spin-labeled biomolecule. Although this demonstrates a certain feasibility for experiments of this type, it also provides strong motivation to go from a single-channel boxcar, as used in references 8 and 9, to a fast multichannel boxcar, thus permitting the collection of substantially more information in the available time.

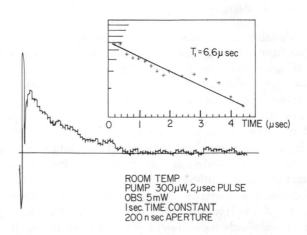

MALEIMIDE LABELED HEMOGLOBIN

$T_1 = 6.6\,\mu$ sec

0   I   2   3   4   TIME (μsec)

ROOM TEMP
PUMP 300μW, 2μsec PULSE
OBS 5mW
I sec. TIME CONSTANT
200 n sec APERTURE

Figure 10 Saturation recovery signal from malemide-labeled hemoglobin.[8] Experimental conditions: $10^{-3}M$, 1 mm capillary; roon temperature; pumping pulse 300 mW and 2 μs duration; observing power 5 mW; 1 s receiver time constant; 200 ns aperture.

7   COMPARISON WITH OTHER TIME DOMAIN TECHNIQUES

It is useful for the reader to be clearly aware of the differences, both experimentally and theoretically, between *step* techniques and *pulse* techniques. Saturation recovery is a step technique. It is assumed that the pumping pulse is on for a time sufficiently long that the spin system has come to equilibrium under saturating conditions. The pumping power is then turned off, or perhaps stepped to a new value, and the dynamic processes by means of which the spin system approaches the new equilibrium conditions are observed.

In contrast, pulse or spin-echo techniques employ pulse durations that are much shorter than the characteristic times for loss of phase coherence, T_2. The entire sequence of pulses, delays, and echoes occurs in a time of the same order of magnitude as T_2. Thus the time response of the equipment, both pumping and observing, must be at least $0.1T_2$.

It is not a simple matter to deliver pulses that have a duration of less than 10^{-7} s both because of the times and because of the magnitudes of rf field intensity required, nor is it a simple matter to observe transient responses shorter than 10^{-7} s. Let us assume that $T_1 = CT_2$ where C is a constant. The saturation recovery technique is feasible while the spin-echo technique is questionable for measurement of spin-lattice relaxation times in the time scale of $T_1/C \leq 10^{-7} \leq T_1$. Köksal and Krüger[34] have measured T_1's as short as 10^{-6} s in a spin-echo liquid-phase experiment but at a very high radical concentration ($\sim 0.1M$).

This distinction that the apparatus must be able to respond in a time of the order of T_1 in saturation recovery and in a time of the order of $0.1T_2$ in spin echo is crucial in the design of experiments. Here T_2 is a characteristic time for loss of phase coherence, a true thermodynamic degradation that cannot be recovered by, say, a 180° pulse as in the 90–180° spin-echo sequence. If one accepts the experimental limit of 10^{-7} s for delivery of a pulse, then the relationship $10^{-7} = 0.1T_2$ leads to a maximum permissible peak-to-peak derivative linewidth of a homogeneous line of 65 mG. Thus spin-echo techniques can hardly be applied to any system of chemical or biological interest in the liquid phase "slow-tumbling" and "very-slow-tumbling" domains. They will not be applicable to any *spin-labeled* biological system. Only for free radicals of relatively low molecular weight in solutions of rather low viscosity can spin-echo experiments in fluids be expected to be successful.

The 65 mG estimate may be too conservative. The factor of 0.1 in the relationship $10^{-7} = 0.1T_2$ might be stretched to 0.3. Spin-echo instruments with response times of 3×10^{-8} s have been reported. Several spin-echo papers on exchange narrowed EPR spectra in the liquid phase have appeared,[24,28-34] indicating a certain experimental feasibility for spin-echo experiments in liquids. But there should be no doubt that spin-echo experiments in the liquid and glassy phases are very difficult, much more difficult than saturation recovery.

In fact, a spectrometer suitable for one type of measurement can easily be used for the other. Referring to Fig. 1, all that must be accomplished is to alter the programmer. The handshake to the transient receiver after the programmer has instructed the exciter is a key element in the design of a flexible time domain EPR spectrometer suitable, for example, for both "pulse" and "step" experiments.

As a specific illustration of this major difference between spin-echo and saturation recovery, Huisjen and Hyde[8] were successful using the saturation recovery technique in measuring the spin-lattice relaxation time of maleimide-labeled hemoglobin in water at room temperature (Fig. 10). Rotational diffusion and nitrogen spin-lattice relaxation results in saturation transfer throughout the entire spectrum and an extremely short time for loss of phase coherence. Brown[29] studied the anthracene radical cation in sulfuric acid glasses using the spin-echo technique. He was also interested in phase-memory times determined by motional effects but was unable to make measurements at temperatures above 180 K where T_2's were perhaps three orders of magnitude longer than that for the labeled hemoglobin. Of course in the fast-tumbling domain, T_2's of free radicals in liquids become longer and spin echo is feasible, but in the gap between very show tumbling and fast tumbling, only saturation recovery has been successful thus far.

If the relaxation times are sufficiently long, as in frozen solutions and single crystals, the spin-echo technique has definite advantages. For example, we have seen that T_2 and T_2^* are folded into the free-induction decays in a complex way. The 90–180° two-pulse sequence is the best available technique for determining the characteristic time for loss of phase coherence in inhomogeneous systems if the measurement is experimentally possible.

Other step techniques besides the saturation recovery method that one might consider include steps in pumping microwave phase, steps in phase accompanied by steps to a new microwave level, and steps in environmental conditions such as incident light level or temperature. These techniques have not been investigated very much thus far in EPR spectroscopy. The author and his colleagues did consider phase steps in some detail from a theoretical point of view, with the hope of devising a step technique that might be competitive with spin echo in measuring times for loss of phase coherence. The calculations were not promising.

Experimentally the literature suggests a certain rivalry between practitioners of step and pulse methods, although the literature also suggests that complementariness is a more appropriate relationship. On the basis of utilization in the solid-state EPR literature, it appears that spin echo is the method of choice to measure times for loss of phase coherence and saturation recovery to measure spin-lattice relaxation times.

Attention is called to an interesting hybrid technique between the step and pulse methods. Dalton et al.[35] and Brown[36] used a method involving a long pumping pulse of duration sufficient for the spin system to reach equilibrium, followed by a delay of variable length and then a 90–180° pulse sequence. This is a method for measuring progression of recovery because of spin-lattice relaxation following the long pumping pulse. The point here is that even if T_2 is sufficiently long to permit spin-echo experiments, the usual sequences to measure T_1 will in fact measure T_1 (effective), that is, the sum of all possible relaxation pathways rather than the relaxation between only the observed levels. By initially saturating all coupled transitions with the long pumping pulse, this complication can largely be avoided.

There is a definite element of uncertainty both in this hybrid step-pulse sequence and in the saturation recovery method when spectral diffusion processes occur on a time scale that is the same as the spin-lattice relaxation time. This is a complexity of not very much significance in NMR where these techniques originated, but it is of commanding importance in EPR spectroscopy. It is intuitively clear that when spectral diffusion is fast, the pumping microwave field saturates the entire spin system and recovery proceeds with the true T_1. And if spectral diffusion is negligible, the true T_1 is measured. It is also fairly obvious that if spectral diffusion and spin-lattice relaxation occur at the same rate, complications can be expected. Freed[37] reaches this same conclusion based on an analysis that includes the full relaxation matrix of a nitroxide radical.

Even when the times are equal, a fairly good number for T_1 can be obtained because diffusion of saturation away from the observed transition tends to be canceled by diffusion of saturation from coupled transitions back to the observed transition. It appears that in the worst possible circumstances, errors in measurement of T_1 that arise from spectral diffusion will not exceed a factor of 2. By use of pumping pulses of various durations, it is likely that information can be gained concerning spectral diffusion that will permit corrections to be made to the data, thus yielding the true T_1. We are not aware, however, of any instances where this suggested procedure has been carried out.

Spectral diffusion in EPR spectroscopy is deserving of further experimental investigation. The optimum technique appears to be the use of

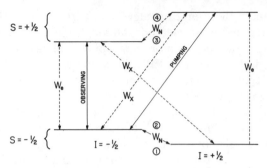

Figure 11   Energy level diagrams for an $S = \frac{1}{2}$, $I = \frac{1}{2}$ system.[38]

long saturating pulses at one part of the spectrum and the direct observation of the time-dependent response at another part of the spectrum with a second rf field. This is the so-called pulse ELDOR method of Nechtschein and Hyde.[38] In Fig. 11 the energy levels of an $S = \frac{1}{2}$, $I = \frac{1}{2}$ system are shown with all relaxation probabilities indicated. Assume, for the purposes of illustration, that W_x, $W_N < W_e$. Application of a long pumping pulse between level 2 and level 4 (a "forbidden" transition) results in a net transfer of spin population between the "allowed" transition manifold consisting of levels 3, 2 and the "allowed" transition manifold consisting of levels 4, 1. Spectral diffusion arises from W_x, W_N processes. Observation of the resulting time dependence of an allowed transition yields a transient signal that measures the relaxation processes connecting the two manifolds. This is, of course, a step-type technique and belongs, conceptually, to the class of methods under consideration in this paper. Many step recovery ELDOR experiments can be devised, and this seems without doubt to be the method of choice in investigating spectral diffusion processes.

We have seen, then, that saturation recovery permits the convenient measurement of spin-lattice relaxation times that are longer than 10^{-7} s, and spin-echo techniques permit the convenient measurement of T_1 and T_2 if $T_2 > 10^{-6}$ s. There is thus no convenient experimental methodology to measure characteristic times for loss of phase coherence in inhomogeneously broadened lines with a signal-to-noise ratio near the theoretical optimum when $T_2 < 10^{-6}$ s. The author views this as a particularly challenging experimental problem. The gap can be effectively narrowed by careful analysis of free-induction decay signals or by isolation of the term in the saturation recovery expressions [Eq. (2) and (3)] that is proportional to z_0. This latter term, it would appear from the equations, can be monitored by suddenly turning on a *nonsaturating* microwave level and observing the resulting transient response. The motivation in calling special attention to this difficult intermediate range of T_2's is that it seems likely to correspond to the T_2's of free radicals in organic crystals and of "immobilized" spin labels in proteins. In these systems fluctuations in the orientation of the radical with respect to its environment determines the T_2 in a manner that depends on the anisotropy of the magnetic interactions.

In summary, the author views time domain EPR spectroscopy to be a central thrust of instrumental development and applications in the next decade. Step-recovery techniques as discussed in this chapter will be the *dominant* instrumental methodology in systems of chemical and biological interest except where temperatures are sufficiently low and immobilization sufficiently great that spin-echo approaches can be used.

ACKNOWLEDGMENT

Preparation of this manuscript was assisted by grants RR-01008 and 5 R01 GM22923 from the National Institutes of Health.

REFERENCES

1. N. Bloembergen, E. M. Purcell, and R. V. Pound, *Phys. Rev.*, **73,** 679 (1948).

2. J. A. Giordmaine, L. E. Alsop, F. R. Nash, and C. H. Townes, *Phys. Rev.*, **109,** 302 (1958).

3. C. F. Davis, Jr., M. W. P. Strandberg, and R. L. Kyhl, *Phys. Rev.*, **111,** 1268 (1958).

4. K. D. Bowers and W. B. Mims, *Phys. Rev.*, **115,** 285 (1959).

5. O. S. Leifson and C. D. Jeffries, *Phys. Rev.*, **122,** 1781 (1961).

6. P. L. Scott and C. D. Jeffries, *Phys. Rev.*, **127,** 32 (1962).

7. M. Huisjen and J. S. Hyde, *J. Chem. Phys.*, **60,** 1682 (1974).

8. M. Huisjen and J. S. Hyde, *Rev. Sci. Instrum.*, **45,** 669 (1974).

9. P. W. Percival and J. S. Hyde, *Rev. Sci. Instrum.*, **46,** 1522 (1975).

10. P. W. Percival and J. S. Hyde, *J. Mag. Res.*, **23,** 249 (1976).

11. J. S. Hyde and T. Sarna, *J. Chem. Phys.* **68,** 4439 (1978).

12. T. Sarna and J. S. Hyde, *J. Chem. Phys.* **69,** 1945 (1978).

13. P. W. Atkins, K. A. McLaughlan, and P. W. Percival, *Mol. Phys.*, **25,** 281 (1973).

14. K. V. Lingam, P. G. Nair, and B. Venkataraman, *Proc. Ind. Acad. Sci.*, **A76,** 207 (1972).

15. S. K. Rengan, M. P. Khakhar, B. S. Brabhananda, and B. Venkataraman, *Pure Appl. Chem.*, **32,** 287 (1972).

16. S. K. Rengan, M. P. Khakhar, B. S. Brabhananda, and B. Venkataraman, *Pramana*, **3,** 95 (1974).

17. S. K. Rengan, M. P. Khakhar, B. S. Brabhananda, and B. Venkataraman, *J. Mag. Res.*, **16,** 35 (1974).

18. J. S. Hwang, R. P. Mason, L. P. Hwang, and J. H. Freed, *J. Phys. Chem.*, **79,** 489 (1975).

19. T. G. Castner, Jr., *Phys. Rev.*, **115,** 1506 (1959).

20. D. Kivelson, *J. Chem. Phys.*, **33,** 1094 (1960).

21. M. D. Smigel, L. R. Dalton, J. S. Hyde, and L. A. Dalton, *Proc. Nat. Acad. Sci. U.S.A.*, **71,** 1925 (1974).

22. J. S. Hyde, M. D. Smigel, L. R. Dalton, and L. A. Dalton, *J. Chem. Phys.*, **62,** 1655 (1975).

23. H. C. Torrey, *Phys. Rev.*, **92,** 962 (1953).

24. G. J. Krüger, *Z. Naturforsch.* **A24,** 560 (1969).

25. J. S. Hyde, in C. H. W. Hirs and S. N. Timasheff, Eds., *Methods in Enzymology*, Vol. 49G, No. 19, Academic, New York, 1978, p. 480.

26. J. S. Hyde and L. R. Dalton, in L. J. Berliner, Ed., *Spin Labeling II. Theory and Applications*, Academic, New York, 1979, p1.

27. J. S. Hyde and L. Dalton, *Chem. Phys. Lett.*, **16,** 568 (1972).

28. A. D. Milov, K. M. Salikhov, and Yu. D. Tsvetkov, *Chem. Phys. Lett.*, **8,** 523 (1971).

29. I. M. Brown, *Chem. Phys. Lett.*, **17,** 404 (1972).

30. I. M. Brown, *J. Chem. Phys.*, **60,** 4930 (1974).

31. A. D. Milov, A. B. Mel'nik, and Yu. D. Tsvetkov, *Theor. Exp. Chem.*, **11,** 656 (1975).

32. G. J. Krüger, *Adv. Mol. Relaxation Proc.*, **3,** 235 (1972).

33. R. Brändle, G. J. Krüger and W. Müller-Warmuth, *Z. Naturforsch.*, **25a,** 1 (1970).

34. K. Köksal and G. J. Krüger, *Z. Naturforsch.*, **30a,** 883 (1975).

35. L. R. Dalton, A. L. Kwiram, and J. A. Cowen, *Chem. Phys. Lett.*, **14,** 77 (1972).

36. I. M. Brown, *J. Chem. Phys.*, **58,** 4242 (1973).

37. J. H. Freed., *J. Phys. Chem.*, **78,** 1155 (1974).

38. M. Nechtschein and J. S. Hyde, *Phys. Rev. Lett.*, **24,** 672 (1970).

2 THEORY OF ESR SATURATION RECOVERY IN LIQUIDS AND RELATED MEDIA

Jack H. Freed

Department of Chemistry
Cornell University
Ithaca, New York

1 INTRODUCTION

This chapter is based on the original paper published in 1974, reference 1. It was motivated by the then growing interest in pulsed electron spin resonance (ESR) experiments on free radicals in liquids and related media, in particular saturation recovery-type experiments.[2] Since then the interest has greatly expanded, as is evidenced by this book. The basis of our theory of time-resolved ESR experiments is a natural outgrowth of our theory of steady-state saturation and multiple resonance behavior. This theory has now been summarized in a chapter in another book, and frequent reference is made to it (reference 3). In fact, it was most interesting to find that the steady-state saturation theory, which has been developed in great detail, could readily be extended, with all its sophistication, to the case of time-dependent, or time-resolved, spectroscopy.

We emphasize saturation recovery in this chapter, but we also include some comments on pulsed electron-electron double resonance (ELDOR),[2b] which, in principle, may be thought of as a saturation recovery, but with observation at a frequency displaced from the high-power pulse frequency. In later chapters we see how such methods may be extended to free-induction decay and echo-type experiments for the free-radical systems, where numerical techniques are useful.[4] We emphasize analytical techniques in the present chapter. This is possible because one finds that over a wide range of types of systems, the saturation recovery experiment is simply interpreted. We, however, give the general expressions that are amenable to computer methods already developed in connection with steady-state problems.[5]

Considerable theoretical and experimental work has focused on ESR spectra in the slow-tumbling region. Such work has (1) extended the range of motional reorientation times over which accurate analyses can be made of these motions and (2) demonstrated that more microscopic features of the motion (i.e., deviations from Brownian motion) could be studied.[2a,5] We also outline how the original steady-state theory may readily be extended to time-resolved experiments with emphasis again on slow-motional saturation recovery (and ELDOR-type) experiments. Again, because of a number of *formal* similarities between a motional narrowing theory and a slow motional theory, we are able to cast both in a single general framework. The analogies that are then established allow us to clarify the more complex slow–motional analysis.

2 GENERAL FORMULATION

We start with the usual density-matrix equation of motion for the spin-density matrix[2a,3,6a], $\sigma(t)$:

$$\dot{\sigma} = -i(\mathcal{H}_0^x + \varepsilon(t)^x + iR)(\sigma - \sigma_{eq}), \tag{1}$$

where $\mathcal{H}_0$ is the zero-order Hamiltonian, $\varepsilon(t)$ is the interaction with the radiation field(s), R is the relaxation matrix, σ_{eq} is the time-independent equilibrium value of $\sigma(t)$, and the superscript x-implies that for two operators A and B, $A^x B \equiv [A, B]$. One speaks of A^x as the superoperator form of the operator A.

This is the usual expression one obtains for the motional narrowing region, where rotational modulation of the perturbing Hamiltonian $\mathcal{H}_1(\Omega)$ is sufficiently rapid that $|\mathcal{H}_1(\Omega)^2| \tau_R^2 \ll 1$, where τ_R is the rotational correlation time. The relaxation matrix R is made up from terms quadratic in matrix elements of $\mathcal{H}_1(\Omega)$, and it contains the linewidths, or in a time-resolved experiment the matrix of the T_2's, as well as the transition probabilities for relaxation from nonequilibrium population distributions. It should be clear, from the form of Eq. (1), that to have stable exponential relaxation, it must be true that the real part of the elements of R be negative (i.e., Re $R < 0$). The explicit appearance of σ_{eq} in Eq. (1) is part of a high-temperature approximation such that

$$\sigma_{eq} = e^{-\hbar \mathcal{H}_0 / kT} / Tr\{e^{-\hbar \mathcal{H}_0 / kT}\}$$
$$\approx \frac{1}{A}\left[1 - \frac{\hbar \mathcal{H}_0}{kT}\right], \tag{2}$$

where A is the total number of spin eigenstates and k is Boltzmann's constant. More generally, we may write a stochastic Liouville expression for $\sigma(\Omega, t)$ wherein the assumption of motional narrowing need not be made:[2a,5,7,8]

$$\dot{\sigma}(\Omega, t) = -i(\mathcal{H}_0^x + \varepsilon(t)^x + \mathcal{H}_1(\Omega)^x + iR' - i\Gamma_\Omega)[\sigma(\Omega, t) - \sigma_0(\Omega)]. \tag{3}$$

Here Γ_Ω is the Markov operator for the rotational tumbling that is modulating $\mathcal{H}_1(\Omega)^x$, R' is that part of the relaxation matrix that is orientation independent. Note that the expression is written for a $\sigma(\Omega, t)$, which is both a spin-density operator and a classical probability function in the values of the random variables Ω. One may recover the ordinary spin-density matrix by averaging over orientations

$$\sigma(t) = \int d\Omega \sigma(\Omega, t) P_{eq}(\Omega) \equiv \langle P_{eq}(\Omega) | \sigma(\Omega, t) | P_{eq}(\Omega) \rangle, \tag{4}$$

where $P_{eq}(\Omega)$ is the equilibrium distribution of orientations, and the convenient bra-ket notation is introduced. When $|\mathcal{H}_1(\Omega)| / |\Gamma_\Omega| \ll 1$ one may recover the motional narrowing limit from Eq. (3) [i.e., one obtains Eq. (1) for $\sigma(t)$].

We note that the Markov operator Γ_Ω has associated with it the expression

$$\frac{\partial}{\partial t} P(\Omega, t) = -\Gamma_\Omega P(\Omega, t), \tag{5a}$$

where $P(\Omega, t)$ is the probability density of finding Ω at a particular value at time t. The process is assumed to be stationary, so that Γ_Ω is time independent and we have

$$\Gamma_\Omega P_{\text{eq}}(\Omega) = 0. \tag{5b}$$

We first study the general approach to the time-dependent solution of Eq. (1); then we generalize to cover Eq. (3). We now introduce for the relevant off-diagonal elements of $\sigma(t)$:

$$\sigma_{\lambda_j}(t) = \sum_{n=-\infty}^{\infty} e^{in\omega t} Z(t)_{\lambda_j}^{(n)}, \tag{6a}$$

and

$$Z(t)_{\lambda_j}^{(n)} = Z(t)_{\lambda_j}^{(n)'} + iZ(t)_j^{(n)''}, \tag{6b}$$

for the λ_jth (ESR) transition. In our notation the matrix element $\sigma_{\alpha\alpha'}$, where α and α' differ by at least the value of electron spin quantum number $M_s = \pm\frac{1}{2}$, may be written as $\sigma_{\lambda_j^- \lambda_j^+} \equiv \sigma_{\lambda_j}$ corresponding to the λ_jth (ESR) transition. The states $\lambda^\pm$ are the $M_s = \pm\frac{1}{2}$ states; and λ_j^+ and λ_j^- have the same nuclear configuration, if one has an allowed ESR transition, or they have different nuclear configurations, if one has a forbidden ESR transition. The matrix elements $\sigma_{\lambda_j}(t)$ are then Fourier-analyzed in harmonics of the frequency ω of the applied radiation field. The $Z(t)_{\lambda_j}^{(n)}$ and $Z''(t)_{\lambda_j}^{(n)}$ are the real and imaginary parts, respectively, of $Z(t)_{\lambda_j}^{(n)}$ and correspond to the dispersive and absorptive modes of a resonant signal.

Similarly, for the diagonal elements, we can let

$$[\sigma(t) - \sigma_{\text{eq}}]_{\lambda_j^\pm} \equiv \chi(t)_{\lambda_j^\pm} \tag{7}$$

and Fourier-analyze the deviations of the diagonal elements of $\sigma(t)$ from their equilibrium values. In the present chapter we are only interested in the case of the $n = 0$ Fourier component of Eq. (7) representing the actual population deviations (i.e., recall that the diagonal elements σ_{aa} gives the population in state a). Also in Eq. (6a, b) we are only interested in the $n = 1$ case, corresponding to the Fourier component rotating with the radiation field.

One can now take matrix elements of Eq. (1) for a general multilevel spin system. The methods have been summarized in reference 3. The result is a set of coupled linear differential equations which may be neatly arranged in matrix notation. That is, we have the matrix differential equation

$$
\begin{bmatrix} \dot{\mathbf{Z}}(t) \\ \dot{\mathbf{Z}}^*(t) \\ \dot{\chi}(t) \end{bmatrix} = \begin{bmatrix} \mathbf{R}-i\mathbf{K} & 0 & +i\mathbf{d} \\ 0 & \mathbf{R}+i\mathbf{K} & -i\mathbf{d} \\ +i\mathbf{d}^{\text{tr}} & -i\mathbf{d}^{\text{tr}} & -\mathbf{W} \end{bmatrix} \begin{bmatrix} \mathbf{Z}(t) \\ \mathbf{Z}^*(t) \\ \chi^{(t)} \end{bmatrix} + \begin{bmatrix} i\mathbf{Q} \\ -i\mathbf{Q} \\ 0 \end{bmatrix}. \tag{8}
$$

The vector $\mathbf{Z}(t)$ is a vector defined in the M-dimensional "space" of all the induced transitions with elements $Z(t)_{\lambda_i}$, whereas $\chi(t)$ is a vector defined in the A-dimensional "space" of all eigenstates with elements $\chi_{\lambda_i^{\pm}}$. $\mathbf{Z}^*(t)$ is the complex conjugate of $\mathbf{Z}(t)$. The matrix $\mathbf{W}$ is the transition probability matrix, whose $\alpha\beta$th element for $\alpha \neq \beta$ is just minus the transition probability from state β to state α. It is defined in "eigenstate space" of dimension A. The width matrix $\mathbf{R}$ contains what in a steady-state experiment are the (coupled) widths of all the induced transitions; in a time-resolved experiment they represent the (coupled)-exponential decays of the off-diagonal density matrix elements. It is defined in "transition space." The coherence matrix $\mathbf{K}$ defined in "transition space" has as its λ_ith diagonal element the deviation of the Larmor frequency of the λ_ith transition from the applied (near)-resonant radiation field. The vector $\mathbf{Q}$, defined in "transition space," results from the driving terms of the radiation field, and is nonzero only for allowed transitions. The transition-moment matrix $\mathbf{d}$ and its transpose $\mathbf{d}^{\text{tr}}$ are in general *not* square matrices. The rows of $\mathbf{d}$ are labeled according to transition space, whereas its columns are labeled according to eigenstate space. Thus it is an $M \times A$ rectangular matrix. Its elements represent the way pairs of eigenstates belonging to the λ_ith transition are coupled by the transitions induced by the radiation field(s). Detailed instructions for writing these matrices down are given in reference 3, and we illustrate with the following examples.

In the case of a simple two-level system between states $a \rightarrow (M_s = \frac{1}{2})$ and $b \rightarrow (M_s = -\frac{1}{2})$ one has $\mathbf{K} \rightarrow \Delta\omega = \omega - \omega_{0'}$ where ω is the frequency of the applied radiation field and ω_0 is the Larmor frequency for the transition. Also $-\mathbf{R} \rightarrow T_2^{-1}, \mathbf{W}$ is the 2×2 matrix:

$$
\begin{bmatrix} +W_{ba} & -W_{ab} \\ -W_{ba} & +W_{ab} \end{bmatrix}
$$

involving the transition probability from state a to b, W_{ba}, and that from state b to a, W_{ab}. They are equal in the high-temperature approximation. Also $\mathbf{d}$ is the 1×2 matrix:

$$
(-d, +d),
$$

where $d = \frac{1}{2}\gamma_e B_1$. This arises from the radiation term

$$\hbar\varepsilon(t) = \frac{1}{2}\hbar\gamma_e B_1[S_+ e^{-i\omega t} + S_- e^{+i\omega t}], \qquad (9a)$$

which is the interaction of the spin $S = +\frac{1}{2}$ with a rotating field:

$$\mathbf{B}_1 = B_1(\mathbf{i}\cos\omega t + \mathbf{j}\sin\omega t). \qquad (9b)$$

Finally, $\mathbf{Q} \to q\omega_0 \mathbf{d}$ with $q = \hbar/2kT$ in this case.

If we now transform our expressions using the definitions

$$\chi^\pm(t) \equiv \frac{1}{\sqrt{2}}(\chi_+ \pm \chi_-), \qquad (10)$$

for this two-level system, then we obtain

$$
\begin{bmatrix} \dot{Z}(t) \\ \dot{Z}^*(t) \\ \dot{\chi}^-(t) \\ \dot{\chi}^+(t) \end{bmatrix} =
\begin{bmatrix}
-T_2^{-1} - i\,\Delta\omega & 0 & -\sqrt{2}\,id & 0 \\
0 & -T_2^{-1} + i\,\Delta\omega & \sqrt{2}\,id & 0 \\
-\sqrt{2}\,id & \sqrt{2}\,id & -T_1^{-1} & 0 \\
0 & 0 & 0 & 0
\end{bmatrix}
$$

$$
\times \begin{bmatrix} Z(t) \\ Z^*(t) \\ \chi^-(t) \\ \chi^+(t) \end{bmatrix} + iq\omega_0 d \begin{bmatrix} 1 \\ -1 \\ 0 \\ 0 \end{bmatrix}, \qquad (11)
$$

where $T_1^{-1} \equiv 2W_{ab}$.

We first of all see that $\chi^+(t)$ is uncoupled to the other density-matrix elements and is time independent. This is because it represents the conservation of probability condition, which more generally is

$$\text{Tr}\,\sigma = \text{Tr}\,\sigma_{eq} = 1, \qquad (12a)$$

or equivalently,

$$\text{Tr}\,\chi = \text{Tr}\,[\sigma - \sigma_{eq}] = 0. \qquad (12b)$$

The remaining portion of Eq. (11) involves a 3×3 symmetric matrix, which, in fact, is nothing more than the well-known Bloch equations[6] describing a single-line spectrum. We need only make the further identifications

$$2d = \omega_1 \equiv \gamma_e B_1, \qquad Z' = \text{Re}\, Z \to \tilde{M}_x, \qquad Z'' = \text{Im}\, Z \to \tilde{M}_y,$$

where $\tilde{M}_x$ and $\tilde{M}_y$ are the x and y components of magnetization in the rotating frame, whereas $(1/\sqrt{2})\chi^- \to (M_{eq} - M_z)$, where M_{eq} is the equilibrium value of the magnetization and M_z is its z-component. These Bloch equations may be solved by standard methods (e.g., Laplace transforms).[6b]

We now return to the general Eq. (8). We wish to transform them in a manner analogous to the type of transformation that led to the simplified form of Eq. (11) for the simple two-level system. We can do that by generalizing the definition Eq. (10) for each eigenstate pair $\chi_{\lambda_i^+}$ and $\chi_{\lambda_i^-}$, and defining new vectors

$$\chi^{\pm} = \frac{1}{\sqrt{2}}(\chi_+ \pm \chi_-), \tag{13}$$

where $\chi_{\pm}$ is the subvector of dimension $A/2$ including all the $\chi_{\lambda_i^{\pm}}$. Thus the vectors $\chi^{\pm}$ are also of dimension $A/2$. This transformation leads to the matrix equations

$$\begin{bmatrix} \dot{\mathbf{Z}}(t) \\ \dot{\mathbf{Z}}^*(t) \\ \dot{\chi}^-(t) \\ \dot{\chi}^+(t) \end{bmatrix} = \begin{bmatrix} \mathbf{R}-i\mathbf{K} & 0 & \sqrt{2}\,i\hat{\mathbf{d}} & \sqrt{2}\,i\tilde{\mathbf{d}} \\ 0 & \mathbf{R}+i\mathbf{K} & -\sqrt{2}i\hat{\mathbf{d}} & -\sqrt{2}\,i\tilde{\mathbf{d}} \\ \sqrt{2}\,i\hat{\mathbf{d}}^{\mathrm{tr}} & -\sqrt{2}\,i\hat{\mathbf{d}}^{\mathrm{tr}} & -\hat{\mathbf{W}} & -\mathscr{W} \\ \sqrt{2}\,i\tilde{\mathbf{d}}^{\mathrm{tr}} & -\sqrt{2}\,i\tilde{\mathbf{d}}^{\mathrm{tr}} & -\mathscr{W}^{\mathrm{tr}} & -\tilde{\mathbf{W}} \end{bmatrix} \begin{bmatrix} \mathbf{Z}(t) \\ \mathbf{Z}^*(t) \\ \chi^-(t) \\ \chi^+(t) \end{bmatrix} + \begin{bmatrix} i\mathbf{Q} \\ -i\mathbf{Q} \\ 0 \\ 0 \end{bmatrix}, \tag{14}$$

where we have introduced the definitions

$$2\hat{\mathbf{d}} \equiv \mathbf{d}_+ - \mathbf{d}_- \tag{15a}$$

$$2\tilde{\mathbf{d}} \equiv \mathbf{d}_+ + \mathbf{d}_-, \tag{15b}$$

and the $\mathbf{d}_{\pm}$ are defined by analogy to $\chi_{\pm}$. That is, we may write $\mathbf{d} = (\mathbf{d}_+, \mathbf{d}_-)$ (i.e., a partitioned matrix where $\mathbf{d}_{\pm}$ represents the couplings to the $M_s = \pm$ eigenstates and is of dimension $M \times A/2$. Also we have

$$2\hat{\mathbf{W}} \equiv (\mathbf{W}_{+,+} + \mathbf{W}_{-,-}) - (\mathbf{W}_{+,-} + \mathbf{W}_{-,+}) \tag{16a}$$

$$2\tilde{\mathbf{W}} \equiv (\mathbf{W}_{+,+} + \mathbf{W}_{-,-} + \mathbf{W}_{+,-} + \mathbf{W}_{-,+}) \tag{16b}$$

$$2\mathscr{W} \equiv (\mathbf{W}_{+,+} - \mathbf{W}_{-,-}) + (\mathbf{W}_{+,-} - \mathbf{W}_{-,+}), \tag{16c}$$

where we have partitioned $\mathbf{W}$ according to the $M_s = \pm$ states as

$$\mathbf{W} = \begin{pmatrix} \mathbf{W}_{+,+} & \mathbf{W}_{+,-} \\ \mathbf{W}_{-,+} & \mathbf{W}_{-,-} \end{pmatrix}. \tag{16d}$$

[Note that in reference 1 we used $\hat{\mathbf{X}}(t) \equiv \sqrt{2}\chi^-(t)$ and $\tilde{\chi}(t) \equiv \sqrt{2}\chi^+(t)$, which led to a slightly more complex form of Eq. (11), but we use both forms below]. The form of Eq. (14) and (16a–d) does indeed appear to be more complex than Eq. (8). However, it is often the case that $\mathscr{W} = 0$. This will be exactly the case [cf. Eq. (16c) and below] when the matrix elements of $\mathbf{W}$ obey

$$W_{\alpha\pm,\beta\mp} = W_{\alpha\mp,\beta\pm}, \tag{17a}$$

$$W_{\alpha\pm,\beta\pm} = W_{\alpha\mp,\beta\mp}, \tag{17b}$$

(where α and β represent any nuclear configurations and the $\pm$ signs refer to M_s), which is a common situation.[3] We further assume that only electron spin resonance (ESR) transitions are excited, in which case one finds that $\tilde{\mathbf{d}} = \mathbf{0}$ [but for electron-nuclear double resonance (ENDOR) $\tilde{\mathbf{d}} \neq \mathbf{0}$] and the $\chi(t)^+$ may now be decoupled from the relevant part of the solution, that is Eq. (14) becomes

$$\begin{bmatrix} \dot{\mathbf{Z}}(t) \\ \dot{\mathbf{Z}}^*(t) \\ \dot{\chi}^-(t) \end{bmatrix} = \begin{bmatrix} \mathbf{R} - i\mathbf{K} & \mathbf{0} & \sqrt{2}\, i\hat{\mathbf{d}} \\ \mathbf{0} & \mathbf{R} + i\mathbf{K} & -\sqrt{2}\, i\hat{\mathbf{d}} \\ \sqrt{2}\, i\hat{\mathbf{d}}^{\mathrm{tr}} & -\sqrt{2}i\hat{\mathbf{d}}^{\mathrm{tr}} & -\hat{\mathbf{W}} \end{bmatrix} \begin{bmatrix} \mathbf{Z}(t) \\ \mathbf{Z}^*(t) \\ \chi^-(t) \end{bmatrix} + \begin{bmatrix} i\mathbf{Q} \\ -i\mathbf{Q} \\ \mathbf{0} \end{bmatrix}. \tag{18}$$

Also we have from Eq. (16a) that

$$\hat{\mathbf{W}} = \mathbf{W}_{+,+} - \mathbf{W}_{+,-}, \tag{19}$$

since from Eq. (15a, b) and the fact that $\mathbf{W}$ is symmetric one has $\mathbf{W}_{+,-} = \mathbf{W}_{-,+}$ and $\mathbf{W}_{-,-} = \mathbf{W}_{+,+}$. Equation (18) is the natural generalization for a multilevel spin system to the Bloch equations for a simple line. It is, however, only valid provided that Eq. (17a) and (17b) are at least approximately correct and that no nuclear magnetic resonance (NMR) transitions are excited. If either or both of these conditions are not valid, then one should use Eq. (18), which is of larger dimension (by the amount of $A/2$). In either case one has a complex symmetric matrix, and one may solve the general case on a computer by diagonalizing this matrix.[8] In Eq. (8) there are one or more eigenvalues of zero corresponding to Eq. (12b) for conservation of total probability, but this does not appear in Eq. (14).

Note that since the $\mathbf{Z}''$ elements are typically detected, we need the unitary transformation

$$\sqrt{2} \begin{bmatrix} \mathbf{Z}' \\ i\mathbf{Z}'' \\ \dfrac{1}{\sqrt{2}}\chi^- \end{bmatrix} = \begin{bmatrix} \dfrac{1}{\sqrt{2}}\mathbb{1} & \dfrac{1}{\sqrt{2}}\mathbb{1} & \mathbf{0} \\ \dfrac{1}{\sqrt{2}}\mathbb{1} & -\dfrac{1}{\sqrt{2}}\mathbb{1} & \mathbf{0} \\ \mathbf{0} & \mathbf{0} & \mathbb{1} \end{bmatrix} \begin{bmatrix} \mathbf{Z} \\ \mathbf{Z}^* \\ \chi^- \end{bmatrix} \equiv \mathbf{u} \begin{bmatrix} \mathbf{Z} \\ \mathbf{Z}^* \\ \chi^- \end{bmatrix}, \tag{20}$$

which transforms Eq. (18) into

$$
\begin{bmatrix} \dot{\mathbf{Z}}'(t) \\ i\dot{\mathbf{Z}}''(t) \\ (2)^{-1/2}\dot{\boldsymbol{\chi}}^-(t) \end{bmatrix} = \begin{bmatrix} \mathbf{R} & -i\mathbf{K} & 0 \\ -i\mathbf{K} & \mathbf{R} & 2i\hat{\mathbf{d}} \\ 0 & 2i\hat{\mathbf{d}}^{\text{tr}} & -\hat{\mathbf{W}} \end{bmatrix} \begin{bmatrix} \mathbf{Z}'(t) \\ i\mathbf{Z}''(t) \\ (2)^{-1/2}\boldsymbol{\chi}^-(t) \end{bmatrix} + \begin{bmatrix} 0 \\ i\mathbf{Q} \\ 0 \end{bmatrix}, \tag{21}
$$

with the new symmetric matrix on the right-hand side of Eq. (21).

We note that the steady-state solutions to Eq. (8) or (18) may be calculated by the methods reviewed in reference 5. In particular, the form of Eq. (8) or (18) is

$$
\Delta\dot{\mathbf{m}}(t) = \mathbf{Lm} + \mathbf{Q}'. \tag{22}
$$

Then the steady-state solution $\mathbf{m}^{ss}$ is formally given as

$$
\mathbf{m}^{ss} = -\mathbf{L}^{-1}\mathbf{Q}' \tag{23}
$$

(ignoring for the moment the singularity of the $\mathbf{W}$ matrix of Eq (8), cf. reference 3). If we define

$$
\Delta\mathbf{Z}(t) \equiv \mathbf{Z}(t) - \mathbf{Z}^{ss}, \text{ and so on,} \tag{24}
$$

or

$$
\Delta\mathbf{m}(t) \equiv \mathbf{m}(t) - \mathbf{m}^{ss}, \tag{25}
$$

then Eq. (22) may be written in terms of these deviations from steady-state value as

$$
\Delta\dot{\mathbf{m}}(t) = \mathbf{L}\Delta\mathbf{m}(t) \tag{26a}
$$

$$
\Delta\mathbf{m}(t) = e^{\mathbf{L}t}\Delta\mathbf{m}(0), \tag{26b}
$$

so as $t \to \infty$, $\Delta\mathbf{m}(t) \to 0$.

It is shown in reference 3 that the solution of Eq. (23) may be simplified in terms of smaller submatrices as

$$
\mathbf{Z}'' = \mathbf{M}^{-1}(-\mathbf{R}^{-1})\mathbf{Q} \tag{27a}
$$

$$
\mathbf{Z}' = (-\mathbf{R}^{-1})\mathbf{KZ}'' \tag{27b}
$$

$$
\tilde{\mathbf{d}}\tilde{\boldsymbol{\chi}} + \hat{\mathbf{d}}\hat{\boldsymbol{\chi}} = -\mathbf{SZ}'', \tag{27c}
$$

where

$$
\mathbf{M} = \mathbb{1} + (\mathbf{R}^{-1}\mathbf{K})^2 + (-\mathbf{R}^{-1})\mathbf{S}. \tag{27d}
$$

Since we are assuming that $\tilde{\mathbf{d}} = 0$ (see above), this means that Eq. (27d) simplifies to

$$
\hat{\mathbf{d}}\hat{\boldsymbol{\chi}} = -\mathbf{SZ}'', \tag{27c'}
$$

and the saturation matrix $\mathbf{S}$ simplifies to

$$\mathbf{S} = 4\hat{\mathbf{d}}(\hat{\mathbf{W}})^{-1}\hat{\mathbf{d}}^{\text{tr}} \tag{28}$$

(and recall that $\hat{\mathbf{W}}$ is nonsingular). It is then only necessary to invert the real symmetric matrices $\mathbf{S}$, $\mathbf{R}$, and then $\mathbf{M}$ to obtain the steady-state solutions. We must proceed differently for convenient solutions to the time-dependent case.

We now particularize the solutions to saturation recovery-type experiments, such that observations are made only for small d. [If we were to consider free-induction decay and spin-echo experiments, then we would be looking for solutions for $d = 0$ after the spins have been prepared by a pulse (90° and/or 180°) of short enough duration that spin relaxation is not yet operative.] We thus wish to develop for present purposes a perturbation scheme to lowest order in d. For this purpose the matrix of Eq. (18) is more satisfactory than that of Eq. (21), since it lacks the two degenerate submatrices ($\mathbf{R}$) along the partitioned diagonal that appear in Eq. (21). Note, however, that for $K_\lambda = 0$ (i.e., λth line is on resonance), if $-R_\lambda \equiv T_{2,\lambda}^{-1} = T_{1,\lambda}^{-1}$, then a triple degeneracy occurs in Eq. (21) with respect to the λth transition that is lifted by $d_\lambda \neq 0$. We must consider the case of $T_{2,\lambda} \neq T_{1,\lambda}$ separately from that for $T_{2,\lambda} = T_{1,\lambda}$. We develop the perturbation scheme by a generalization of the Van Vleck transformation procedure.[9] We first introduce the partitioned matrices

$$\mathbf{A} = \begin{bmatrix} \mathbf{R} - i\mathbf{K} & 0 & 0 \\ 0 & \mathbf{R} + i\mathbf{K} & 0 \\ 0 & 0 & -\hat{\mathbf{W}} \end{bmatrix}, \quad (29a) \qquad \mathbf{B} = i\sqrt{2}\begin{bmatrix} 0 & 0 & \hat{\mathbf{d}} \\ 0 & 0 & -\hat{\mathbf{d}} \\ \hat{\mathbf{d}}^{\text{tr}} & -\hat{\mathbf{d}}^{\text{tr}} & 0 \end{bmatrix}, \tag{29b}$$

where $\mathbf{L} = \mathbf{A} + \mathbf{B}$ and consider a vector $\mathbf{m}$ [cf. Eq. (22)] and solve for

$$\mathbf{OAO}^{-1}(\mathbf{Om}) + \mathbf{OB0}^{-1}(\mathbf{Om}) = (\mathbf{A} + \mathbf{b})\mathbf{m}', \tag{30}$$

where the partitioned matrix $\mathbf{B}$ is transformed approximately to be block diagonal [i.e., partitioned matrices along the diagonal as is $\mathbf{A}$ in eq. (29a)] by the (complex) orthogonal transformation $\mathbf{O}$ to lowest order in d. That is, we let

$$\mathbf{m}' = e^{i\mathbf{s}}\mathbf{m} \cong (\mathbb{1} + i\mathbf{s})\mathbf{m}, \tag{31}$$

where $\mathbf{s}$ is found to be the (complex) antisymmetric operator

$$\mathbf{s} = +i(\mathbf{A}^\infty)^{-1}\mathbf{B}, \tag{32}$$

and

$$\mathbf{b} = i\tfrac{1}{2}\mathbf{B}^\infty\mathbf{s} = -\tfrac{1}{2}\mathbf{B}^\infty[(\mathbf{A}^\infty)^{-1}\mathbf{B}]. \tag{33}$$

That is

$$
\mathbf{s} = \begin{bmatrix} \mathbf{0} & \mathbf{0} & -(\mathbf{A}^{o,d^{\infty}})^{-1}\sqrt{2}\hat{\mathbf{d}} \\ \mathbf{0} & \mathbf{0} & (\mathbf{A}^{c,d^{\infty}})^{-1}\sqrt{2}\hat{\mathbf{d}} \\ -(\mathbf{A}^{d,o^{\infty}})^{-1}\sqrt{2}\hat{\mathbf{d}}^{\text{tr}} & (\mathbf{A}^{d,c^{\infty}})^{-1}\sqrt{2}\hat{\mathbf{d}}^{\text{tr}} & \mathbf{0} \end{bmatrix} . \quad (34)
$$

[Our present use of the symbols **s** and **m** replaces **S** and **M** in reference 1 in order not to have these symbols confused with the saturation and **M** matrices; cf. Eq. (27).]

Here the inverse operator $(\mathbf{A}^{j,k^{\infty}})^{-1}$ for $j, k = o, c, d$ may be conveniently defined by the prescription

$$
(\mathbf{A}^{j,d^{\infty}})^{-1}\hat{\mathbf{d}} = \lim_{\varepsilon \to 0_{+}} -\int_{0}^{\infty} d\tau \exp\left[-\varepsilon\tau\right]\exp\left[\mathbf{A}^{j}\tau\right]\hat{\mathbf{d}}\exp\left[-\mathbf{A}^{d}\tau\right],
$$

and
$$
j = o \quad \text{or} \quad c \quad (35a)
$$

$$
(\mathbf{A}^{d,j^{\infty}})^{-1}\hat{\mathbf{d}}^{\text{tr}} = \lim_{\varepsilon \to 0_{+}} -\int_{0}^{\infty} d\tau \exp\left[-\varepsilon\tau\right]\exp\left[\mathbf{A}^{d}\tau\right]\hat{\mathbf{d}}^{\text{tr}}\exp\left[-\mathbf{A}^{j}\tau\right],
$$

$$
j = o \quad \text{or} \quad c \quad (35b)
$$

where

$$
\mathbf{A}^{o} = \mathbf{R} - i\mathbf{K}; \qquad \mathbf{A}^{c} = \mathbf{R} + i\mathbf{K} = \mathbf{A}^{o*}; \qquad \mathbf{A}^{d} = -\hat{\mathbf{W}}. \quad (36)
$$

(The convergence factor $\varepsilon > 0$ is always taken as large enough to guarantee vanishing of the integrand as $\tau \to \infty$, and the limit is taken only after preforming the integration.) Thus in a M-dimensional basis set $a, b, \ldots$ in which $\mathbf{A}^{o}$ (or $\mathbf{A}^{c}$) is diagonal and an $A/2$ dimensional basis set $\alpha, \beta \ldots$ in which $\mathbf{A}^{d}$ is diagonal, one has, for example

$$
[(\mathbf{A}^{o,d^{\infty}})^{-1}]_{a\alpha b\beta} = \frac{\delta_{ab}\delta_{\alpha\beta}}{A_{aa}^{o} - A_{\alpha\alpha}^{d}} = \frac{\delta_{ab}\delta_{\alpha\beta}}{(-R_{aa}) - \hat{W}_{\alpha\alpha} + iK_{aa}} . \quad (37)
$$

Thus the expansion is in terms of

$$
\frac{|\sqrt{2}\hat{d}_{a\alpha}|}{|(-R_{aa}) - \hat{W}_{\alpha\alpha} + iK_{aa}|} \ll 1, \quad (37b)
$$

for any nonvanishing $\hat{d}_{a\alpha}$, or more simply for a simple line:

$$
\frac{|(\sqrt{2}/2)\omega_{1}|}{|T_{2}^{-1} - T_{1}^{-1} + i\,\Delta\omega|} \ll 1. \quad (37c)
$$

One finds, utilizing the fact that $\mathbf{A}^{j}$ are symmetric matrices, that

$$
[(\mathbf{A}^{j,d^{\infty}})^{-1}\hat{\mathbf{d}}]^{\text{tr}} = -[(\mathbf{A}^{d,j^{\infty}})^{-1}\hat{\mathbf{d}}^{\text{tr}}], \qquad j = o, c \quad (37d)
$$

from which it follows that **s** is antisymmetric, as required. Also we have

$$
\mathbf{b} = \begin{bmatrix} \mathbf{C} + \mathbf{C}^{tr} & -(\mathbf{C}^{tr} + \mathbf{C}^*) & 0 \\ -(\mathbf{C} + \mathbf{C}^{tr*}) & \mathbf{C}^* + \mathbf{C}^{tr*} & 0 \\ 0 & 0 & \mathbf{E} + \mathbf{E}^{tr} \end{bmatrix},
\tag{38}
$$

where

$$
\mathbf{C} = \hat{\mathbf{d}}(A^{d,o^{\approx}})^{-1}\hat{\mathbf{d}}^{tr},
\tag{39a}
$$

and

$$
\mathbf{E} = 2\hat{\mathbf{d}}^{tr}\,\mathrm{Re}\,[(\mathbf{A}^{o,d^{\approx}})^{-1}]\hat{\mathbf{d}}.
\tag{39b}
$$

When the transformation of Eq. (18) is utilized, then in the basis of **Z′**, i**Z″**, $(1/\sqrt{2})$**X**⁻ one has

$$
\mathbf{s} = 2 \begin{bmatrix} 0 & 0 & -i[\mathrm{Im}(\mathbf{A}^{o,d^{\approx}})^{-1}]\hat{\mathbf{d}} \\ 0 & 0 & -[\mathrm{Re}\,(\mathbf{A}^{o,d^{\approx}})^{-1}]\hat{\mathbf{d}} \\ -i[\mathrm{Im}(\mathbf{A}^{d,o^{\approx}})^{-1}\hat{\mathbf{d}}^{tr} & -[\mathrm{Re}\,(\mathbf{A}^{d,o^{\approx}})^{-1}]\hat{\mathbf{d}}^{tr} & 0 \end{bmatrix},
\tag{40}
$$

whereas

$$
\mathbf{b} = 2 \begin{bmatrix} 0 & 2i\,\mathrm{Im}\,\mathbf{C}^{tr} & 0 \\ 2i\,\mathrm{Im}\,\mathbf{C} & 2\,\mathrm{Re}\,(\mathbf{C} + \mathbf{C}^{tr}) & 0 \\ 0 & 0 & \mathbf{E} + \mathbf{E}^{tr} \end{bmatrix}.
\tag{41}
$$

Thus one may solve either Eq. (18) or (21) in the approximations used as

$$
\Delta\dot{\mathbf{m}}'(t) \cong (\mathbf{A} + \mathbf{b})\,\Delta\mathbf{m}'(t)
\tag{42a}
$$

so

$$
\Delta\mathbf{m}'(t) \cong \exp\,[+(\mathbf{A} + \mathbf{b})t]\,\Delta\mathbf{m}'(0),
\tag{42b}
$$

and

$$
\Delta\mathbf{m}'(t) \cong (1 - i\mathbf{s})\,\exp\,[+(\mathbf{A} + \mathbf{b})t](1 + i\mathbf{s})\Delta\mathbf{m}(0).
\tag{42c}
$$

Note that $\mathbf{A} + \mathbf{b}$ given either by Eq. (29a) plus Eq. (38) [in the representation of Eq. (18)] or by $\mathbf{u}\mathbf{A}\mathbf{u}^{tr}$ plus Eq. (41) [in the representation of Eq. (21)] have the eigenstate-pair space (represented by superscript d) approximately uncoupled from the transition space (o and c superscripts) so $-\hat{\mathbf{W}} + (\mathbf{E} + \mathbf{E}^{tr})$ may be diagonalized separately. However,

$$
\begin{bmatrix} \mathbf{R} & -i(\mathbf{K} - 4\,\mathrm{Im}\,\mathbf{C}^{tr}) \\ -i(\mathbf{K} - 4\,\mathrm{Im}\,\mathbf{C}) & \mathbf{R} + 4\,\mathrm{Re}\,(\mathbf{C} + \mathbf{C}^{tr}) \end{bmatrix}
$$

will in general couple **Z′** to **Z″**. [Alternatively, the coupling can be written for **Z** and **Z″** from Eq. (29a) and Eq. (38).]

We note here that it is always possible to choose basis sets $a, b \cdots$ for transition space and $\alpha, \beta \cdots$ for eigenstate-pair space such that $\hat{\mathbf{d}}$ has a simple structure with $\hat{d}_{i,\hat{j}} = \hat{d}_{i,\hat{i}} \delta_{i,\hat{j}}$ where $\hat{j}$ refers to the eigenstate pair corresponding to the jth ESR transition. Several examples appear below.[24a] However, this choice does not, in general, simultaneously diagonalize $\mathbf{A}^o$ and $\mathbf{A}^d$. In those cases where it does, and if $\hat{d}_{i,i} = d$ independent of i, it then follows from the preceding definitions that $\mathbf{U}^d = \mathbf{U}^o$ (see below), $\mathbf{C}^{tr} = \mathbf{C}$ and $\mathbf{E}^{tr} = \mathbf{E}$. Also the mixing of the $\mathbf{Z}$ and $\mathbf{Z}^*$ components by the terms in $\mathbf{b}$ is in general not easily simplified. This mixing becomes important as the elements $K_{i,j} \to 0$ representing exact resonances.

2.1 Simple One-Line Case

We illustrate the preceding formalism for the simple one-line case, which is otherwise well known, in preparation for the more complex cases given below. In this case we have $\mathbf{C} = \mathbf{C}^{tr}$, $\mathbf{E} = \mathbf{E}^{tr}$ and

$$\mathbf{s} = \frac{+\omega_1}{(T_2^{-1} - T_1^{-1})^2 + \Delta\omega^2} \begin{bmatrix} 0 & 0 & -i\,\Delta\omega \\ 0 & 0 & (T_2^{-1} - T_1^{-1}) \\ i\,\Delta\omega & -(T_2^{-1} - T_1^{-1}) & 0 \end{bmatrix} \quad (43a)$$

and

$$\mathbf{b} = \frac{\omega_1^2}{(T_2^{-1} - T_1^{-1})^2 + \Delta\omega^2} \begin{bmatrix} 0 & -i\,\Delta\omega & 0 \\ -i\,\Delta\omega & (T_2^{-1} - T_1^{-1}) & 0 \\ 0 & 0 & -(T_2^{-1} - T_1^{-1}) \end{bmatrix}$$
$$(43b)$$

in the $\mathbf{Z}'$, $i\mathbf{Z}''$, $(1/\sqrt{2})\mathbf{X}^-$ representation. When we neglect terms of order $\omega_1^2/[(T_2^{-1} - T_1^{-1})^2 + \Delta\omega^2]$ compared to unity, one has

$$\mathbf{A} + \mathbf{b} \cong \begin{bmatrix} -T_2^{-1} & -i\,\Delta\omega & 0 \\ -i\,\Delta\omega & -T_2^{-1} + \delta & 0 \\ 0 & 0 & -T_1^{-1} - \delta \end{bmatrix}, \quad (44)$$

where

$$\delta = \frac{\omega_1^2 (T_2^{-1} - T_1^{-1})}{[(T_2^{-1} - T_1^{-1})^2 + \Delta\omega^2]}. \quad (44a)$$

The 2×2 submatrix may be diagonalized by the orthogonal transformation $\mathbf{U}$

$$\mathbf{U} = \begin{bmatrix} [1 - a_+]^{-1/2} & [1 - a_+^{-1}]^{-1/2} \\ [1 - a_-]^{-1/2} & -[1 - a_-^{-1}]^{-1/2} \end{bmatrix}, \quad (45a)$$

such that

$$U[A+b]U^{-1} = \begin{bmatrix} E_+ & 0 \\ 0 & E_- \end{bmatrix}, \tag{45b}$$

where

$$a_\pm = \frac{(\delta \pm \sqrt{\delta^2 - 4\,\Delta\omega^2})^2}{4\,\Delta\omega^2}, \tag{45c}$$

and

$$E_\pm = -T_2^{-1} + \delta\sqrt{2} \pm \tfrac{1}{2}\sqrt{\delta^2 - 4\,\Delta\omega^2}. \tag{45d}$$

For $|\delta^2/\Delta\omega^2| \ll 1$, that is, a line off resonance, one has complex eigenvalues of eq. (44) of $\lambda = -T_2^{-1} + \delta/2 \mp i\,\Delta\omega$ corresponding to the eigensolutions $(1/\sqrt{2})Z$ and $(1/\sqrt{2})Z^*$ [cf. Eq. (29a)]; whereas for $|\delta^2/\Delta\omega^2| \gg 1$, that is, a line close to resonance, one has simple decaying solutions $\lambda = -T_2^{-1}$ and $-T_2^{-1} + \delta$ for eigensolutions Z' and iZ'', respectively [cf. Eq. (44)]. It then follows from the preceding equations that the complete solution is

$$\begin{bmatrix} \Delta Z'(t) \\ \Delta Z''(t) \\ \frac{1}{\sqrt{2}}\Delta\chi^-(t) \end{bmatrix} = (1-is)U^{tr} \begin{bmatrix} e^{-E_+ t} & & \\ & e^{-E_- t} & \\ & & e^{-(T_1^{-1}+\delta)t} \end{bmatrix} U(1+is) \begin{bmatrix} \Delta Z'(0) \\ \Delta Z''(0) \\ \frac{1}{\sqrt{2}}\Delta\chi^-(0) \end{bmatrix},$$

$$\tag{46}$$

where only terms linear in s are kept. Some simple and well-known limiting cases are[10]

CASE 1 $\Delta\omega = 0$.

Then

$$\Delta Z''(t) = e^{-t(T_2^{-1}-\delta)}\Delta Z''(0) + \frac{\omega_1}{(T_2^{-1}-T_1^{-1})}(e^{-t[T_2^{-1}-\delta]} - e^{-t[T_1^{-1}+\delta]})\frac{1}{\sqrt{2}}\Delta\chi^-(0), \tag{47}$$

CASE 2 $T_2^{-1} \gg T_1^{-1}$.

Then for $t > T_2^{-1}$

$$\Delta Z''(t) = \frac{-\omega_1 T_2^{-1}}{T_2^{-2} + \Delta\omega^2} e^{-t[T_1^{-1}+\delta]} \left[\frac{1}{\sqrt{2}}\Delta\chi^-(0)\right]. \tag{48}$$

If we use conditions of partial saturation such that $M_z(0) = \alpha M_0$, $0 \leq \alpha \leq 1$, with M_0 the equilibrium magnetization, then

$$\tilde{M}_x(0) = \alpha \, \Delta\omega\omega_1 T_2^2 M_0$$
$$\tilde{M}_y(0) = \alpha\omega_1 T_2 M_0$$
$$M_z(0) = \alpha M_0, \tag{49}$$

and

$$\Delta Z'(0) = \Delta\tilde{M}_x(0) = \Delta\omega T_2 \, \Delta\tilde{M}_y(0) = \Delta\omega T_2^2 \omega_1 (\alpha - 1) M_0$$
$$\Delta Z''(0) = \Delta\tilde{M}_y(0) = (\alpha - 1)\omega_1 T_2 M_0$$

$$-\frac{1}{\sqrt{2}} \Delta\chi^- = \Delta M_z(0) = (\alpha - 1) M_0. \tag{50}$$

Then for Case 1 we have

$$\Delta Z''(t) = \frac{-(1-\alpha)\omega_1 T_2 M_0}{(T_2^{-1} - T_1^{-1})} \left[T_2^{-1} e^{-t(T_1^{-1}+\delta)} - T_1^{-1} e^{-t(T_1^{-1}-\delta)} \right], \tag{51}$$

whereas for Case 2 we have

$$\Delta Z''(t) \cong \frac{-(1-\alpha)\omega_1 T_2^{-1} M_0}{T_2^{-2} + \Delta\omega^2} e^{-t[T_1^{-1}+\delta]}. \tag{52}$$

[Note that to achieve a (partial) saturation condition it is necessary to apply a strong microwave field over a time $t > T_1$, so the spins can properly respond to the saturating field. This is of course different from the use of 90° and 180° pulses in free-induction decay and spin-echo experiments.[6]]

2.2 General Case for $T_2 \ll T_1$

The preceding formalism permits the solution of a variety of situations involving saturation recovery for which Eq. (46) is immediately generalized, and the general expression of Eq. (8) or (18) may be used for more general cases. We now, however, particularize our solutions to the case for $T_2 \ll T_1$ or, more generally, $|\mathbf{R}| \gg |\mathbf{W}|$. This is a useful case, especially in the slow-tumbling region, and also one for which some relatively simple analytic solutions may be obtained even for spectra that otherwise appear complex. In this case we have from Eq. (18), (21), (29), (40), and (41) that for

$$|\mathbf{R}|t > 1, \tag{53}$$

$$\Delta Z''(t) \cong [\mathrm{Re}\,(\mathbf{A}^{o,d^-})^{-1} 2\mathbf{d}] e^{-\hat{\mathbf{W}}t} \frac{1}{\sqrt{2}} \Delta\boldsymbol{\chi}^-(0). \tag{54}$$

In Eq. (54) we have dropped the small correction $\mathbf{E} + \mathbf{E}^{tr}$ of Eq. (44) to $\hat{\mathbf{W}}$. In the simple line case, this is just the neglect of δ of Eq. (44a) compared to T_1^{-1}, which is valid since for $T_2^{-1} > T_1^{-1}$:

$$\delta T_1 \cong \frac{\omega_1^2 T_1 T_2^*}{1 + T_2^{*2} \Delta\omega^2} \lesssim \omega_1^2 T_1 T_2^* \ll 1 \tag{55}$$

(where $T_2^{*-1} = T_2^{-1} - T_1^{-1}$). The last inequality is a consequence of the no-saturation condition during the recovery. Now if $\mathbf{U}_o$, $\mathbf{U}_c$, and $\mathbf{U}_d$ are the orthogonal transformations that diagonalize $\mathbf{A}^o$, $\mathbf{A}^c$, and $\mathbf{A}^d$, respectively, we may rewrite Eq. (54) as

$$\Delta\mathbf{Z}''(t) \cong -\int_0^\infty d\tau \{ \mathbf{U}_o^{tr} \exp [\tau \mathbf{U}_o (\mathbf{R} - i\mathbf{K}) \mathbf{U}_o^{tr}] \mathbf{U}_o$$
$$+ \mathbf{U}_c^{tr} \exp [\tau \mathbf{U}_c (\mathbf{R} + i\mathbf{K}) \mathbf{U}_c^{tr}] \mathbf{U}_c \} \hat{\mathbf{d}} \mathbf{U}_d^{tr}$$
$$\times \exp [\tau \mathbf{U}_d (+\hat{\mathbf{W}}) \mathbf{U}_d^{tr}] \exp [\mathbf{U}_d^{tr} (-\hat{\mathbf{W}} t) \mathbf{U}_d^{tr}] \mathbf{U}_d \frac{1}{\sqrt{2}} \Delta\boldsymbol{\chi}^-(0). \tag{56}$$

[The convergence factor has been dropped in Eq. (56) since $|\mathbf{R}| > |\mathbf{W}|$ implies satisfactory behavior of the integrals.] Note, however, by the functional properties $\mathbf{U}_o = \mathbf{U}_o\{\mathbf{R}, -i\mathbf{K}\}$ and $\mathbf{U}_c = \mathbf{U}_c\{\mathbf{R}, +i\mathbf{K}\}$ it follows that $\mathbf{U}_c = \mathbf{U}_o^*$. Then if we let

$$\mathbf{r} - i\mathbf{k} \equiv \mathbf{U}_o (\mathbf{R} - i\mathbf{K}) \mathbf{U}_o^{tr}, \tag{57a}$$

and

$$\mathbf{w} \equiv \mathbf{U}_d (\hat{\mathbf{W}}) \mathbf{U}_d^{tr}, \tag{57b}$$

Eq. (56) may be written more simply as

$$\Delta\mathbf{Z}''(t) \cong -\int_0^\infty d\tau \, \mathrm{Re} \, \{ \mathbf{U}_o^{tr} \exp [\tau (\mathbf{r} - i\mathbf{k}] \mathbf{U}_o \} (2\hat{\mathbf{d}}) \mathbf{U}_d^{tr}$$
$$\times \exp [(\tau - t)\mathbf{w}] \mathbf{U}_d \frac{1}{\sqrt{2}} \Delta\boldsymbol{\chi}^-(0). \tag{58}$$

We consider specific examples in the next section.

3　MOTIONAL NARROWING EXAMPLES

3.1　Well-Separated Hyperfine Lines (Nitroxide)

We first illustrate the application of our expression to a nitroxide (^{14}N) in the motional narrowing region when the three Lorentzian hyperfine lines

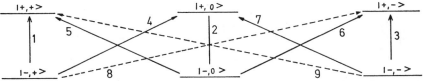

Figure 1 Energy levels and transitions for a nitroxide in high fields. Here $S = \frac{1}{2}$ and $I = 1$ and the notation is $|M_S,M_I\rangle$. The forbidden transitions are 4–9, and the allowed transitions are 1–3.

are well separated. If we consider just pure electron spin-flip transitions with rate W_e; pure nuclear spin-flip transition rates, which arise from the pseudosecular contributions of the electron-nuclear dipolar (END) interactions, and are, for a spin $I = 1$, given by $2W_n$ (where detailed expressions for W_n are given in reference 3), and spin exchange with rate ω_{HE}, we have

$$\mathbf{W}_{+,+} = \mathbf{W}_{-,-} = W_e \begin{bmatrix} 1+4b''+2b & -2b''-2b & -2b'' \\ -2b''-2b & 1+4b''+4b & -2b''-2b \\ -2b'' & -2b'' & 1+4b''+2b \end{bmatrix}, \quad (59a)$$

and

$$\mathbf{W}_{+,-} = \mathbf{W}_{-,+} = W_e \mathbb{1}, \quad (59b)$$

where we have introduced the dimensionless parameters $b \equiv W_n/W_e$ and $b'' = \omega_{HE}/AW_e$. Then from Eq. (19) we have

$$\hat{\mathbf{W}} = 2W_e \begin{bmatrix} 1+2b''+b & -b''-b & -b'' \\ -b''-b & 1+2b''+2b & -b''-b \\ -b'' & -b''-b & 1+2b''+b \end{bmatrix}. \quad (60)$$

Now we may write (by symmetry considerations, see below)

$$\mathbf{U}_d = \begin{bmatrix} \dfrac{1}{\sqrt{3}} & \dfrac{1}{\sqrt{3}} & \dfrac{1}{\sqrt{3}} \\ \dfrac{1}{\sqrt{2}} & 0 & \dfrac{-1}{\sqrt{2}} \\ \dfrac{1}{\sqrt{6}} & \dfrac{-2}{\sqrt{6}} & \dfrac{1}{\sqrt{6}} \end{bmatrix}, \quad (61a)$$

so that

$$\mathbf{w} = 2W_e \begin{bmatrix} 1 & 0 & 0 \\ 0 & 1+3b''+b & 0 \\ 0 & 0 & 1+3b''+3b \end{bmatrix}. \quad (61b)$$

We also have

$$-R_{i,j} = T_{2,i}'^{-1}\delta_{i,j} + (\hat{W}_{i,j} - W_e\,\delta_{i,j}), \quad (62)$$

where the $T'^{-1}_{2,i}$ give the purely secular contributions to the ith hyperfine line and the terms involving transition probabilities add up to the Heisenberg uncertainty in lifetime contributions to the linewidths (cf. reference 3). Also

$$K_{i,j} = \Delta\omega_i \, \delta_{i,j}, \tag{63}$$

and

$$-\hat{d}_{i,j} = \tfrac{1}{2}\omega_1 \, \delta_{i,j} = \tfrac{1}{2}\gamma_e B_1, \tag{63a}$$

where j refers to the eigenstate pair associated with the jth transition.

For the general spectrum of well-separated lines we have for $i \neq j$ (cf. 3):

$$|-R_{i,j}| = |\hat{W}_{i,j}| \ll |\omega_i - \omega_j|, \qquad i \neq j. \tag{64}$$

Thus $\mathbf{R} \pm i\mathbf{K} = \mathbf{r} \pm i\mathbf{k}$ is diagonal in the basis of the separate transitions, and $\mathbf{U}_o = \mathbb{1}$ [except for higher-order terms in $2W_e b''$ and $2W_e b$ vs. $(\omega_i - \omega_j)$]. Then elements of Eq. (58) are just

$$\Delta Z''(t) = -\sum_{j,\beta} \int_0^\infty d\tau \, \text{Re} \exp\left[\tau(r_{i,i} - ik_{i,i})\right]\omega_1 (U_d^{\text{tr}})_{i,\beta}$$

$$\times \exp\left[(\tau - t)w_{\beta\beta}\right](U_d)_{\beta\hat{j}} \frac{1}{\sqrt{2}} \Delta\chi_{\hat{j}}^-(0)$$

$$= +\omega_1 \, \text{Re} \sum_{j,\beta} \frac{e^{-w_{\beta\beta}t}}{r_{i,i} + w_{\beta\beta} - i\,\Delta\omega_i} (U_d^{\text{tr}})_{i\beta}(U_d)_{\beta\hat{j}} \frac{1}{\sqrt{2}} \Delta\chi_{\hat{j}}^-(0). \tag{65}$$

To complete the solution we must specify the initial condition

$$\frac{1}{\sqrt{2}} \chi_{\hat{j}}^-(0) = -(M_{z,\hat{j}}(0) - M_{0,\hat{j}}) = (1 - \alpha_{\hat{j}})M_{0,\hat{j}}, \tag{66}$$

or

$$\alpha_{\hat{j}} = \frac{M_{z,\hat{j}}(0)}{M_{0,\hat{j}}}. \tag{66a}$$

It is now convenient to consider two limiting cases depending on whether $b, b'' \ll 1$ or $\gg 1$.

CASE 1   Uncoupled relaxation: $b, b'' \ll 1$. For this case a saturating pulse on the jth transition leading to $\alpha_{\hat{j}} \neq 0$ will not appreciably affect the $i \neq j$ lines (except for terms higher order in b, b'', see below). Furthermore,

$$w_{\beta\beta} \cong 2W_e = T_1^{-1}, \quad \text{all } \beta. \tag{67}$$

Then, since $\sum_\beta (U_d^{\text{tr}})_{i\beta}(U_d)_{\beta\hat{j}} = \delta_{i,j}$, Eq. (65) becomes

$$\Delta Z_i''(t) \cong -\text{Re} \frac{\omega_1(1 - \alpha_{\hat{j}})M_{0,\hat{j}}}{T'^{-1}_{2i,j} + i\,\Delta\omega_i} e^{-t/T_1} \, \delta_{i,j}, \tag{68}$$

which is just Eq. (41) for each line.

CASE 2 Coupled relaxation $b', b'' \gg 1$. For this case a saturating pulse on the jth transition will have its effects transmitted equally to all the eigenstate pairs so that $\alpha_{\hat{i}} = \alpha_{\hat{j}} = \alpha = \frac{1}{3}\alpha_{\text{Total}} \neq 0$. Then since $\sum_j (U_d)_{\beta\hat{j}} = \sqrt{3}\,\delta_{\beta,1}$, Eq. (65) becomes

$$\Delta Z_i''(t) = -\text{Re } \omega_1 (1-\alpha) M_{0,\hat{i}} \frac{e^{-t/T_1}}{(-R_{i,i})^{-1} - T_1^{-1} + i\,\Delta\omega_i}, \tag{69}$$

for $i = 1, 2$, or 3 (corresponding to transitions with nuclear spin of $-1, 0$, and $+1$). Thus only one of the eigenvalues of $\mathbf{w}$ (i.e., $2W_e = T_1^{-1}$) is seen.

If it were possible to saturate one of the lines relative to the other two, then one could obtain a superposition of three decay terms each decaying by one of the eigenvalues of $\mathbf{w}$. Such would be the case if b and/or b'' is of order of magnitude unity. But then the three eigenvalues of $\mathbf{w}$ would not be much different, so that the superposition of three decay terms would not differ much from a single average exponential decay. A rigorous solution of this intermediate region would require a calculation from Eq. (18) of the values of $\boldsymbol{\chi}^-$ resulting from a pulse of finite duration $\Delta t'$. However, if $\Delta t' > w_{\beta\beta}^{-1} \leq T_1 = (2W_e)^{-1}$, then one may use as the ratios $\alpha_{\hat{j}}/\alpha_{\hat{i}} = \chi_{\hat{j}}^{-ss}/\chi_{\hat{i}}^{-ss}$ (i.e., the steady-state values obtained in the presence of the saturating field). Thus

CASE 3 The steady-state approximation on the pulse duration is

$$\hat{\boldsymbol{\chi}}^{\text{sat'd}} = -2\hat{\mathbf{W}}^{-1}\hat{\mathbf{d}}_{\text{sat'd}}^{\text{tr}}\mathbf{Z}''^{\text{sat'd}}, \tag{70}$$

with $\mathbf{Z}''^{\text{sat'd}}$ calculated by standard means (cf. reference 3). Then we can use Eq. (28) for the saturation matrix elements $S_{i,j}$ or the saturation parameters $\Omega_{\hat{i},\hat{j}}$ defined by

$$S_{i,j} = \hat{d}_{i,\hat{i}}\Omega_{\hat{i},\hat{j}}\hat{d}_{j,\hat{j}}^{\text{tr}}, \tag{71a}$$

that is,

$$4(\hat{\mathbf{W}}^{-1})_{\hat{i},\hat{j}} = \Omega_{\hat{i},\hat{j}},$$

to rewrite Eq. (65) as (with $\hat{d}_{i,\hat{i}\text{sat'd}} \equiv \frac{1}{2}\omega_1^s$). Thus

$$\Delta Z_i''(t) = \omega_1 \,\text{Re} \sum_{\substack{j,\beta,\\k}} \frac{e^{-w_{\beta\beta}t}}{r_{i,i} + w_{\beta\beta} - i\,\Delta\omega_i} (U_d)_{i\beta}^{\text{tr}}(U_d)_{\beta\hat{j}} \frac{\Omega_{\hat{j},\hat{k}}}{4} \frac{Z_k''^{\text{sat'd}}}{2} \omega_1^s$$

$$= \omega_1 \,\text{Re} \sum_{k,\beta} \frac{e^{-w_{\beta\beta}t}}{r_{i,i} + w_{\beta\beta} - i\,\Delta\omega_i} \frac{1}{w_{\beta\beta}} (U_d)_{i\beta}^{\text{tr}}(U_d)_{\beta k}^{\text{tr}} \frac{Z_k''^{\text{sat'd}}}{2} \omega_1^s, \tag{72}$$

where the second equality follows because

$$\mathbf{w}^{-1} = \mathbf{U}_d\hat{\mathbf{W}}^{-1}\mathbf{U}_d^{\text{tr}}. \tag{73}$$

Case 1 is obtained from the second form of Eq. (72) by setting only one $Z_k''^{\text{sat'd}}$ unequal to zero and then using Eq. (67). Case 2 is obtained from the first form of Eq. (72) by recognizing that for b and/or $b'' \gg 1$, $\Omega_{j,k}$ becomes independent of j and k, that is $\Omega_{j,k} \to 2/(A/2)W_e$. (See reference 3.) (These $\Omega_{j,k}$ are given explicitly for the nitroxide case in Table V of reference 11.) Then one may use $\sum_j (U_d)_{\beta \hat{j}} = \delta_{\beta,1}\sqrt{A/2}$. (Recall, however, that our original derivations of Cases 1 and 2 did not require the "steady-state pulse" approximation.) Cases intermediate between 1 and 2 exhibiting effects of all three decay constants are also obtained from Eq. (72). [Note that eq. (72) also covers ELDOR-type situations.] It follows from Eq. (72) that the exponential decays of larger $w_{\beta\beta}$ have the weaker amplitudes.

Now let us assume that the $w_{\beta\beta}$ are nearly equal, because $b, b'' \ll 1$. Then if the line observed is $i = 1$, while $k = 1$ has been saturated (simple saturation recovery), one obtains the following from Eq. (72):

$$\Delta Z_1''(t) \cong T_1 \omega_1 \, \text{Re} \, (r_{1,1} + T_1^{-1} - i \, \Delta\omega_1)^{-1} e^{-t/T_1}\left[1 - (2b'' + b)\left(\frac{1+t}{T_1}\right)\right] Z_1''^{\text{sat'd}}\omega_1^s.$$

(74)

However, if we let $k = 2$ (an ELDOR case),

$$\Delta Z_1''(t) = \frac{b+b''}{1+b+b''} e^{-t/T_1} T_1 \omega_1 \, \text{Re} \, (r_{1,1} + T_1^{-1} - i \, \Delta\omega_1)^{-1}\left(\frac{1+t}{T_1}\right) z_2''^{\text{sat'd}}\omega_1^s.$$

(75)

This emphasizes how the relaxation is dominated by T_1 and how an ELDOR effect would be weak (but potentially noticeable) compared to the direct saturation recovery effect for this case. When the $w_{\beta\beta}$ are very different (e.g., b and/or $b'' \gg 1$, such that $w_{22}, w_{33} > w_{11}$), other steady-state approximations appropriate to pulses of duration Δt fulfilling $w_{22}^{-1}, w_{33}^{-1} \ll \Delta t \ll w_{11}^{-1} = 2T_1$ may be used by solving for the steady-state solutions appropriate for $W_e \approx 0$ but W_n and/or $\omega_{HE} \neq 0$. This steady-state solution yields equal degrees of saturation of all the eigenstate pairs, and thus gives comparable results to that for Case 2. [Note, however, that for a steady-state approximation to apply here, $\Delta t \gtrsim 2(T_1^{-1} + T_2^{-1})^{-1}$ and $(\omega_1^s)^2 \gtrsim \frac{1}{4}(T_1^{-1} - T_2^{-1})^{2.6}$.]

3.2 Single Average Hyperfine Line (Nitroxide)

Here we assume the opposite of Eq. (64), that is,

$$|-R_{i,j}| = |\hat{W}_{i,j}| \gg |\omega_i - \omega_j|, \qquad i \neq j, \tag{76}$$

or

$$2W_e b \text{ and/or } 2W_e b'' \gg |\omega_i - \omega_j| \sim a_N, \tag{76a}$$

so the original three-line spectrum has collapsed into a single average Lorentzian. If we also assume $|T_{2,i}^{-1} - T_{2,j}^{-1}| \ll 2W_e b$ and/or $2W_e b''$ then $U_o \cong U_d$ of Eq. (60) and Eq. (58) becomes

$$\sum_{i=1}^{3} \Delta Z_i''(t) = - \sum_{i,\alpha,k,m,\beta} \int_0^\infty d\tau \, \text{Re} \, \{U_{oi,\alpha}^{tr} \exp[\tau(r_{\alpha,\alpha} - k_{\alpha,\alpha})] U_{o\alpha k}(2\hat{d})_{k,\hat{k}} U_{\hat{k},\beta}^{tr}\}$$

$$\times \exp[(\tau - t)w_{\beta\beta}](U_d)_{\beta\hat{m}} \frac{1}{\sqrt{2}} \Delta \chi_{\hat{m}}^-(0). \tag{77}$$

But since $\hat{d}_{k,\hat{k}}$ is independent of k and $\sum_k U_{o\alpha,k} U_{dk,\beta}^{tr} \cong \delta_{\alpha,\beta}$, whereas $\sum_i U_{oi,\alpha}^{tr} = \sum_i U_{o\alpha,i} = \sqrt{3} \, \delta_{\alpha,1}$ and $\chi_{\hat{m}}^-(0) = \alpha$ independent of m, one has

$$\sum_{i=1}^{3} \Delta Z_i''(t) \cong 3\omega_1 \, \text{Re} \frac{e^{-w_{11}t}}{r_{11} + w_{11} - i \, \Delta\omega_{11}} \alpha, \tag{78}$$

where $w_{11} = 2W_e$,

$$r_{11} = T_{2,av}'^{-1} + 2W_e, \quad \text{with} \quad T_{2,av}'^{-1} = \frac{1}{3} \sum_{i=1}^{3} T_{2,(i)}'^{-1},$$

and

$$\Delta\omega_{11} = \frac{1}{3} \sum_{i=1}^{3} \Delta\omega_i. \tag{79}$$

Corrections due to the incomplete averaging of effects of the b and b'' terms can be obtained by perturbations methods in the usual fashion. Again the relaxation is dominated by $T_1 = 1/2W_e$. [Note that Eq. (72) does not violate the validity of the perturbation approach as long as $T_2^{-1} > 2W_e$.]

3.3 General Case

The preceding discussion, given for the example of a nitroxide in the motional narrowing region, is seen to apply quite generally to the case of any hyperfine spectrum in the motional narrowing region. That is, Eq. (65) is still applicable in the well-resolved spectral region, as are the discussions and conclusions of cases (1), (2), and (3); Eq. (77) also applies in the limit of a single average hyperfine line. When there are degenerate hyperfine lines, it is only necessary to replace the vectors (e.g., $\mathbf{Z}, \mathbf{\chi}^-$) and matrices (e.g., $\mathbf{R}, \hat{\mathbf{W}}$) by their appropriate symmetrized forms as discussed in reference 3, which then properly include the degeneracy factors. [Care must be exercised in describing the (coupled) relaxation of the components of the degenerate line, but the methods previously described are applicable.]

Note that in the diagonalization of $\hat{\mathbf{W}}$ (and $\mathbf{R}$), one can take advantage of the symmetries of these matrices. Thus the feature of spin exchange for *nondegenerate* transitions—viz., that it leads to equal transition probabilities among *all* the eigenstate pairs (i.e., $-W_{i,j}^{ex} = 2W_e b''$, $\hat{i} \neq \hat{j}$)[3]— means that $\hat{\mathbf{W}}$ in the presence of exchange [but absence of electron-nuclear dipolar (END) terms] is invariant to all permutations of the $A/2$ nondegenerate eigenstate pairs (i.e., it commutes with the permutation group $P_{A/2}$). It is then a simple matter to show from the properties of this group that there is one eigenvalue

$$w_{11} = T_1^{-1} = 2W_e, \tag{80}$$

corresponding to $U_{i1} = 1/(\sqrt{A/2}$ (for all i) or the totally symmetric linear combination of eigenstate pairs. Furthermore, all the other eigenvalues are found to be degenerate [belonging to an $(A/2) - 1$ dimensional representation of $P_{A/2}$] and equal to

$$w_{ii} = 2W_e\left(1 + \frac{A}{2} b''\right), \qquad i \neq 1. \tag{81}$$

The END interaction shows less symmetry. However, for the eigenstate pairs of a single nucleus of I (or for the $J^{(\kappa)}$th set of eigenstate pairs corresponding to n completely equivalent nuclei with $J = \sum_{i=r} I_i$ and κ referring to a particular partner[3]) the $W_{i,j}^{END}$ are symmetric in the quantum number M. Thus the only symmetry operation involves $W_{M,M\pm1} \rightarrow W_{-M,-M\mp1}$ and $\chi_M^- \rightarrow \chi_{-M}^-$. However, one may also take advantage of the structure of W (see reference 3):

$$\hat{W}_{i,j} = 2W_e \, \delta_{i,j} + \hat{W}(END)_{i,j}, \tag{82}$$

so only $\hat{W}(END)_{i,j}$ the END contribution, needs to be diagonalized. Then since[3]

$$\hat{W}(END)_{ii} = -\sum_{j\neq i} \hat{W}(END)_{ij} = -\sum_{j\neq i} \hat{W}(END)_{ji}, \tag{83}$$

the matrix $\hat{\mathbf{W}}(END)$ must have a single eigenvalue of zero corresponding to the eigenvector $\sum_i \chi_i^-$ (by analogy with the equivalent property of symmetric $\mathbf{W}$ matrices corresponding to the conservation of probability). Thus one again has

$$w_{11} = T_1^{-1} + 2W_e,$$

corresponding to $U_{i1} = 1/\sqrt{A/2}$ with $w_{ii} = 2W_e[1 + f_i(b)] > 2W_e$ for $i \neq 1$ where the function $f_i(b)$ is of form seen in Eq. (61). (The preceding symmetry considerations are sufficient to determine the w_{ii} for the nitroxide.)

When both END and exchange are present, the lower symmetry of the END interaction is to be used. Also, if the hyperfine pattern is degenerate with different degeneracies for the different lines, then the $\hat{\mathbf{W}}^s$ matrix (the symmetrized form; see reference 3), in the presence of exchange only, one no longer has $P_{A/2}$ symmetry but usually symmetry like $\hat{\mathbf{W}}(\text{END})$, since, $D(\lambda)$ the degeneracy of the λth transition is symmetric about the center of the spectrum.

One can further generalize the problem to include a $\hat{\mathbf{W}}$ that depends on M (i.e., effects of the cross term between g- and dipolar tensors). This will, however, destroy the symmetries discussed. When cross transitions are not negligible, the $\mathbf{W}$ matrix is nonzero and one must return to Eq. (8) but perturbation methods comparable to those of Section 2 may still be employed.

3.4 On Contributions of T_2-Type Decays

We now wish to discuss the validity of the neglect of the terms appropriate when $|\mathbf{R}| \gg |\hat{\mathbf{W}}|$. [See Eq. (53) and (54).] Such an approximation is valid, for example, for dilute solutions of semiquinones where the secular g-tensor broadening dominates the widths, except at higher temperatures when spin rotation is most important and $T_1 \approx T_2$.[12] In the latter case, each hyperfine line is uncoupled from the others, and one treats each such line separately. This latter case is also the case for the nitroxides at low viscosity; at higher viscosity the secular g-tensor broadening is, however, not dominant, and $T_{2,i}^{-1} \sim W_n$ in Eq. (62), so $T_{2,i}^{-1}$ is only somewhat larger than W_n. However, $b = W_n/W_e$ is then usually substantially greater than unity. Thus, although $w_{11} = 2W_e \ll T_{2,i}^{-1}$ in this case, the w_{22} and w_{33} of Eq. (61) are of comparable order of magnitude to $T_{2,i}^{-1}$. We have already seen that for $b \gg 1$ we can neglect the effects of w_{22} and w_{33} in the saturation recovery, and for the same reason of rapid decay we can neglect terms decaying with time constant of order T_2. A similar argument applies when exchange makes a major contribution to the widths. However, in that region where W_e and W_n (or $\omega_{\text{EX}} \equiv Ab''W_e$) are of the same order, and $T_{2,i}^{-1}$ is not large, then the complicating effects of the decay of T_2-type terms from the complete solution of Eq. (42) might become important. Note further that in the well-resolved region, where only a single hyperfine line is observed, this solution may be achieved fairly simply utilizing the techniques given earlier, since $\mathbf{U}_o = \mathbb{1}$ (even though $\mathbf{U}_d$ is not so simple).

4 SLOW-TUMBLING EXAMPLES

4.1 Simple Line

By means of the eigenfunction expansion method of Freed, Bruno, and Polnaszek (FBP),[7] one obtains the following from Eq. (3):

$$\frac{1}{\sqrt{2}}\dot{C}_m(t) = \sum_n (R_{m,n} - iK_{m,n})\frac{1}{\sqrt{2}}C_n(t) + i\sqrt{2}\hat{d}_{m,\hat{m}}(\tfrac{1}{2}b_{\hat{m}}(t)) + i\frac{1}{\sqrt{2}}Q_m,$$

(84)

as well as the complex-conjugate form of Eq. (84), and

$$\tfrac{1}{2}\dot{b}_{\hat{m}}(t) = i\sqrt{2}\hat{d}_{m,n}^{\text{tr}}\left(\frac{1}{\sqrt{2}}C_m(t) - \frac{1}{\sqrt{2}}C_m^*(t)\right) - \sum_n \hat{W}_{\hat{m},\hat{n}}(\tfrac{1}{2}b_{\hat{n}}(t)).$$

(85)

Equations (84) and (85) are obtained by expanding the orientation-dependent matrix elements of $\sigma(\Omega, t)$ of Eq. (3) as

$$Z(\Omega, t)_{\lambda_j} = \sum_m C_m(t)\,|G_m(\Omega)\rangle,$$

(86a)

and

$$\hat{\chi}(\Omega, t)_{\lambda_j} = \sum b_{\hat{m}}(t)\,|G_{\hat{m}}(\Omega)\rangle,$$

(86b)

where $Z(\Omega, t)$ and $\chi(\Omega, t)$ are defined by analogy with Eq. (6) and (7) and the $G_m(\Omega)$ are eigenfunctions of the Markov operator Γ_Ω,

$$\Gamma_\Omega G_m(\Omega) = E_m G_m(\Omega).$$

(87)

After the expansions of Eq. (86a) are performed in Eq. (5) one multiplies through by $\langle G_{m'}(\Omega)\,|$ and uses the orthonormality property of the $G_m(\Omega)$ to obtain Eq. (84) and (85). (In this section we drop the λ_j subscript, since only a simple line is being considered. We also let $\hat{m} \to m$.) These eigenfunctions $G_n(\Omega)$ may be written for Brownian rotation in isotropic liquids as the normalized Wigner rotation matrices

$$G_n(\Omega) \to G_{KM}^L(\Omega) = \sqrt{\frac{(2L+1)}{8\pi^2}}\,\mathscr{D}_{KM}^L(\Omega),$$

(87a)

with eigenvalues E_n for isotropic motion

$$E_n \to E_{L,K,M} = L(L+1)\mathscr{R},$$

(87b)

where $\mathscr{R}$ is the rotational diffusion coefficient.[2a,5,7] For models involving reorientation by appreciable jumps, it is found that the functions of Eq. (87a) are still good eigenfunctions, and Eq. (87b) becomes

$$E_n \to E_{K,M}^L = B_L(L+1)\mathscr{R},$$

(87c)

where the model parameter B_L ranges from unity for Brownian motion to $B_L = [L(L+1)]^{-1}$, $L \neq 0$ (and $B_L = 1$ for $L = 0$) for a strong collision model. It is discussed in detail elsewhere.[5,8,13-15] (We note that there is a simple analogue between Brownian rotational diffusion with an END mechanism, on the one hand, and strong jump diffusion with a Heisenberg exchange mechanism, on the other hand. The former pair have significant "selection rules", the latter have none.)

Note that the probability function of Eq. (5a) is itself representable by the eigenfunction expansion[5,9]

$$P(\Omega, t) = \sum_{L,K,M} a_{KM}^L(t) G_{KM}^L(\Omega). \tag{88a}$$

In particular, the conditional probability distribution $P(\Omega_0; \Omega, t)$ defined as the probability density of finding Ω at a particular value at time t provided it had the value Ω_0 *at time* $t = 0$, obeys

$$P(\Omega_0; \Omega, t) = \sum_{L,K,M} |G_{KM}^L(\Omega)\rangle \, e^{-E_{K,M}^L t} \, \langle G_{KM}^L(\Omega_0)|, \tag{88b}$$

where we again use bracket notation. In addition, for an isotropic liquid we have [cf. Eq. (5b)]

$$P_{eq}(\Omega) = \frac{1}{8\pi^2} \mathscr{D}_{0,0}^0(\Omega) = \frac{1}{8\pi^2}. \tag{88c}$$

Now one observes an average over the orientations according to the prescription of Eq. (4). It then follows that the absorption is given by

$$Z''(t) = \langle P_{eq}(\Omega)| Z(\Omega, t) |P_{eq}(\Omega)\rangle \propto C_{0,0}^0(t), \tag{89}$$

and the other coefficients $C_{KM}^L(t)$ and $b_{KM}^L(t)$ are coupled into the problem by Eq. (84) and (85).

In particular, if we assume the orientation-dependent perturbation in Eq. (3) is an axially symmetric g-tensor, one finds only the $C_{0,0}^L(t)$ and the $b_{0,0}^L(t)$ for L even affect the observed signals [cf. Eq. (91b)]. For this case the terms in Eq. (84) and (85) are[7]

$$R_{L,L'} = r_{L,L} \, \delta_{L,L'} = -(T_2^{-1} + E_L) \, \delta_{L,L'} \tag{90}$$

$$K_{L,L'} = \Delta\omega + \kappa_{L,L'}, \tag{91a}$$

with

$$\kappa_{L,L'} = [(2L+1)(2L'+1)]^{1/2} \begin{pmatrix} L & 2 & L' \\ 0 & 0 & 0 \end{pmatrix}^2 \mathscr{F}, \tag{91b}$$

and

$$\mathscr{F} = \tfrac{2}{3} \hbar^{-1} \beta_e B_0 (g_\parallel - g_\perp). \tag{91c}$$

These expressions include only the secular contribution of the axially symmetric g-tensor with parallel and perpendicular components $g_\parallel$ and $g_\perp$, respectively. Also β_e is the Bohr magneton and B_0 is the dc magnetic-field strength, and $\begin{pmatrix} L & 2 & L' \\ 0 & 0 & 0 \end{pmatrix}$ is a 3-j symbol for which $L' = L$ or $L \pm 2$.[16] [The nonsecular contributions have been omitted in Eq. (84) and (85) (see FBP, Section IIIB1).] Also

$$\hat{W}_{L,L'} = w_{L,L} \, \delta_{L,L'} = (2W_e + E_L) \, \delta_{L,L'}, \tag{92}$$

and

$$Q_L = q\omega_\lambda \, d_\lambda \, \delta_{L,0}. \tag{93}$$

We have introduced an orientation-independent width T_2^{-1} and $T_1^{-1} = 2W_e$ into Eq. (90) and (92), respectively. Equation (84), its complex conjugate, and Eq. (85) are seen to be of the same matrix form as Eq. (18) (with the matrices $\hat{\mathbf{d}}$ and $\hat{\mathbf{d}}^{tr}$ which by Eq. (84) and (85) only couple $C_{0,0}^L$ with $b_{0,0}^L$). So provided the inequality of Eq. (36) for the present case applies; then the same perturbation treatment in d_λ utilized for solving Eq. (18) may be utilized for the present case.

We wish to point out at this stage that the eigenfunction expansion method immediately yields $R_{m,n}$ and $W_{m,n}$ in *diagonal* form. Thus when $E_m/\mathscr{F} \gg 1$ corresponding to motional narrowing the $\mathbf{R} \pm i\mathbf{K}$ is approximately diagonal in this representation. However, κ, which arises from $\mathscr{H}_1(\Omega)$, is diagonal in the space of orientational unit vectors $|\delta(\Omega - \Omega_0)\rangle$.[2a] We note that from the representation of the δ-function:

$$|\delta(\Omega - \Omega_0)\rangle = \sum_n G_n^*(\Omega_0) \, |G_n(\Omega)\rangle, \tag{94a}$$

where, here, $|G_n(\Omega)\rangle$ are any complete orthonormal set of functions, one has

$$|G_n(\Omega)\rangle = \int d\Omega_0 \, G_n(\Omega_0) \, |\delta(\Omega - \Omega_0)\rangle, \tag{94b}$$

and if they are also eigenfunctions of Γ_Ω, then

$$\langle \delta(\Omega - \Omega_1)| \, \Gamma_\Omega \, |\delta(\Omega - \Omega_0)\rangle = \sum_n G_n^*(\Omega_0) G_n(\Omega_1) E_n. \tag{95}$$

[Equation (95) also follows from evaluating $\partial P(\Omega_0, \Omega, t)/\partial t$ at $t = 0$, since this is the same as Eq. (94a).]

Equations (94a) and (94b) define the unitary transformation

$$U_{n,\Omega_o} = G_n(\Omega_0) \tag{96a}$$

$$(U^{-1})_{\Omega_0,n} = U_{n,\Omega_o}^* = G_n^*(\Omega_0), \tag{96b}$$

between the two sets of basis vectors. It is often the case, however, that the real linear combinations of the $G_n(\Omega)$ can be used so $\mathbf{U}$ becomes an orthogonal transformation.

When $E_L/\mathscr{F} \ll 1$, corresponding to the very slowly tumbling region, then $\mathbf{R} \pm i\mathbf{K}$ is approximately diagonal in the $|\delta(\Omega - \Omega_0)\rangle$ representation, with

$$K_{\Omega_i,\Omega_i} \equiv k(\Omega_j) = \Delta\omega_\lambda - \omega'(\Omega_j), \tag{97a}$$

where

$$\omega'(\Omega_j) = \mathscr{F}\mathscr{D}_{0,0}^2(\Omega_j) = \mathscr{F}P_2(\beta_j), \tag{97b}$$

with $P_2(\beta)$ the second-rank Legendre polynomial. (Of course, actual calculations are performed using finite grid points on the unit sphere.)

Again the solution may be written in the form of Eq. (58), with an equation like Eq. (65) appropriate when $E_L/\mathscr{F} \lll 1$, except that $\mathbf{U}_o \cong (\mathbf{U}^{-1})$ are defined by Eq. (96b), whereas $\mathbf{U}_d = \mathbb{1}$, (since the initial basis sets are the eigenfunctions of Γ and not the individual orientational component; whereas in the motional-narrowing case of Section 3 the individual hyperfine components are utilized). When $E_L/\mathscr{F} < 1$, defining the slow-tumbling region where the spectrum is intermediate between the motional narrowing and rigid limit ones, the matrix $\mathbf{U}_o$ may be obtained by diagonalizing $(\mathbf{R} - i\mathbf{K})$ following methods already well described[5] while $\mathbf{U}_d = \mathbb{1}$. Thus we may write from Eq. (58)

$$\Delta C_0''(t, \omega) = \omega_1 \operatorname{Re} \sum_{L,j} \frac{e^{-w_{LL}t}}{(R - iK)_{jj} + w_{LL}} (U_0^{tr})_{0,j}(U_0)_{j,L} \tfrac{1}{2}\Delta b_L(0), \tag{98}$$

and when $E_L \ll \mathscr{F}$, this may be rewritten as

$$\Omega C_0''(t, \omega) \cong \frac{\omega_1 \operatorname{Re}}{\sqrt{8\pi^2}} \sum_L \int d\Omega_j \frac{e^{-w_{LL}t}}{r(\Omega_j) - ik(\Omega_j) + w_{LL}} G_L(\Omega_j) \tfrac{1}{2}\Delta b_L(0). \tag{99}$$

In both cases it follows from Eq. (91b) that only even values of L are required.

Note that in Eq. (99) it is never really necessary to take an infinite sum over L. This is because in the integration over Ω_i, the $G_L(\Omega_i)$ for large L have rapid oscillations compared to the rest of the integrand, so that they average to zero, and for large enough L, one usually has $W_{LL} \gg W_{00}$. That is, we do not need values of L so large that $G_L(\Omega_i)$ varies much faster in Ω_i than $[r(\Omega_i) - ik(\Omega_i) + w_{LL}]^{-1}$. The effect of a large T_2^{-1} in $r(\Omega_i)$ is to broaden out the features of the near-rigid spectrum, thus decreasing the maximum value of L required. The approximate equality of Eq. (99) reflects the fact that we have taken $\mathbf{R} - i\mathbf{K}$ as diagonal in the $|\delta(\Omega - \Omega_0)\rangle$ representation, with the dominant part of $f(\Omega_i)$ being $-T_2^{-1}$(with any

small residual motional broadening calculated using the correct representation that diagonalizes $\mathbf{R} - i\mathbf{K}$, which for practical purposes, involves
difference methods).

We can also rewrite Eq. (99) in a form more closely resembling Eq.
(65):

$$\Delta Z''(\Omega_i, t, \omega) = \omega_1 \operatorname{Re} \sum_{L \text{ even}} \frac{e^{-w_{LL}t}}{r_{ii} - ik_{i,i} + w_{LL}} \int d\Omega_j G_L(\Omega_i) G_L^*(\Omega_j) \tfrac{1}{2} \Delta \hat\chi(\Omega_j, 0).$$

(100)

First suppose that $E_L \ll W_e$, so that $w_{LL} \cong 2W_e$ for all values of L that
contribute appreciably to the sum [since as already noted the sum may be
truncated, but also the $b_L(0)$ may be negligible for large L], then Eq.
(100) becomes

$$\Delta Z''(\Omega_i, t) \approx \omega_1 e^{-t/T_1} \frac{1}{r_{i,i} - ik_{i,i} + 2W_e} \tfrac{1}{2} \Delta \hat\chi(\Omega_j, 0),$$

(101)

representing the fact that the spin packet at Ω_i is uncoupled to the other
orientations [cf. Eq. (60). Then one may use the general relation

$$\Delta C_0''(t, \omega) = \frac{1}{\sqrt{8\pi^2}} \int d\Omega_i \, \Delta Z''(\Omega_i, t, \omega)$$

(102)

to calculate $\Delta C_0''(t, \omega)$, which is observed in an experiment. Now suppose
that $E_L \gg W_e$ for $L > 0$ such that $w_{LL} \gg 2W_e$ for $L > 0$. Then a saturating
pulse will have its effects transmitted by the rotational diffusion equally to
all parts of the line; that is, only $b_0(0)$ is normally saturated, so only
$\tfrac{1}{2}\Delta b_0 \neq 0$, and Eq. (99) becomes [with an equivalent form for the more
general equation (98)]

$$\Delta C_0''(t, \omega) = \frac{\omega_1}{8\pi^2} e^{-t/T_1} \operatorname{Re} \left[\int d\Omega_i \frac{1}{r(\Omega_i) - ik(\Omega_i, \omega) + 2W_e} \right] \tfrac{1}{2} \Delta b_0(0),$$

(103)

again giving relaxation with a simple $T_1 = \tfrac{1}{2}W_e$ [cf. Eq. (69)].

We can, again, introduce the "steady-state approximation on the pulse
duration" and the analogue of Eq. (70) becomes

$$\tfrac{1}{2}\hat\chi(\Omega_i)^{\text{sat'd}} = -\int d\Omega_j \hat W_{\Omega_i \Omega_i}^{-1} \hat d_{\Omega_i, \Omega_i}^{\text{tr sat'd}} Z''(\Omega_j)^{\text{sat'd}},$$

(104)

where

$$\hat W_{\Omega_i \Omega_i}^{-1} = \sum_m G_m(\Omega_i) w_{mm}^{-1} G_m^*(\Omega_j) = \sum_{L,K,M} \left(\frac{2L+1}{8\pi^2} \right) \mathscr{D}_{KM}^L(\Omega_i) w_{KM}^L \mathscr{D}_{KM}^{L*}(\Omega_j),$$

(104a)

and for an orientation-independent transition moment we may write $d_{\Omega_i,\Omega_i}^{\text{tr sat'd}} = -\frac{1}{2}\omega_1^s\,\delta_{i,j}$. Although, in principle, the sum in Eq. (104a) includes a complete sum over the orthonormal set, the nature of $Z''(\Omega_j)$ for the present case, as determined by Eq. (91), again means that only-the restricted sum of L even and $K = M = 0$ need be used in Eq. (104). An alternative form of Eq. (104) is

$$b_L^{\text{sat'd}} = \omega_1^s w_{LL}^{-1} C_L''^{(\text{sat'd})}. \tag{104b}$$

In general, the $b_L^{\text{sat'd}}$ will be nonnegligible only for those L such that the $C''^{(\text{sat'd})}$ are strongly coupled into the problem by the term in $\mathcal{F}$ of Eq. (97b) *and* for which $4(d^{(\text{sat'd})})^2 w_{LL}^{-1} \gtrsim (-R_{L,L})$, that is, the $C''^{(\text{sat'd})}$ are indeed being saturated. It is usually the latter condition that is limiting, since one has

$$\frac{T_2^{-1}}{E_L},\frac{W_e}{E_L} \ll \frac{|\mathcal{F}|}{E_L},$$

and usually

$$(\omega_1^s)^2 \gtrsim (2W_e)^{-1}T_2$$

(but not very much greater). Note also that the $C^{\text{sat'd}}$ are obtained from Eq. (84)–(91) once $\dot{C}_L(t)$ and $\dot{b}_L(t)$ are set equal to zero.

We now obtain from Eq. (98):

$$\Delta C_0''(t) = +\omega_1 \operatorname{Re}\sum_{L,j} \frac{e^{-w_{LL}t}}{(R - iK)_{j,j} + w_{LL}}\left(\frac{1}{w_{LL}}\right)(U_o^{\text{tr}})_{0,j}(U_o)_{j,m}C_L''^{(\text{sat'd})}\tfrac{1}{2}\omega_1^s, \tag{105}$$

with the obvious modification when Eq. (99) is appropriate. Equation (100) may be rewritten as

$$\Delta Z''(\Omega_i, \omega, t) = \omega_1 \operatorname{Re}\sum_{\text{even } L}\frac{e^{-w_{LL}t}}{r(\Omega_i) - ik(\Omega_i) + w_{L,L}}$$
$$\times \int \frac{G_L(\Omega_i)G_L^*(\Omega_j)}{w_{L,L}}Z''^{\text{sat'd}}(\Omega_j)\frac{\omega_1^s}{2}\,d\Omega_j. \tag{106}$$

where it is again clear that only $w_{0,0}$ and those $w_{L,L}$ comparable to $w_{0,0}$ would contribute substantially. And the saturation recovery spectrum given by $\Delta C''(\omega, t)$ is obtained by integrating Eq. (106) over Ω_i and utilizing $\langle G_L(\Omega_j)\,|\,Z''(\Omega_j)\rangle = C_L$. The case of $w_{L,L}$ comparable to $w_{0,0}$ for all L contributing appreciably to Eq. (106) may be dealt with in the manner of Eq. (72) to yield

$$\Delta Z''(\Omega_i, \omega, t) = \omega_1 \operatorname{Re}\frac{e^{-w_{0,0}t}}{r(\Omega_i) - ik(\Omega_i) + w_{0,0}}$$
$$\times \int d\Omega_j[\delta(\Omega_i - \Omega_j) - \overline{\gamma(\Omega_i, \Omega_j)}(1 + w_{0,0}t)]w_{0,0}^{-1}Z''(\Omega_j)^{\text{sat'd}}(\tfrac{1}{2}\omega_1^s), \tag{107}$$

where

$$\overline{\gamma(\Omega_i, \Omega_j)} = \sideset{}{'}\sum_{L \neq 0} \gamma_L G_L^*(\Omega_i) G_L(\Omega_j), \tag{107a}$$

and γ_L is defined by $w_{LL} = w_{00}\gamma_L$ with $\gamma_L \ll 1$ for all values of L contributing appreciably to Eq. (106). [The prime on Eq. (107a) indicates it may be calculated over this restricted set of L values. Note also that $\sum_L' G_L(\Omega_i) G_L^*(\Omega_j) \approx \delta(\Omega_i - \Omega_j)$ compared to the much slower variation of $Z''(\Omega_j)$ with Ω_j. For Brownian diffusion $\gamma_L = L(L+1)\mathcal{R}/2W_e$; whereas for the limit of strong jumps

$$\overline{\gamma(\Omega_i, \Omega_j)} = \frac{[\delta(\Omega_i - \Omega_j) - 1/8\pi^2]\mathcal{R}}{2W_e}. \tag{108}$$

Equation (107) again emphasizes how the dominant relaxation is via $T_1 = 1/2W_e$ and it shows how one may calculate the magnitude of the weak recovery signal for an ELDOR experiment in this case when Ω_i and Ω_j are quite different. Equation (108) substituted into Eq. (107) gives the simple orientation-independent result expected for a strong-collision model (for $\Omega_i \neq \Omega_j$). For this model, the more general expression, Eq. (106) takes on the simpler form

$$\Delta Z''(\Omega_i, \omega, t) = \omega_1 \operatorname{Re} \frac{e^{-w_{0,0}t}}{r(\Omega_i) - ik(\Omega_i) + w_{0,0}} \left(\frac{C''^{(\text{sat'd})}}{w_{0,0}\sqrt{8\pi^2}} \right)$$

$$+ \frac{e^{-w't}}{r(\Omega_i) - ik(\Omega_i) + w'} \left(\frac{1}{w'} \right) \left[Z''(\Omega_i)^{\text{sat'd}} - \frac{C_0''^{(\text{sat'd})}}{\sqrt{8\pi^2}} \right] \tfrac{1}{2}\omega_1^s, \tag{109}$$

where $w' = w_{0,0} + \mathcal{R}$.

One may speculate, for slow tumbling, whether other initial conditions may be created before observing the recovery. In particular, one could try initially to saturate a single orientation. In the case of $E_L \ll W_e$ (i.e., $w_{LL} \cong 2W_e$ for all L) this is essentially the same as Eq. (109). The opposite limit of $w_{LL} \gg w_{0,0}$ for $L > 0$, a normal saturating pulse is transmitted to all parts of the line. However, it may be possible by such techniques as the use of a relatively weak 180° pulse (i.e., $\omega_1 \tau = \pi$ where τ is the pulse duration, such that $\tau \ll r_{LL}^{-1}, w_{LL}^{-1}$ for values of L contribute to the spectrum, and $r_{L,L}, w_{L,L} \ll \omega_1 = \pi/\tau < \kappa_{L,L}$) initially to saturate a small range of orientations, and still allow for $w_{L,L} > 2W_e$ for $L > 0$ (but not $w_{L,L} \gg 2W_e$). This situation could then show interesting effects from several decay constants: $w_{L,L}$.

4.2 Other Aspects of a Simple Line

4.2.1 ORIENTATION-DEPENDENT T_1 AND T_2

Near the rigid limit it is possible that orientation-dependent effects of T_1 and T_2 begin to show up. We can examine such effects by introducing terms: $T_{1,\Omega}^{-1}\mathscr{D}_{0,0}^2(\Omega)$ and $T_{2,\Omega}^{-1}\mathscr{D}_{0,0}^2(\Omega)$. Then

$$R_{L,L'} = -(T_2^{-1} + E_L)\,\delta_{L,L'} - T_{2,\Omega}^{-1}[(2L+1)(2L'+1)]^{1/2}\begin{pmatrix} L & 2 & L' \\ 0 & 0 & 0 \end{pmatrix}^2,$$

and
$$\tag{110}$$

$$W_{L,L'} = (T_1^{-L} + E_L)\,\delta_{L,L'} + T_{1,\Omega}^{-1}[(2L+1)(2L'+1)]^{1/2}\begin{pmatrix} L & 2 & L' \\ 0 & 0 & 0 \end{pmatrix}^2.$$

$$\tag{111}$$

When Eq. (110) is compared with Eq. (90) and (91), it is seen that its only effect on the previous results is to change $\mathscr{F} \to \mathscr{F} - iT_{2,\Omega}^{-1}$, but Eq. (111) renders $\hat{W}_{L,L'}$ nondiagonal. When $T_{1,\Omega}E_L \gg 1$ (or more precisely $T_{1,\Omega}E_2 \gg 1$) then these orientation-dependent effects may be neglected, but for very slow motions it would be necessary to diagonalize $\hat{W}_{L,L}$ [where, in the limit $E_L \to 0$, one would obtain the $|\delta(\Omega - \Omega_i)\rangle$ representation].

4.2.2 ASYMMETRIC g-TENSOR

The correct expressions may be obtained for this case by direct comparison of the preceding expressions with the steady-state case given by FBP. The main feature to note is that the $\mathscr{D}_{K,0}^L(\Omega)$ for even L and nonzero K appear in the problem, so effects of anisotropic rotational diffusion can appear.[5] Otherwise the discussion is analogous to that given for symmetric g-tensors.

4.2.3 CONTRIBUTION FROM T_2-TYPE DECAYING TERMS

In general, one finds that T_2^{-1} is significantly larger than $2W_e$ in the slow-motional region, so the T_2-type decaying terms should decay much faster. However, it is possible for $E_L = \mathscr{R}L(L+1)$ to play a dominant role for large L in Eq. (90) and Eq. (92), that is, $\mathscr{R}L(1+1) \gg T_1^{-1}$ and T_2^{-2}. But this is the case where these terms of large L decay too rapidly in $e^{-\hat{W}t}$ to be important compared to the $w_{0,0}$ case, and similar comments would apply to the T_2-type decaying terms. Again one can return to the complete Eq. (42) for a detailed examination of such effects. We note that in the slow-tumbling region, the $K_{L,L'}$ of Eq. (91a) will result in contributions from slightly off-resonant components of the line, and their T_2-type decay (but not T_1-type decays) will have some oscillatory character [cf. Eq. (44) and (45)]. Our preceding analysis can be further refined by distinguishing between that portion of the line broadening which is homogeneous, and that which is inhomogeneous (assumed Lorentzian). This is unimportant for unsaturated effects but is important when considering saturation.[5,6,14]

4.3 Complex Spectra: Nitroxides

Very often a slow-tumbling spectrum is not just describable as a simple line but is, rather, a complex one involving the coupling of the different transitions. Methods for solving the steady-state spectra in such cases are given in detail elsewhere.[5,7-9,13,14] However, the partitioned-matrix concept of Eq. (18) may again be applied in a manner analogous to the simple line case treated in the previous section. The important generalizations are just to regard each of the coefficients $C_{K,M}^L(t)$ and $b_{K,M}^L(t)$ as vectors in spin space,[17] such that $C_{K,M}^L(t, i)$ refers to the component representing the ith ESR transition. For the particular case of nitroxides, one need only consider the three allowed transitions ($i = 1$, 2, or 3) and three linear combinations of the six forbidden transitions (cf. Fig. 1).

These forbidden transitions are coupled into the (high-field motional narrowing) allowed transitions by the pseudosecular terms in $\mathscr{H}_1(\Omega)$ that induce nuclear spin flips, and as the motion slows, they also affect the actual resonance frequencies. For very slow motions, one achieves the rigid-limit resonance frequencies, which may, to a good approximation, be described by three allowed transitions for each orientation.

Similarly, there are six components $b_{KM}^L(t, i)$ representing population differences between pairs of eigenstates, that is, three for the allowed transitions ($i = 1$, 2, or 3) and three that are really (mixed) nuclear magnetic resonance (NMR) transitions. These latter arise from the pseudosecular terms in $\mathscr{H}_1(\Omega)$, and thus play a role closely analogous to the three (mixed) forbidden transitions for the $C_{KM}^L(i)$. As the rigid limit is approached (in particular, for $|\mathscr{H}_1(\Omega)|/\mathscr{R} \gg 1$) their inclusion becomes equivalent to representing the diagonalized eigenstates characteristic of the rigid limit.

One may thus generalize all our preceding procedures to such cases wherein the vector spaces of Eq. (18) include the product space of the $C_{KM}^L(t)$ for the different L, K, M (or alternatively, the $|\delta(\Omega - \Omega_i)\rangle$ representation) with the appropriate spin space as just described. In the slow-tumbling region where

$$|\mathscr{H}_1(\Omega)|\,\mathscr{R} > 1,$$

that is, the unsaturated slow-tumbling spectra still show important motional effects, the detailed diagonalizations required for Eq. (58), for example, are complex although tractable.[5,8,21] However, one usually (but not always) has the condition

$$2W_e \ll \mathcal{R},$$

and any saturation effects are transmitted throughout the spectrum. In this event we again have a case where the $\hat{\mathbf{W}}$ matrix is characterized by a $w_{0,0} = (2W_e)$ and $w_{L,L} \gg w_{0,0}$ for $L \neq 0$, and the dominant (slow) decay will again give just $T_1^{-1} = 2W_e$.

A more careful analysis of the slow-tumbling region shows that two types of saturation transmission effects are operative: (1) the motional effect, which contributes terms of type $B_L L(L+1)\mathcal{R}$ to the $w_{L,L}$, and (2) nuclear spin-flip processes, which in the fast-motional case depend on b. (See Section 3.) For $|\mathcal{H}_1(\Omega)|/\mathcal{R} \lesssim 1$ (i.e., $\tau_R \lesssim 10^{-9}$ sec) one gets values of $b \sim 10$–40 representing strong coupling of the hyperfine lines.[14] But when $|\mathcal{H}_1(\Omega)|/\mathcal{R} \gg 1$, and only residual motional effects are important, one must examine their effects more carefully.

In Section 1 a hypothetical case called the *quasi-nitroxide* case, such that $\sqrt{\frac{3}{10}} |A_\parallel - a_N| \ll |a_N|$ (where a_N is the isotropic hyperfine splitting and $A_\parallel$ is the parallel component of an axially symmetric hyperfine tensor), was considered. [In reality, $a_N \approx 15\,G$ and $\sqrt{\frac{3}{10}}(A_\parallel - a_N) \approx 11\,G.$] The quasi-nitroxide case allows one to use (van Vleck) perturbation theory on the pseudosecular terms to decouple the $C_{K,M}^L(t, i)$ and the $b_{K,M}^L(t, i)$ for the three allowed transitions and the three eigenstate pairs from the forbidden ESR transitions and the three NMR-type transitions. In particular, one obtains a simplified approximate nuclear spin-flip-induced transition probability (i.e., a W_n for $L = 0$) given by $2W_n' = (D^2/5)[\tau_R/(1 + \tilde{b}_2^2\tau_R^2)]$, where $\tau_R = 1/6\mathcal{R}$, $D = -\frac{3}{2}(1/\sqrt{6})|\gamma_e|(A_\parallel - a_N)$ and $\tilde{b}_2 = -\frac{1}{2}|\gamma_e|[a_N + \frac{1}{7}(A_\parallel - a_N)]$, which is essentially the fast-motional result, but is correct for slow motion (for the quasi-nitroxide) where its asymptotic behavior goes as $(D^2/5)\tilde{b}_2^{-2}\tau_R^{-1}$. This W_n term couples the relaxation of all three allowed transitions (or more precisely eigenstate pairs) for $i = 1$, 2, and 3. [More generally, for $L \neq 0$, these W_n-type terms will couple the three allowed eigenstate pairs and will also couple coefficients $b_{K,M}^L(i)$ of different L values.]

In the limit of slow motion we can then write $b = W_n'/W_e$ as

$$b = \frac{\frac{1}{2}(D^2/5)\tilde{b}_2^{-2}\tau_R^{-1}}{W_e},$$

which for $\mathcal{R}/W_e \gg 1$ will still allow $b \gtrsim 1$, and nuclear spin flips are an important part of the problem. For $\mathcal{R}/W_e \lesssim 1$, the contribution of nuclear spin flips is less important relative to effects of W_e.

The quasi-nitroxide model immediately leads to

$$\frac{W_n'}{6\mathscr{R}} = \frac{1}{5}\left(\frac{D}{\bar{b}_2}\right)^2 \ll 1,$$

indicating that motional effects spread the saturation within a hyperfine line much faster than nuclear spinflips occur. But for a true nitroxide, where D^2/b_2^2 is *not* much smaller than unity, this is no longer true, and one may expect the nuclear spin flips to play an almost comparable role. Thus we would not expect $b \ll 1$ until $\mathscr{R}/W_e \ll 1$, so that the importance of nuclear spin-flip terms should persist to almost as slow motions as does the direct motional effects.

5 SUMMARY

One may conclude that the general methods developed for steady-state saturation experiments in both the motional narrowing and slow-tumbling region may be applied to time-dependent experiments such as saturation recovery. The solution is again dependent on the same matrix representations. The complex coupled differential equations are most effectively solved in terms of separate diagonalizations in transition space (in which the relaxation and coherence matrices are defined) and in eigenstate (or eigenstate-pair) space (in which the transition probability matrix is defined). The saturation recovery-type experiment (which also includes pulsed-ELDOR) may be readily handled by a general procedure, based on having a weak nonsaturating observing mode.

One finds from the analysis of spectra with hyperfine lines exhibiting coupled relaxation (with a nitroxide being a particular example) that, quite generally, the saturation recovery signal is dominated by a single exponential decay of time constant $T_1 = (2W_e)^{-1}$ despite the complexities of coupled relaxation that may exist. Simply stated, this is because when W_n or ω_{HE} are much greater than W_e, so as strongly to couple the relaxation of the eigenstate pairs, then the whole spectrum first rapidly adjusts to a common level of saturation with time constants $\sim W_n^{-1}$ or ω_{HE}^{-1} and then proceeds to relax to equilibrium more slowly with T_1, which is the slow decay observed experimentally. When a steady-state pulse approximation (i.e., the saturating pulse is on for times > the T_1's) is applicable, then one finds the fast decays all have much weaker amplitudes (proportional to their decay time constants). For W_n, $\omega_{HE} \ll W_e$, the lines are essentially uncoupled and all decays are $\sim T_1$. However, when W_n, $\omega_{HE} \sim W_e$, then more complex behavior may be seen with several (not very different) decay constants, which are weighted differently for ELDOR versus direct observation. Thus ELDOR would be helpful in deciphering the different decays. When W_n, $\omega_{HE} \ll W_e$, it is still possible to observe ELDOR-recovery effects with time constant $\approx T_1$, but with a signal attenuated by factors of the order of $b = W_n/W_e$ or $Ab'' = \omega_{HE}/W_e$.

In the slow-motional case, for a simple line, the important comparison is between $\mathcal{R}$ versus W_e, where $\mathcal{R}$ is the rotational diffusion constant. For $\mathcal{R} \gg W_e$ (a frequent situation even in the slow-motional region), the rotational reorientation spreads the saturation over the whole spectrum, and the observed slow decay is again given by $T_1 = (2W_e)^{-1}$. For $\mathcal{R} \ll W_e$ each orientational component of the spectrum is separately saturated and it relaxes with $T_1 = (2W_e)^{-1}$. The region of $6\mathcal{R} \sim W_e$ allows for the superposition of several decays of comparable order of magnitude that might be effectively explored by a combination of direct and ELDOR-observational techniques. The multiple-line (e.g., the nitroxide) case involves a combination of reorientational and nuclear spin-flip effects.

ACKNOWLEDGMENTS

We wish to thank the NSF for partial support of this work through Grant no. CHE 77-26996. This chapter was completed while the author was a visiting professor of Physics at Delft Technical University, and he greatly appreciates the facilities made available to him.

REFERENCES

1. J. H. Freed, *J. Phys. Chem.*, **78**, 1155 (1974).

2. This is reviewed by (a) J. H. Freed, *Ann. Rev. Phys. Chem.*, **23**, 265 (1972); and (b) J. S. Hyde, *ibid.*, **25**, 407 (1974).

3. J. H. Freed, in M. Dorio and J. H. Freed, Eds., *Multiple Electron-Spin Resonance*, Plenum, New York, 1979, Chap. 3.

4. M. D. Smigel, L. A. Dalton, L. R. Dalton, and A. L. Kwiram, *Chem. Phys.* **6**, 183 (1974) discussed computer simulations of saturation recovery.

5. J. H. Freed, in L. J. Berliner, Ed., *Spin Labeling: Theory and Applications*, Academic, New York, 1976, Chap. 3.

6. (a)A. Abragam, *The Principles of Nuclear Magnetism*, Oxford University Press, New York, 1961; (b) H. C. Torrey, *Phys. Rev.*, **76**, 1059 (1949).

7. J. H. Freed, G. V. Bruno, and C. F. Polnaszek, *J. Phys. Chem.*, **75**, 3385 (1971), referred to as FBP.

8. G. V. Bruno, Ph.D. Thesis, Cornell University, 1973.

9. C. F. Polnaszek, G. V. Bruno, and J. H. Freed, *J. Chem. Phys.*, **58**, 3185 (1973).

10. The special case of $\Delta\omega = 0$ and $T_1 = T_2 \equiv T$, which as noted, represents a breakdown of

the expansion equation (2.35a), results in $\Delta Z'(t)$ decaying as $e^{-t/T}$ while the coupled modes $[\Delta Z''(t) \pm \frac{1}{2}\Delta\hat{\chi}(t)]$ decay as $e^{-t(T^{-1}\pm i\omega_1)}$. Since we are assuming $\omega_1^2 T_1 T_2 = (\omega_1 T)^2 \ll 1$, it follows that $e^{-t(T^{-1}\pm i\omega_1)} \cong e^{-t/T}(1 \pm i\omega_1 t)$, essentially an exponential decay in T_1.

11. J. S. Hyde, J. C. W. Chien, and J. H. Freed, *J. Chem. Phys.*, **48**, 4211 (1968).

12. D. S. Leniart, H. D. Connor, and J. H. Freed, *J. Chem. Phys.*, **63**, 165 (1975).

13. S. A. Goldman, G. V. Bruno, C. F. Polnaszek, and J. H. Freed, *J. Chem. Phys.* **56**, 716 (1972).

14. S. A. Goldman, G. V. Bruno, and J. H. Freed, *J. Chem. Phys.*, **59**, 3071 (1973).

15. J. S. Hwang, R. P. Mason, L. P. Hwang, and J. H. Freed, *J. Phys. Chem.*, **79**, 489 (1975).

16. A. R. Edmonds, *Angular Momentum in Quantum Mechanics*, Princeton University Press, Princeton, N.J., 1957.

3 ELECTRON SPIN-LATTICE RELAXATION IN NONIONIC SOLIDS

Michael K. Bowman

Chemistry Division
Argonne National Laboratory
Argonne, Illinois

Larry Kevan

Department of Chemistry
Wayne State University
Detroit, Michigan

1 INTRODUCTION

1.1 Spin-Lattice Relaxation

In all forms of linear spectroscopy it is necessary to have some relaxation pathway connecting the two levels that are being excited. If this were not the case, the transition being observed would saturate or bleach during the course of the observation and only transient measurements would be possible. In magnetic resonance this relaxation process is called spin-lattice relaxation. It is the conversion of the energy absorbed by the spin system into thermal energy in the sample.

One relaxation process that is often important in optical spectroscopy is radiative relaxation in which the excited states in the sample under observation emit a photon and return to the ground state. This is usually observed as fluorescence or phosphorescence. Because of the small energy difference between the spin states, a calculation of the radiative relaxation rate or its inverse, the radiative lifetime, must consider both spontaneous emission and stimulated absorption and emission due to black body radiation in the sample.[1] For a system of two levels, a and b, with a transition frequency in angular units of ω, the transition rate from level a to b is $W_{ab} = B\rho_{em}$. The transition rate in the opposite direction is $W_{ba} = A + B\rho_{em} = B\rho_{em} \exp(\hbar\omega/kT)$ where A and B are the Einstein coefficients for spontaneous and stimulated emission, respectively, and ρ_{em} is the thermal equilibrium density of electromagnetic radiation of energy $\hbar\omega$. The radiative relaxation rate is $W_{ab} + W_{ba} = B\rho_{em}(1 + \exp(\hbar\omega/kT))$. The Einstein coefficient for stimulated emission or absorption is simply $B = (\frac{2}{3})\pi^2\gamma^2$ where γ is the magnetogyric ratio of the electron for the case of a magnetic dipole transition. The radiation density with energy $\hbar\omega$ is given by $\rho_{em} = (\hbar\omega^3/\pi^2 c^3)(\exp(\hbar\omega/kT) - 1)^{-1}$, where c is the velocity of light in that medium.[1] Thus for an $S = \frac{1}{2}$ spin system at 4.2 K, a 9 GHz electron paramagnetic resonance (EPR) transition would have a radiative decay rate of

$$W_{ab} + W_{ba} = \frac{2\gamma^2\hbar\omega^3}{(3c^3)} \coth\left(\frac{\hbar\omega}{kT}\right) \approx 10^{-10}\,\text{s}^{-1}. \tag{1}$$

In more comprehensible terms the relaxation time due to radiative decay is about three centuries at 4.2 K and is still several years at room temperature—many orders of magnitude slower than what is observed.

One reason for this discrepancy is that there are many more lattice phonons or vibrations with energy $\hbar\omega$ than there are of photons with the same energy. At frequencies below the Debye cutoff, the relative density of phonons and electromagnetic photons of the same frequency is given as $\rho_{ph}/\rho_{em} = (c/v)^3 \approx 10^{15}$ where v is the velocity of sound in the solid.[1] Provided that the phonons are at least 10^{-15} times as effective in stimulating spin transitions as are photons, these phonons will be dominant over photons in spin-lattice relaxation. In solids in which the Debye model is not a good description of the dynamics of the lattice, other vibrations can couple to the spin system and increase the relaxation rate still further.

From this brief consideration of spin-lattice relaxation in solids, it is apparent that at least two things can be investigated by studying spin-lattice relaxation—the phonon structure of the solid in the region of the EPR frequency and the interaction between the electron spin and the motion of the lattice. Actually there is much more that can be learned. Spin-lattice relaxation studies can yield information about the entire vibrational spectrum of the solid and even about the vibrational modes of the free radical itself, as well as revealing the presence and energies of electronically excited states of the radical.

Aside from the information that spin-lattice relaxation studies can provide about the radical and the solid, a good understanding of spin-lattice relaxation can be very helpful in the design and execution of other magnetic resonance experiments. Many of the more exotic EPR techniques depend very strongly on the magnitudes and the ratios of various relaxation rates in the spin system. The knowledge of how to manipulate even one of these relaxation rates in a well-defined manner may be the *sine qua non* for the success of an electron-electron double resonance (ELDOR) or an electron-nuclear double resonance (ENDOR) experiment on a difficult system.[2]

1.2 Spin-Lattice Relaxation Rate Measurement

The simplest method for measurement of the spin-lattice relaxation rate for a paramagnetic sample using commercial EPR spectrometers is one of the various progressive saturation techniques. One records an EPR signal from the sample and progressively increases the power used for the observation of the signal over a range in which the microwave-induced transition rate changes from far less than the spin-lattice relaxation rate to far more. Assuming a simple enough model for the response of the spin system at the experimental conditions, the spin-lattice relaxation rate can be derived from such experiments. Whether one uses the simple Bloch equations or more sophisticated models for the response of the spin system to saturation [3–11] there is still some uncertainty about the validity of such measurements in solids. Nearly all these theories are quite sensitive to the assumed inhomogeneous broadening distribution function and ignore the presence of forbidden spin-flip satellite lines[12] that must be present if there are any $I \neq 0$ nuclei in or near the radical. In addition, the Bloch equations are always assumed implicitly to be valid in solids, whereas spin-echo studies[13,14] usually find that they do not describe the response of a spin system in solids.

In principle, the picture is even more complicated for radicals in disordered solids. Implicit in the discussion so far has been the assumption that like radicals have identical spin-lattice relaxation rates. This is not necessarily the case in disordered solids. If the radicals experience either a different strength of interaction with the surrounding lattice or if the lattice exhibits a different dynamic behavior around individual radicals, the sample may exhibit a distribution of spin-lattice relaxation times.

1.3 Time Domain Relaxation Measurements

In view of the problems in obtaining meaningful and accurate measurements of the spin-lattice relaxation rate in solids using the continuous wave methods mentioned in Section 1.2, there have been many time domain spectrometers constructed to measure these rates directly. With one exception the techniques used for the measurements of spin-lattice relaxation times have consisted of two parts. The first part is the preparation of the spin system in some nonequilibrium configuration from which it relaxes toward an equilibrium determined by the sample temperature. The second part of the technique is the observation of the spin system's magnetization as it relaxes toward the equilibrium value.

Since these two parts of the measurement are in some sense independent of each other, they are described independently. In view of the excellent descriptions of the techniques in several books and articles,[15-26] the descriptions here are brief.

The exception to the rule that all time domain measurements of the spin-lattice relaxation rate involve the preparation of a nonequilibrium state of the spin system lies in those spin-echo methods wherein the transitions between spin states of a radical system in thermal equilibrium with the lattice influences some relaxation parameter of another spin system.[17]

1.3.1 PREPARATION OF A NONEQUILIBRIUM POPULATION

The simplest method to create a nonequilibrium spin population is to saturate or partially saturate an EPR transition. This can be accomplished by a short microwave pulse of sufficiently high microwave power. This procedure tends to equalize the populations of the two spin states and decreases the EPR signal or magnetization in the direction of the external magnetic field.

This simple procedure is inadequate for some cases. Unless very large microwave powers are used, only a small part of the EPR line is saturated; the line is then said to have a hole burned in it. When this happens the EPR signal at that part of the EPR line may not recover with a rate governed by the spin-lattice relaxation rate. Instead the hole in the line often tends to become broader and shallower as a process called spectral diffusion changes the resonant frequency of those spins that are saturated. In those cases the recovery, if only part of the EPR line is observed, consists of the concurrent processes of spectral diffusion and spin-lattice relaxation.[27-29]

The effects of spectral diffusion can sometimes be eliminated in experimental measurements by the use of a very long saturation pulse. In this case spectral diffusion during the saturating pulse saturates the whole EPR line and the whole line relaxes together toward equilibrium. Another way of eliminating spectral diffusion in certain cases is to apply a series of short saturating pulses over a long period of time. This is sometimes called the "picket fence" technique for suppressing spectral diffusion. The technique of using pulses of microwaves to saturate an EPR transition before observing its recovery to thermal equilibrium with the lattice is called the saturation recovery method.

Another pulsed microwave method of preparing the sample for observing relaxation consists of inverting the spin state populations instead of equalizing them. This inversion can be accomplished in one of two ways. Either a single pulse of microwaves with an intensity and duration sufficient to invert the spin populations is used, or an adiabatic, rapid passage inversion is used. When a single inverting pulse is used in a system with broad EPR lines, a hole is burned in the line and the recovery can be complicated by the effects of spectral diffusion. Adiabatic, rapid passage inversion, which is accomplished by applying an intense pulse of microwaves while the magnetic field is swept through resonance avoids the effects of spectral diffusion, since the whole EPR line or the whole spin system is inverted. Although an adiabatic, rapid passage inversion cannot be performed very rapidly, since the magnetic field must be swept, spectral diffusion is often a problem only when the spin-lattice relaxation rate is slow enough for this technique to be used.

The third method of preparing a sample for observation with a non-equilibrium population is rapidly to generate radicals that have initial electron spin polarization in the sample. This can be done either using light or ionizing radiation to produce the radicals.[30] This technique has been used with photoexcited triplet states in solids where, in addition to the spin system relaxing toward equilibrium, the number of triplets is changing and the analysis of the kinetics in order to obtain the spin-lattice relaxation rate is complicated. With triplets, microwave pulses can be used in addition to light pulses to create a variety of nonequilibrium states for observation.

1.3.2 OBSERVATION OF RELAXATION

The obvious method for monitoring the recovery of the spin system toward equilibrium is the observation of the EPR signal. This method has been used for many years very successfully and shows no sign of being replaced by another means. For systems with slow spin-lattice relaxation rates, the microwave power used to observe the EPR signal must sometimes be so low as to make detection of the signal impossible. In that case, sweeping the magnetic field through resonance periodically and momentarily observing the signal is often necessary.

The EPR signal or the sample magnetization along the magnetic field can be measured by generating a spin echo. Since the spin echo leaves the spin system in a saturated condition, it can be used as its own saturating pulse. However, it is possible to produce an echo only once before the spin system must be prepared for observation again.

The last method used to observe the relaxation of the spin system toward equilibrium is by optical detection of magnetic resonance. This can take one of a number of forms[31] and has been applied only to the measurement of spin-lattice relaxation in triplets.

In addition to the techniques available for the measurement of spin-lattice relaxation rates in solids using EPR, there are a number of non-EPR methods that are outside the scope of this article.[32]

2 KINETICS OF SPIN-LATTICE RELAXATION

2.1 Kinetics for Identical $S = \frac{1}{2}$ Spins

For a two-level spin system with levels called a and b, the equilibrium populations of these levels N_a and N_b are related to each other by $N_b/N_a = \exp(-\hbar\omega/kT)$. The equilibrium population difference is $n_0 = N_a - N_b$. The instantaneous populations of the two levels that we shall call n_a and n_b, with the instantaneous difference n, are not necessarily the equilibrium values for those quantities. The total number of spins in the system is $N = N_a + N_b = n_a + n_b$.

The transition rate in the absence of a microwave magnetic field or in the presence of a negligible one from level a to level b is W_{ab} and the reverse rate is W_{ba}. For a spin system in a solid whose phonon, vibrational, or electromagnetic radiation fields can be described by the temperature T, the relation between W_{ab} and W_{ba} is $W_{ba}/W_{ab} = \exp(\hbar\omega/kT)$. The rate of change in the populations in levels a and b is

$$\frac{dn_a}{dt} = -\frac{dn_b}{dt} = -W_{ab}n_a + W_{ba}n_b. \tag{2}$$

This equation can be rearranged into a more convenient form as follows:

$$\frac{dn}{dt} = \frac{dn_a}{dt} - \frac{dn_b}{dt} = 2(-W_{ab}n_a + W_{ba}n_b) \tag{3}$$
$$= (W_{ba} - W_{ab})N - (W_{ba} + W_{ab})n.$$

Making use of the relationship between W_{ab}, W_{ba}, N, and n_0, this equation can be reduced to

$$\frac{dn}{dt} = (n_0 - n)(W_{ab} + W_{ba}).$$ (4)

This differential equation for the population difference is easily solved and gives

$$n(t) = [n_0 - n(t=0)] \exp\left(-\frac{t}{T_1}\right),$$ (5)

where the time dependence of the population difference n is emphasized by writing it as $n(t)$. The relaxation time of the spin system T_1, is the spin-lattice relaxation time and is given by $T_1^{-1} = W_{ab} + W_{ba}$. Thus the recovery of the spin system toward its equilibrium configuration is exponential with a rate constant equal to the spin-lattice relaxation rate.

2.2 Kinetics for Identical $S = 1$ Spins

The measurement of the spin-lattice relaxation rate for a triplet spin system with three spin levels, a, b, and c is more difficult than that for a two-level system. This is because spin-lattice relaxation connects all three levels together with rates that are not necessarily very different. For a triplet produced by photoexcitation the problem is even more difficult because the population and decay rates for each of the triplet levels must be considered. In the most general case the time dependence of the population of the triplet spin state a is described by the differential equation

$$\frac{dn_a}{dt} = (-K_a - W_{ab} - W_{ac})n_a + P_a + W_{ca}n_c + W_{ba}n_b,$$ (6)

where P_a and K_a are the populating and decay rates, respectively, for triplet spin level a. There are analogous equations for the other two spin levels.[18-26] This set of three coupled differential equations has no simple analytical solution. For certain simplifying assumptions that one or another rate constant is infinite or zero, analytical solutions yielding the spin-lattice relaxation rates are possible. Solutions for these coupled differential equations have been generated by computer with the rate constants treated as variables and adjusted until the response of the spin system is sufficiently well modeled.

In those systems where multiple optical detection of magnetic resonance (ODMR) techniques are feasible, the kinetic equations can be simplified by saturating the EPR transition between two of the levels.

When this is done, the three-level system can be formally treated as a two-level system and analytical solutions of the rate equations can be obtained fairly easily.

In those experiments where the signal from the spin system is the usual EPR signal, one often uses square-wave modulated light to produce the triplets and to let the total triplet population decay to zero during the dark periods. In this way the initial populations of the spin levels are known to be zero and the populating rates are parameters to be determined from the response of the system while the light is on as well as while it is off. If only a short, intense light pulse is used to generate the triplets, the response of the system in the dark when the populating rates are zero is observed. However, the initial populations of the three different spin states then become parameters to be evaluated from the response of the spin system in time.

Although ground-state triplet systems are far less common than are excited-state triplets, their spin-lattice relaxation rates can be measured with more ease than can those of excited-state triplets. There are again three coupled differential equations describing the time response of the system; however, the populating and decay rates for the triplet molecules are all zero. Thus the response of the system is described by the sum of three exponential functions for the most general case. The amplitudes of the three exponential components are determined in part by the initial state in which the spin system is prepared. Thus the exact recovery of the spin system toward equilibrium will be partly determined by the experimental conditions. However, the recovery or spin-lattice relaxation rates will be independent of the experimental conditions used to measure them.

2.3 Multiphasic Kinetics

In Section 2.1 it was shown that, for a system of identical doublets, the spin system will always approach equilibrium exponentially with a rate equal to the spin-lattice relaxation rate. In many measurements of the spin-lattice relaxation rate the response of the spin system is often not described by a single exponential function, particularly in saturation recovery experiments. The response can sometimes be described as the sum of several exponential functions with different amplitudes and sometimes the responses seem to be very nonexponential.[33]

There are a number of possible reasons for this behavior. Perhaps the simplest one is that the spin system under observation does not have a single spin-lattice relaxation rate. When the system contains more than one radical with overlapping spectra, the recovery of the EPR signal can be the sum of the recoveries of the individual radicals that appears as the sum of two or more exponentials. Multiple exponential recoveries may be observed when one radical is stabilized in more than one site in the lattice or when overlapping components of the EPR signal have different spin-lattice relaxation rates because of the orientational anisotropy or the hyperfine component dependence of the spin-lattice relaxation rate.

The recovery of the EPR signal detected during a saturation recovery experiment or an inversion recovery experiment may not be a single exponential for a number of reasons. The most prevalent reason would seem to be the existence of spectral diffusion processes during the experiment. Spectral diffusion will alter the kinetics of the recovery signal because the radicals that were saturated or inverted at the start of the experiment are not the ones being observed at the end of the experiment. When spectral diffusion takes place, the EPR signal being observed can increase in intensity, not because of spin-lattice relaxation but because spins that were saturated are no longer in resonance and spins that were not saturated begin to contribute to the EPR signal being observed.

Cross relaxation or cross saturation is another process that can result in a multiexponential recovery. Cross relaxation or cross saturation is when the electron spins of two different radicals exchange their spin angular momentum with each other. This can lead to a transfer of energy or saturation from one spin system to another or even from one hyperfine component to another. Although the concepts of spectral diffusion and cross relaxation are somewhat overlapping, one is not a subset of the other. When cross relaxation is present in a sample, the recovery of the EPR signal is often characterized by a recovery of the EPR signal of the radical being observed as it comes into thermodynamic equilibrium with the other radical, and then another recovery as the combined spin system comes into thermal equilibrium with the lattice. In the event that the second spin system has a very fast spin-lattice relaxation rate so that during cross relaxation it remains in thermal equilibrium with the lattice, the recovery of the EPR signal is determined only by the cross relaxation kinetics and not by any spin-lattice relaxation rate constant.

2.4 Exp $(-(at)^{1/2})$ Kinetics

A dilute solid solution of a radical is by definition a set of nonidentical radicals if only because the radicals differ from one another on the basis of the distribution of radicals about themselves. If there are several radical species in the sample, this can be important in the kinetics of spin-lattice relaxation.

If a radical is surrounded by a random distribution of other radicals that always remain in thermal equilibrium with the lattice, cross relaxation similar to dipole-dipole energy transfer[34] takes place. The kinetics of the recovery of the observed radical have been calculated for this case.[35,36] The recovery of the signal at long times approaches $\exp\left(-(at)^{1/2}\right)$. The rate constant a in this equation is determined by the spectral properties of both radicals and by the spin-lattice and spin-spin relaxation times of the other radical.

This same kinetic form has also been attributed to the recovery of a single isolated radical species—hydrogen atoms in quartz and sulfuric acid glass.[37] The authors attributed this to a distribution of spin-lattice relaxation rates caused by hydrogen atoms in traps that had different interactions with their surroundings. Why the kinetics of the recovery of the signal takes on this mathematical form for hydrogen atoms in these solids is not clear, since only one distribution function for the spin-lattice relaxation rates in the sample yields kinetics of this form.

2.5 Determination of Rate Constants of Recovery

Once the kinetic form of the recovery of the EPR signal or sample magnetization is known, the determination of accurate spin-lattice relaxation rate constants is not very difficult.

For exponential recovery the simplest method is to plot the logarithm of the recovery as a function of time. This should yield a straight line whose slope is the relaxation rate. This procedure can lead to an experimental bias because the slope of the line is dependent on the value used for the baseline, the unsaturated magnetization of the sample. It is well known that an error of 1% of the original amplitude of the exponential in the determination of the baseline will result in a 3% error in the relaxation rate constant.[38,39]

The problem is even worse in multiexponential recoveries where the component due to spin-lattice relaxation may be only a few percent of the initial transient signal. In this case it may be necessary to determine the baseline to a few parts in 10^4 of the initial transient signal. A similar problem must exist for the kinetics of Section 2.4, but these kinetics have not been invoked often and the influence of baseline error on the rate constant is not known.

When the recovery is composed of two or more exponentials, the best general method of determining the spin-lattice relaxation rate is to fit an exponential function to the very tail of the recovery, where, it is hoped, only one exponential contributes to the recovery. When the two rate constants are vastly different in a biexponential recovery, the whole recovery can be used to determine both rate constants. When the two rate constants differ by a factor of 3 or less, the uncertainty in their determination is tremendous.

2.6   Determination of Kinetic Form of Recovery

The determination of the form of the kinetics of the recovery of a signal in the presence of noise is equivalent to the problem of determining the form over a limited range of time. The simplest method of determining whether a recovery is exponential, is multiexponential, or has the form given in Section 2.4 is by making comparisons over a large time range.

One of the greatest problems in the determination of the kinetic form of the recovery is the uncertainty in the value of the baseline. Over a limited range in time, an exponential and a biexponential function are indistinguishable if a constant can be added or subtracted from them. This is illustrated in Fig. 1. It is noteworthy that when an exponential recovery,

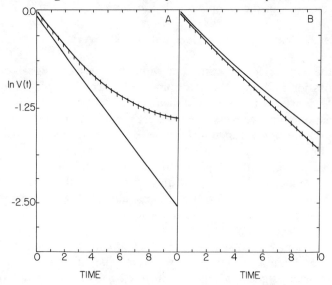

Figure 1   Plots of the natural logarithm of $V(t)$ versus t for four functions, $V(t)$, commonly encountered in spin-lattice relaxation measurements. In section A the solid line is $V(t) = \exp(-t/0.4)$ and yields a straight line whose inverse slope or time constant is 0.4. The hatched line is the function $V(t) = \exp(-t/0.4) + 0.1$ and appears to be the sum of two exponential decays. Tangents to the initial and final portions of the curve have time constants of 0.44 and 2.0, respectively. Only the initial portion of the curve has a slope approaching the slope of the true exponential. In section B the solid line is the function $V(t) = \exp(-t/0.4) + \exp(-t/0.8)$ and appears to be curved in this plot. The hatched line that seems to be a straight line with an inverse slope or time constant of 0.54 is actually a plot of the function $V(t) = \exp(-t/0.4) + \exp(-t/0.8) - 0.045$. These plots illustrate the difficulty in determining the functional form of spin-lattice relaxation kinetics when recorded over a limited time range.

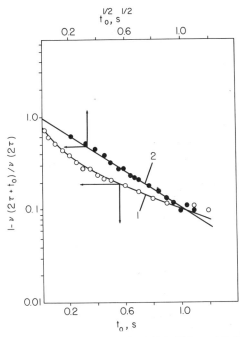

Figure 2 Plots of spin-lattice relaxation data for V^{4+} in TiO_2 at 4 K. [From R. L. Saunders and L. G. Rowan, *Phys. Rev. Lett.*, **21**, 140 (1968).] Curve 1 is a plot of the data to a biexponential recovery. Curve 2 is a plot of the data to $\exp(-t^{1/2})$ kinetics characteristic of cross relaxation between radicals. Plots are taken from reference 17, p. 101. Note that both functional forms fit the data equally well.

through the mischoice of the baseline, is treated as a biexponential, the longer-lived portion has a rate constant that is not closely related to the rate constant of the true recovery.

When the spin-lattice relaxation rate is measured from a small-intensity tail in a recovery curve described by bi- or multiexponential kinetics, the value of the baseline used may have an order of magnitude influence on the measured rate constant.

Reference 17 illustrates the difficulty of discriminating between a biexponential recovery and one described in Section 2.4. Figure 2 shows an actual set of data with both functional forms fit to it. It appears to be impossible to determine visually which is better.

The safest, although the most time-consuming, method of determining the kinetic form and the spin-lattice relaxation rates is to record the data with the greatest signal-to-noise ratio over the longest range in time as is practical, to fit the various kinetic forms to the experimental data using nonlinear least-squares techniques, and to choose the best rates and kinetic form on the basis of statistical tests. It must be remembered that the kinetic form of the recovery may change with the temperature of the sample, with the concentration of the radical itself or of any other paramagnetic species present, and with other experimental conditions.

3   RELAXATION PROCESSES

Electron spin-lattice relaxation can be described in terms of the phonons or lattice vibrations taking part or in terms of the coupling mechanism between the lattice and the electron spin. The former is called the spin-lattice relaxation process and the latter is called the mechanism.

There are basically three different kinds of processes. There is the "direct" process, in which the energy required or released by the electron spin flip is compensated for by the creation or annihilation of one phonon of that energy. This is analogous to single photon absorption or emission in optical spectroscopy. There is also a Raman process, similar to Raman spectroscopy, in which the difference in energy between two phonons is used to conserve the energy of the entire system. In addition, there are higher-order processes involving local vibrational or excited states that are somewhat similar to resonant Raman optical spectroscopy.

The various optical effects that are similar to the spin-lattice relaxation processes have different dependencies on the intensity of light used; moreover, the different spin-lattice relaxation processes depend on the phonon densities that are determined by the temperature of the sample. Thus it is possible to distinguish between the different relaxation processes on the basis of their temperature dependencies.

In this section the temperature dependencies of the relaxation processes are derived; in the next section the mechanisms associated with the different processes that are important in disordered solids are discussed.

3.1   The Direct Process

The probability of a transition conserving energy for a spin in a solid is

$$P_{\substack{ab \\ ba}} = \frac{2\pi}{\hbar} \left| \left\langle \begin{matrix} b \\ a \end{matrix} \right| V \left| \begin{matrix} a \\ b \end{matrix} \right\rangle \right|^2, \tag{7}$$

where $|a\rangle$ and $|b\rangle$ are eigenfunctions of the entire system of both spin and phonons and V is the time-dependent perturbation connecting these two eigenstates of the system. It is implicitly assumed that the energies of states a and b are equal. The perturbation V must contain operators that operate not only on the spin system but also on the phonon system. This can be written in the form

$$V = \sum_n \frac{\partial \mathcal{H}_s}{\partial \varepsilon_n} \varepsilon_n + \sum_{n,m} \frac{\partial^2 \mathcal{H}_s}{\partial \varepsilon_n \, \partial \varepsilon_m} \varepsilon_m \varepsilon_n + \cdots, \tag{8}$$

where $\mathcal{H}_s$ is the spin Hamiltonian for the system and $\varepsilon_n(t)$ is the relative time-dependent displacement or strain of nucleus n with respect to the radical from its equilibrium position.

Writing the first term in the time-dependent perturbation as $v'\varepsilon$, the transition probabilities between states with different electron spins is given as

$$P_{ab \atop ba} = \frac{2\pi}{\hbar} |\langle s_1| v' |s_2 \rangle|^2 |\langle n| \, \varepsilon \, |n \pm 1 \rangle|^2, \tag{9}$$

where $|s_1\rangle$ and $|s_2\rangle$ are the electron spin states involved and $|n\rangle$ and $|n \pm 1\rangle$ are the eigenfunctions of the phonon system with n and $n \pm 1$ phonons. Since the total energy of the system is constant, the energy difference between electron spin states exactly equals the energy of one phonon. That means that the phonon frequency equals the EPR frequency.

The first integral in the transition rate expression contains terms that depend only on the spin system and is independent of the temperature of the lattice. The exact form of this integral and the perturbation v' depend on the spin-lattice relaxation mechanism.

The second integral depends on the lattice vibrations only and provides the temperature dependence for the transition probability. These integrals have been calculated for a Debye lattice[40] that should approximate nonlocalized vibrations of disordered solids. These integrals are

$$\langle n| \, \varepsilon \, |n-1 \rangle = \frac{\omega}{v} \left(\frac{\hbar}{2\rho\omega} \right)^{1/2} n^{1/2}$$

$$\langle n| \, \varepsilon \, |n+1 \rangle = \frac{\omega}{v} \left(\frac{\hbar}{2\rho\omega} \right)^{1/2} (n+1)^{1/2}, \tag{10}$$

where v is the velocity of sound in the solid, ρ is the density of the solid, and n is the occupation number of phonons. For a phonon system that is in thermal equilibrium with a temperature T, n can be replaced with the thermal occupation number

$$n = \left(\exp\left(\frac{\hbar\omega}{kT} \right) - 1 \right)^{-1}. \tag{11}$$

The transition rate is equal to the transition probability times the phonon density given by[1]

$$W_{ab} = P_{ab}\rho_{ph} = P_{ab} \frac{3\omega^2}{2\pi^2 v^3}. \tag{12}$$

The spin-lattice relaxation rate from Eq. (9)–(12) is then

$$T_1^{-1} = \frac{3\omega^2}{2\pi^2 v^3}(P_{ab}+P_{ba}) = \frac{3\omega^3|\langle s_1|\,v'\,|s_2\rangle|^2}{2\pi^2 v^5 \rho}\left(1+\frac{2}{\exp(\hbar\omega/kT)-1}\right)$$

$$= \frac{3\omega^3|\langle s_1|\,v'\,|s_2\rangle|^2}{2\pi^2 v^5 \rho}\coth\left(\frac{\hbar\omega}{kT}\right). \tag{13}$$

When kT is much larger than the spin-state energy difference, $\coth(\hbar\omega/kT) = 2kT/\hbar\omega$ and the spin-lattice relaxation rate is proportional to the temperature. Thus for EPR experiments performed at frequencies near 9 GHz, the direct process exhibits a linear temperature dependence above 1 or 2 K. In this high-temperature limit the relaxation rate from the direct process is proportional to $\omega^2|\langle s_1|\,v'\,|s_2\rangle|^2$. The microwave frequency dependence of the relaxation rate reveals the frequency dependence of the integral involving the spin states, which, in turn, can reveal the mechanism responsible for the spin-lattice coupling for the direct process.

3.2 The Raman Process

In the derivation of the direct process in Section 3.1, the time-dependent perturbation responsible for the spin-lattice relaxation was taken to first order in the nuclear displacements and was treated with first-order perturbation theory. It is possible and necessary in some cases, however, to use a more complete treatment. One can treat the perturbation to second order in the nuclear displacements using first-order perturbation theory or one can use second-order perturbation theory with the perturbation carried to first order in the nuclear displacement.

Writing the time-dependent perturbation to second order in the nuclear displacements as

$$\sum_{m,n}\frac{\partial^2\mathcal{H}_s}{\partial\varepsilon_m\,\partial\varepsilon_n}\varepsilon_m\varepsilon_n = v''\varepsilon_1\varepsilon_2, \tag{14}$$

the transition probability for a transition from state a to b is

$$P_{ab}_{ba} = \sum_{m,n}|\langle s_1|\,v''\,|s_2\rangle|^2\,|\langle n_n|\,\varepsilon_1\,|n_n\pm 1\rangle|^2\,|\langle n_m|\,\varepsilon_2\,|n_m\mp 1\rangle|^2$$

$$\times\,\delta(\hbar\omega\pm E_n\mp E_m), \tag{15}$$

where E_n and E_m are the energies of phonons with frequencies ω_n and ω_m. The spin-lattice relaxation rate may be calculated as before by multiplying the transition probabilities by the phonon densities and performing the summation. Since this second-order process is expected to be important only when the temperatures and phonon occupation numbers are high and since the phonon density is greatest at high phonon energies, it is usual to approximate $\omega_m = \omega_n$ in the calculation. Converting the summation in the expression for the transition probability to an integration over phonon frequency, the spin-lattice relaxation rate can be written in the form

$$T_1^{-1} = \int_0^{\omega_D} |\langle s_1| \, v'' \, |s_2\rangle|^2 \frac{9\omega_n^4}{8\pi^3 v^{10}\rho^2} 2n_n(n_n + 1) \, d\omega_n, \qquad (16)$$

where ω_D is the Debye cutoff frequency for the solid. A disordered solid is not expected to resemble the Debye model when the phonon wavelength approaches the intermolecular distance which occurs as the phonon frequency approaches the Debye frequency. Unfortunately, there are no better models that are general enough to be used here. Nevertheless this integral should approximate the temperature dependence of the Raman process when $\hbar\omega_D/kT$ is large. Using the thermal occupation numbers for the phonons involved, the spin-lattice relaxation rate can be written as

$$T_1^{-1} = \int_0^{\omega_D} |\langle s_1| \, v'' \, |s_2\rangle|^2 \frac{9\hbar}{8\pi^3 v^{10}\rho^2} \frac{\omega_n^6 \exp(\hbar\omega_n/kT)}{(\exp(\hbar\omega_n/kT) - 1)^2} \, d\omega_n$$

$$= |\langle s_1| \, v_0'' \, |s_2\rangle|^2 \frac{9\hbar}{8\pi^3 v^{10}\rho^2} \left(\frac{kT}{\hbar}\right)^{7+d} \int_0^{\hbar\omega_D/kT} \frac{x^{6+d} \exp(x)}{(\exp(x) - 1)^2} \, dx, \qquad (17)$$

where the last integral is the well-known $6+d$ transport integral and where $v'' = v_0'' \omega_n^d$ with v_0'' independent of the phonon frequency. The transport integrals have been tabulated and there are several numerical approximations to them.[41] For thermal energies below the Debye frequency, the $6+d$ transport integral is approximately $(6+d)!$ Depending on the mechanism responsible for spin-lattice relaxation, d can take on the values of zero or ± 2.

The Raman process just described is due to the nonlinear mixing of phonons in the solid producing perturbations on the electron spin that are at the frequency difference between the two phonons. The second possible Raman process involves transitions to virtual states with their instantaneous emission of another phonon. However, the temperature dependence is the same as the Raman processes described here and the mathematical formalism is the same as that used to describe the Orbach–Aminov process in Section 3.3.

3.3 The Orbach–Aminov Process

It is not sufficient to consider spin-lattice relaxation using only first-order perturbation theory. The phonons themselves can mix excited states into the zero-order spin phonon states, which will significantly increase spin-lattice relaxation. This must be treated by second-order perturbation theory using one time-dependent perturbation to introduce admixtures of excited states and a second time-dependent perturbation to cause spin-lattice relaxation transitions.

The transition probability from state a to state b is

$$P_{ab} = \frac{2\pi}{\hbar} \left| \left\langle b \left| \sum_c \frac{v' \varepsilon_1 |c\rangle\langle c| v' \varepsilon_2}{\Delta E - \hbar \omega_n} \right| a \right\rangle \right|^2 \delta(\hbar\omega - \hbar\omega_n + \hbar\omega_m), \qquad (18)$$

where $|c\rangle$ is an excited state and ΔE is its separation from the states being observed by EPR. If ΔE is much larger than $\hbar\omega_D$, the $\hbar\omega_n$ term in the denominator can be ignored and the spin-lattice relaxation rate can be calculated in the same manner as in Section 3.2, with $v''\varepsilon_1\varepsilon_2$ replaced by

$$\sum_c \frac{v'\varepsilon_1 |c\rangle\langle c| v'\varepsilon_2}{\Delta E}. \qquad (19)$$

The temperature dependence and field dependence of this Raman process are the same as those of the Raman process described in Section 3.2. Because the nature of the excited state c, has not been specified yet, an exact expression for the spin-lattice relaxation rate in terms of parameters of the solid and of the ground state of the radical cannot be written.

When the energy of the excited state c lies within the energy spectrum of the phonon band, the integration used in Section 3.2 to calculate the spin-lattice relaxation rate is not possible, since the energy term in the denominator of the transition probability is zero at some point. This problem was solved by Orbach[42] and Aminov,[43] who independently included the finite lifetime of the excited state in the calculation. This allows the integral to be evaluated.

Orbach and Aminov considered the approach toward equilibrium of the population difference between the two lowest states in their calculations. They found that the spin-lattice relaxation rate for this measured quantity is

$$T_1^{-1} \propto \frac{\Delta E^3}{\rho v^5} \left[\exp\left(\frac{\Delta E}{kT}\right) - 1 \right]^{-1}. \qquad (20)$$

However, Murphy[44] has pointed out that in the case of doublet radicals, the excited states are likely to be doublets also with the same EPR frequency as the ground state of the doublet. If this is the case, then a typical saturation or inversion recovery experiment that monitors the EPR signal measures not only the difference in population between the spin states in the ground doublet, but also the population difference between the spin states in the excited-state doublet. Murphy calculated that the spin-lattice relaxation rate in this case is $T_1^{-1} \propto \operatorname{csch}(\Delta E/kT)$.

There is a slight but noticeable difference between these two functional forms in the region when $\Delta E/kT$ is about unity. However, they both have the same high and low temperature limits. At low temperature, $T_1^{-1} \propto \exp(-\Delta E/kT)$ and at high temperatures, $T_1^{-1} \propto T$.

When the lifetime of the phonons in the sample approaches the inverse of the EPR frequency, or when the motion involved becomes very anharmonic, the Debye model and the concept of lattice phonons is no longer useful in the calculation of the temperature dependence of the spin-lattice relaxation rate. In this limit it is more useful to consider the nuclear motion as described by a correlation function with a correlation time τ_c that is temperature dependent. The relaxation transition probability becomes

$$P_{ab} = \hbar^{-2}|\langle s_1| v' \Delta\varepsilon |s_2\rangle|^2 \frac{2\tau_c}{1+\omega^2\tau_c^2}, \tag{21}$$

where the operator in the integral is the interaction that is being modulated by the motion having a correlation time τ_c. The spin-lattice relaxation rate then has a temperature dependence determined by the temperature dependence of τ_c.

An example of such a treatment is the description of the relaxation of free radicals in glassy solids by Bowman and Kevan.[45] These authors propose that in a glassy solid, the tunneling of nuclei or groups of nuclei produces a modulation of the electron nuclear dipolar interaction between those nuclei and the free radical that can produce electron spin-lattice relaxation at low temperatures. As the matrix nuclei tunnel between positions in the matrix that they can occupy with little change in their energy, the electron dipolar interaction with the free radical changes. Since the interaction includes the operators $S_\pm$, the electron spin can change its spin quantum number and thus relax. This process is merely another Orbach–Aminov process in which the excited state $|c\rangle$ is the electron spin state in the tunneling state of higher energy. In this case the evaluation of the temperature dependence of the spin-lattice relaxation rate would be rather cumbersome, since the form of the low-energy phonon spectrum in glasses and its coupling to nuclear tunneling are somewhat uncertain.

However, low-temperature measurements of the relaxation time of these low-energy tunneling modes have been made in polyethylene by Phillips.[46] The relaxation times can be simply related to the correlation times for these modes, thus allowing a phenomenological treatment of tunneling in the calculation of the spin-lattice relaxation. The correlation time for these tunneling modes has a temperature dependence given as

$$\tau_c^{-1} \propto E \operatorname{csch}\left(\frac{2E}{kT}\right), \tag{22}$$

where E is the energy difference between the two positions in the lattice that the nuclei may occupy. Furthermore, the correlation time, at least in polyethylene and probably in most glasses at low temperatures, is much greater than the inverse of the EPR frequency. The spin-lattice relaxation rate can then be written as

$$T_1^{-1} = \frac{2}{\hbar^2} |\langle s_1| \, v' \, \Delta \varepsilon \, |s_2\rangle|^2 \frac{2\tau_c}{1 + \omega^2 \tau_c^2} = \frac{2}{\hbar^2} |\langle s_1| \, v' \, \Delta \varepsilon \, |s_2\rangle|^2 \frac{2}{\omega^2} \tau_c^{-1},$$

and

$$T_1^{-1} \propto |\langle s_1| \, v' \, \Delta \varepsilon \, |s_2\rangle|^2 \frac{E}{\omega^2} \operatorname{csch} \left(\frac{2E}{kT}\right). \tag{23}$$

This formulation contains the interaction $v' \, \Delta \varepsilon$ between the nuclei and the electron spin, the microwave frequency dependence, and the temperature dependence of this relaxation process. As expected, the temperature dependence is that predicted by Murphy[44] for an Orbach–Aminov process in which the EPR spectrum of the excited state is the same as that of the ground state. It is also the temperature dependence observed for the relaxation of an electron spin by nearby tunneling methyl groups in a single crystal.[47]

3.4   Other Relaxation Processes

Since the publication of the original papers, which have been collected by Manenkov and Orbach,[48] deriving the direct, Raman, and Orbach–Aminov relaxation processes, many refinements have been made. Among these have been the consideration of anharmonic lattices, point crystal defects, the optical phonon branch, and non-Debye phonon spectra. In addition, Orbach–Aminov processes involving a wide variety of different excited states have been described and many describe more than one relaxation process.[41–75] In all the relaxation processes described, the temperature dependence of the spin-lattice relaxation rate is identical to or nearly identical to that of the direct, Raman, and Orbach–Aminov processes discussed here in Section 3.

4   SPIN-LATTICE RELAXATION MECHANISMS

Electron spin-lattice relaxation is a transition between electron spin levels with different spin quantum numbers. Since the electron spin states are orthogonal to each other, there is no transition between them unless there is a perturbation that contains an operation on the spin quantum number. This means that the v' or v'' portion of the time-dependent perturbation must contain the operator S_+ or S_- where the z-axis is taken to be the electron spin quantization axis in the absence of the time-dependent perturbation.

In the case of the direct and Raman relaxation processes described using first-order perturbation theory in Sections 3.1 and 3.2, the operators v' and v'' must contain, at the very least, the spin operators S_+ and S_-. However, in the case of the Orbach–Aminov processes described using second-order perturbation theory in section 3.3, it is only necessary that $v'|c\rangle\langle c|v'$ function as one of these operators. This can come about because v' contains such an operator or because the electron spin function in $|c\rangle$ is not identical to the spin functions in either $|a\rangle$ or $|b\rangle$. The latter occurs if the excited state has a different g-factor or hyperfine structure or axis of quantization than the ground state. Nevertheless there is some element in the spin Hamiltonian necessary to describe the excited state that is different from that in the ground state.

Whatever interaction is modulated by the lattice phonons or by excitation into an excited state and affects the electron spin is considered to be the mechanism of the spin-lattice relaxation.

4.1 Mechanism in Ionic Solids

Paramagnetic metal ions have been studied extensively for many years. Among these studies have been many on the spin-lattice relaxation of ions in ionic solids. In most cases the spin-lattice relaxation mechanism involves the spin-orbit interaction of the ion. Vibrations in the lattice of an ionic crystal produce relatively large time-dependent changes in the crystal field acting on the paramagnetic ion. This in turn produces time-dependent changes in the orbital of the unpaired electron. Changes in the orbital angular momentum of the electron act on the electron spin through the spin-orbit coupling of the ion. This crystal field-orbit spin-orbit mechanism for spin-lattice relaxation is the dominant mechanism in the majority of ions in ionic solids. A major exception to this would be $S = \frac{1}{2}$ s-electron paramagnets and ions with low-lying excited electronic states where the Orbach–Aminov or Raman processes can be important.

4.2 Rotational Oscillations

In paramagnetic systems the EPR frequency is often a function of the angle between the radical and the external magnetic field applied to the sample. This can be due to the zero-field splittings of an $S > \frac{1}{2}$ system, an anisotropic hyperfine interaction, an anisotropic g-tensor, or an anisotropic exchange interaction. In these cases the electron spin is not aligned along the external magnetic field for most orientations of the radical relative to the external field. The exact direction of the spin quantization axis is a function of the orientation of the radical.

Some of the vibrations and phonons in a solid act on a radical within the solid as a small rotation. This small rotational perturbation of the radical is translated by the anisotropic interactions into a small change in the direction of the quantization axis of the electron spin. One description of such a rotation is in terms of the electron spin operators S_+ and S_- that produce changes in the electron spin state.

Since the exact quantization direction of the electron depends on the orientation of the radical with respect to the external magnetic field, this relaxation mechanism is characterized by a large anisotropy of the electron spin-lattice relaxation rate. It is most easily identified in single crystals rather than in disordered systems, although it should be no less important in the spin-lattice relaxation of radicals in disordered systems.[55,76]

4.3 Modulation of Hyperfine Interaction

Another interaction that can be modulated by nuclear motion or vibrations is the hyperfine interaction. Changes in either the electron wavefunction or the position of a nucleus can change the isotropic hyperfine interaction of the nucleus. Since the isotropic hyperfine interaction contains the operators $S_x I_x$ and $S_y I_y$, which can be rewritten in terms of S_+ and S_-, modulation of the isotropic hyperfine interaction can produce spin-lattice relaxation.

Likewise, modulation of the anisotropic hyperfine interaction can also produce spin-lattice relaxation, whether the modulation is a modulation of the direction of the principal axes or of the magnitude of the principal values of the interaction. It is important to note that, although the terms in the hyperfine interaction containing S_x and S_y are usually unimportant in the determination of the EPR spectrum, they can be more important than any other interaction in determining the electron spin-lattice relaxation rate.

Spin-lattice relaxation rates have been calculated for several systems based on this electron spin-lattice relaxation mechanism.[55,57,58,60,61,77-79]

4.4 Modulation of the Zero-Field Splitting Tensor

The electron spin-electron spin interaction can be relatively large in triplet states and radical pairs. Since this interaction contains operators S_+

and S_-, a modulation of this interaction can be a mechanism for electron spin-lattice relaxation in these systems. This is the mechanism originally proposed for the relaxation of transition metal ions in ionic crystals by Waller in 1932[49] and expanded by Al'tshuler.[80,81] In radical clusters or pairs, modulation of the distance between radicals or their orientation with respect to the external magnetic field will modulate the electron spin-spin interaction and cause electron spin-lattice relaxation.[82-84]

In molecular triplets, modulation of the electron spin-spin interaction or zero-field splitting tensor can be produced in other ways. Since a triplet state is a non-Kramers system, an electric field can perturb the spin system. This can be from the motion of ions or electric dipoles that are external to the triplet, or it can be from the motion of charges (nuclei) that make up the triplet. This is equivalent to the modulation of the crystal field around a paramagnetic ion.

The zero-field splitting tensor of a triplet in a solid can also be modulated by the exchange interactions with molecules in its environment. These exchange interactions delocalize a small amount of the triplet spin onto the neighboring molecules, producing a state that has a small component of a triplet exciton. This exciton component has a different zero-field splitting tensor from that of the triplet molecule, and the tensor of the triplet in the solid is a weighted average of the two tensors. Since the exchange interaction between the triplet and the other molecules near it can be modulated by modulating the distance between them, the tensor for the triplet in the solid will be modulated also. This may, of course, produce spin-lattice relaxation.[85,86]

4.5 Electric Field Modulation

An electric field can, of course, affect a non-Kramers spin that has an integral electron spin. An electric field, for a triplet, alters the zero-field splittings. In a half-integral spin system such as a doublet free radical, an electric field is not able to influence the spin. However, an applied magnetic field mixes other spin states with the ground state doublet spin states in unequal amounts. One result of this mixing is that an electric field becomes capable of affecting the spin system in an amount proportional to the applied magnetic field. Thus a time-dependent electric field acting on the radical can produce spin-lattice relaxation, but the perturbation operator will be proportional to the applied magnetic field. This field dependence will result in a different dependence of the relaxation rate on EPR frequency than will perturbations that are independent of magnetic field.

5 RELAXATION MECHANISMS AND PROCESSES IN NONIONIC SOLIDS

An attempt is made in this section to identify the different spin-lattice relaxation mechanisms and processes that are actually involved in the relaxation of paramagnetic centers in nonionic solids. The relaxation of paramagnetic ions in ionic solids has received a great deal of study both before and after the advent of EPR spectroscopy. Since the mechanisms and processes involved are fairly well known and have been described many times,[48,87–90] they are not discussed here.

As far as possible, all time domain EPR measurements of electron spin-lattice relaxation rates are discussed as well as selected measurements using other techniques that are particularly relevant to the determination of the mechanisms and processes involved.

5.1 s-Electron Atoms and Ions

Atoms and ions with the unpaired electron in an s-orbital generally have no low-lying electronic excited states, very little if any g-tensor anisotropy, and no internal vibrations. This greatly restricts both the mechanisms and processes available for relaxation.

5.1.1 HYDROGEN ATOMS

Hydrogen atoms in solids appear to be relatively unperturbed electronically by their surroundings. The hyperfine coupling is very isotropic and is nearly identical with the coupling in the gas phase, which is about 505 G. The g-factor is also very isotropic and is nearly equal to the free electron g-factor. In solids containing nuclei with nonzero nuclear moments, the EPR spectrum of the hydrogen atom often exhibits some hyperfine coupling with the nearest nuclei.

There have been three experiments in which the spin-lattice relaxation rate of trapped hydrogen atoms was measured in the presence and absence of oxygen. In both silica gel[91,92] and HD[93] the hydrogen atoms were found to relax faster in the presence of oxygen. Hydrogen atoms in glassy sulfuric acid were also observed to have spin-lattice relaxation rates that could be altered by the addition of paramagnetic ions. This was attributed to cross relaxation and spin-lattice relaxation induced by the paramagnetic ions. From these studies the spin-lattice relaxation rates of the paramagnetic ions were deduced.[36,94]

A careful study of the hydrogen and deuterium atoms in single crystals of CaF_2 was performed by Castle, Feldman, and Murphy.[95] The atoms are located in interstitial sites surrounded by fluorine ions and exhibit rather large fluorine hyperfine coupling with their nearest neighbors. Between 2.1 and 165 K the hydrogen atoms showed electron spin-lattice relaxation rates that are characteristic of a direct process, a Raman process and an Orbach–Aminov process with an excited state with an energy corresponding to 850 K. Deuterium atoms showed the same processes with the same rate coefficients—except for the excited-state energy, which was 640 K. The Orbach–Aminov process was postulated to involve a local vibrational mode of the lattice that involves the interstitial hydrogen atom. The absence of a hydrogen isotope effect on the direct and Raman processes implies that the relaxation mechanism does not involve the hydrogen hyperfine interaction.

Other experiments by Feldman, Castle, and Wagner[96] examined hydrogen atoms in fused silica. Aside from a low-abundance isotope of silicon there are no nuclear moments in the sample other than those of the occasional proton, and the EPR lines of the hydrogen atom are quite narrow, in contrast to hydrogen atoms in CaF_2. The interpretation of the relaxation rate is rather complex. A direct process is observed that depends on the sample history. It is interpreted as indicating a distribution of spin-lattice relaxation rates for hydrogen atoms in slightly different environments. They also observed an Orbach–Aminov process. The exact nature of the excited states involved is not clear. One interpretation proposed by Murphy[44] involves a single excited state with an energy 16 K above the ground state involving tunneling of the hydrogen atom in a double potential well with no other thermally accessible states. A second interpretation involves motion of the H atom in a one-dimensional square well with the first excited state 14 K above the ground state. In the square-well model several excited states can contribute to the relaxation via magnetic dipole coupling from nearby magnetic lattice nuclei or via electronic field coupling from nearby lattice ions. The square-well model fits the data slightly better than the tunneling model.

Actually the relaxation can occur with either the paramagnetic species or a lattice magnetic nucleus or ion undergoing the restricted tunneling motion or constrained to the square well. The identity of the species undergoing either motion in a box or tunneling can only be conjectured, since no experiments on the relaxation of deuterium atoms in fused silica were reported.

Measurements of the spin-lattice relaxation rate of hydrogen atoms trapped in concentrated sulfuric acid glass at 77 K have been made using spin-echo techniques.[37] At low hydrogen atom concentrations the spin-lattice relaxation rate of the hydrogen atoms appeared to be composed of the independent relaxation of a collection of hydrogen atoms with different spin-lattice relaxation rates. At concentrations of hydrogen atoms above 10^{17} cm^{-1}, the hydrogen atoms underwent cross relaxation with each other and the spin-lattice relaxation took the form of a single exponential decay. Hydrogen atoms in quartz were also examined and equivalent results obtained. The conclusions were that the hydrogen atoms are trapped in nonequivalent traps and that nothing inconsistent with the earlier results of Feldman, Castle, and Wagner[96] was observed.

A study of hydrogen and deuterium atoms in 85 mass % H_3PO_4 in H_2O and in 85 mass % D_3PO_4 in D_2O by Bowman and Kevan[97,98] found extensive cross relaxation to other spin species at low temperatures, a linearly temperature dependent process at intermediate temperatures and another process with a different temperature dependence at higher temperatures. The linearly temperature dependent process showed no deuterium isotope effect. This process for hydrogen atoms is an order of magnitude more effective in phosphoric acid than is the direct process in fused silica or in CaF_2. On the basis of this and by analogy with the relaxation mechanisms for trapped electrons in glassy solids,[45,97] Bowman and Kevan proposed that the linearly temperature dependent process is the high-temperature region of an Orbach–Aminov process involving the tunneling of either the hydrogen atom or a phosphate group with an extremely small splitting between the tunneling levels so that $\Delta E \ll kT$. The details of this process were sketched in Section 3.4. It is somewhat different from that described by Murphy[44] in that the potential for the tunneling particle is asymmetric and the tunneling rate is rapid enough to be treated as a random process.[45]

The relaxation mechanisms involved in the relaxation of hydrogen atoms in these systems are largely unknown. The modulation of hyperfine interaction with neighboring phosphorus nuclei[97,98] is a conjecture at this stage. The direct process in fused silica and in CaF_2 has identical magnitudes, and it is difficult to imagine why these two solids should have the same effect on the hydrogen atom. The mechanism for the Raman process in CaF_2 is unknown also. The Orbach–Aminov process in fused silica can be explained by the modulation of the electron spin by an electric field during the motion of a particle in a box. Although electric fields do not affect a spin-$\frac{1}{2}$ particle in zero magnetic field, they may do so in a finite magnetic field, so this mechanism could be tested by a study of the magnetic field dependence of this relaxation process.

5.1.2 SILVER ATOMS

Michalik and Kevan[99] have examined the electron spin-lattice relaxation rate of silver atoms in a variety of disordered solids between 6 and 80 K.

They examined silver atoms in H_2O, D_2O, CH_3OH, CH_3OD, CD_3OD, ethanol, propylene carbonate, and 2-methyltetrahydrofuran. They interpreted their results in terms of a general $\exp(\Delta E/kT)$ process at high temperatures and in terms of an Orbach–Aminov process involving tunneling states of the lattice molecules at low temperature.[45] The relaxation seems to be dominated by the tunneling process below 40 K. Deuteration experiments show that the tunneling nuclei are protons and that in methanol the methyl protons have more tunneling modes available than the hydroxyl protons.

In polycrystalline ice matrices it is possible to stabilize silver atoms with two different orientations of surrounding water molecules.[100] The fact that silver atoms can exist in a variety of different sites in polycrystalline ice and that these sites readily intraconvert by thermal and optical excitation[100] is probably related to the apparent existence of many proton tunneling modes near silver atoms in ice matrices. It was found that the coefficient of the tunneling mode relaxation process differs for silver atoms in different sites in ice matrices. The two sites studied are designated A and D. Silver atoms in site A are hydrogen bonded to their first solvation waters and should have a stronger nuclear dipolar interaction with the nearest protons than silver atoms in site D, in which the first solvation shell waters are hydrogen bonded more efficiently to the bulk water. If the densities of the proton tunneling modes are similar in the two cases, one would expect site A to show a stronger tunneling relaxation process. It was in fact found that site A relaxes about five times more efficiently by tunneling mode interactions than site D. It was also noted that there was a rough correlation between the strength of the tunneling mode relaxation interaction and the static dielectric constant of the various matrices in which the silver atoms were trapped. This trend probably reflects a change in the strength of the electron-nuclear dipolar interaction that is modulated by the tunneling modes. This study demonstrated that the tunneling relaxation process is quite sensitive to the immediate environment of the trapped paramagnetic species and can differentiate small geometrical differences as well as indicate the rigidity or orderedness of the immediate environment.

5.1.3 E'-Centers in Quartz

Castle et al.[75,101] have examined the relaxation of E'_1- and E'_2-centers in quartz. A direct process was observed in the case of the E'_1-center and a variety of Raman processes with unusual temperature dependencies in both samples. The unusual Raman processes are due to the departure of the lattice from the Debye model due to the E'-centers themselves. These centers are defects in the structure of the solid and have various vibrational modes associated with them that produce the strange temperature dependencies observed. The distinction between these Raman processes and Orbach–Aminov processes is somewhat arbitrary as elements of both processes are involved. The relaxation mechanisms are not known for this system.

5.1.4   F-Centers and Trapped Electrons

The relaxation of F-centers in a variety of ionic crystals has been extensively studied. We note only three studies on F-centers in alkali halides that report the relaxation mechanisms and processes involved.

Feldman, Warren, and Castle[102,103] made very careful studies of the electron spin-lattice relaxation of F-centers in alkali halides. They were able to control the concentrations of F-center aggregates and obtained the spin-lattice relaxation properties of both isolated F-centers and aggregates. They were able to obtain not only the temperature dependence of the relaxation rate but also the field dependence. The isolated F-centers exhibited a direct and a Raman relaxation process. The direct process was proportional to the square of the magnetic field and the Raman process was independent of field. This indicates a time-dependent perturbation responsible for relaxation that is independent of magnetic field. Modulation of the electron nuclear hyperfine interaction with surrounding nuclei was chosen as the most probable mechanism for both processes.

Panepucci and Mollenauer[104] investigated F-centers in KBr and KI crystals at 1.6 K between 0 and 50 kG. They measured the magnetic field dependence of the direct process and discovered not only a contribution from modulation of the electron nuclear hyperfine interaction proportional to H^2, but also, at very high magnetic fields, a contribution from electric field modulation proportional to H^4.

Bowman and Kevan[45,97,98] have examined trapped electrons in a number of aqueous and organic glassy frozen matrices. The trapped electron in these systems is somewhat similar to F-centers in that the electron is localized in a vacancy in the matrix and exhibits some hyperfine interaction with nearby nuclei. The electric field of the trapped electron orients the molecules nearest it.

The spin-lattice relaxation rate in the different matrices showed the presence of some high-order relaxation process and an Orbach–Aminov process involving low-energy tunneling modes.

The Orbach–Aminov process involved the tunneling of nuclei in the environment of the trapped electron. This tunneling motion modulates the electron nuclear hyperfine interaction and causes electron spin-lattice relaxation. By the selective use of isotopic substitutions, the nuclei and the functional groups involved in the tunneling were identified for a few matrices. In $10M$ NaOH glasses, the protons and ^{17}O nuclei have no observable effect on the Orbach–Aminov process, leaving the sodium nuclei as the only candidate for the tunneling species. Because the sodium ions are charged, they can produce a modulation of their electric field at the trapped electron as well as a modulation of their hyperfine interaction. In methanol and ethanol glasses protons on the methyl groups were much more effective in electron spin-lattice relaxation than the hydroxyl protons, even though there is a much larger hyperfine interaction with the hydroxyl protons. This was interpreted as indicating that the methyl protons could tunnel much more easily than the hydroxyl protons. Combined with the data from $10M$ NaOH glass, it was suggested that the lack of participation of the nearest protons with the largest hyperfine interaction in the relaxation of the trapped electron indicates that the electric field of the trapped electron strongly restricts motion of the nearest nuclei that it has ordered in its immediate environment.

5.2 Organic Free Radicals

The first measurement of the spin-lattice relaxation time of an organic free radical was reported in 1957 by Weissman, Feher, and Gere.[105] They found the relaxation time of the triphenylmethyl radical in a single crystal to be 1 msec at 300 K, 5 msec at 77 K, 4 min at 4.2 K, and 20 min at 1.2 K. They did not attempt at that time to explain the magnitude or temperature dependence of the relaxation time of this radical. Since then there have been many studies of the relaxation rate of free radicals and much discussion of the mechanisms and processes involved. Unfortunately, many of the measurements have used the progressive saturation technique and yielded results of dubious validity. However, some of these results are included in this discussion when they shed some light on the problem of identifying the processes and mechanisms involved in the electron spin-lattice relaxation of free radicals in solids.

5.2.1 CONSTANT TEMPERATURE STUDIES

A series of substituted acetylene radicals with different g-factors was studied at 4.2 and 77 K by saturation recovery methods.[106] The spin-lattice relaxation rate was found to be independent of Δg at 4.2 K but dependent on Δg at 77 K. Hence the contribution of spin-orbit coupling to the relaxation appears to increase with temperature. Radicals in γ-irradiated linear dicarboxylic acids at 77 K had spin-lattice relaxation rates uncorrelated with chain length.[107] Similarly, the spin-lattice relaxation rate of radicals in γ-irradiated amino acids at 4.2 K did not seem to be correlated with any radical property.[108]

At 77 K radicals formed by irradiation of single crystals of long-chain alkanes had much faster spin-lattice relaxation rates when the radical was located next to the terminal methyl group than when it was located between methylene groups.[109] It was suggested that methyl rotation and the resultant modulation of the hyperfine interaction was an important relaxation mechanism at this temperature. Curious deuteration effects were noted in this system.

The major radical in irradiated malonic acid was examined by pulsed ELDOR techniques at room temperature.[110] The allowed hyperfine lines in the EPR spectrum had a sevenfold faster spin-lattice relaxation rate than the forbidden hyperfine lines. According to the theory developed by Simizu,[111] this indicates that the modulation of the hyperfine interaction of the CH proton in the radical is not an important relaxation mechanism.

The spin-lattice relaxation of diphenyl nitroxide has been examined at 80 K and 300 K in a benzophenone crystal.[112] The relaxation rate is concentration dependent and anisotropic. The spin-lattice relaxation rate of vitreous carbon that had been carbonized at various temperatures has been measured at 78 and 298 K in the presence and the absence of molecular oxygen.[113] The spin-lattice relaxation rate of the paramagnetic centers increases as the carbonization temperature increases, and the rate is also increased in the presence of molecular oxygen.

5.2.2   Temperature-Dependent Studies

A number of peroxide radicals formed by radiation have been studied between 1.8 and 4.2 K in several crystalline halopolyethylenes.[114] A direct relaxation process was observed whose magnitude increased with increased spin-orbit coupling. The spin-lattice relaxation rate of α,α'-diphenyl-β-picryl hydrazyl (DPPH) in polystyrene has been examined between 1 and 300 K.[115,116] The rate is proportional to the temperature throughout the whole range. It is rather unlikely that this is a direct process at the higher temperatures. It is more likely that the linear temperature dependence is due to an Orbach–Aminov process in the high-temperature limit where kT is much larger than the energy of the excited state involved. The most likely excited state that has an energy less than 1 K is that of a particle tunneling between two positions in the polystyrene lattice.

The nitroxide radical Tanol has been examined in a number of frozen solutions between 20 and 120 K.[117-119] The spin-lattice relaxation rate was found to be proportional to $T^{2.2}$. This was interpreted as indicating that the relaxation was produced by small hindered rotational motions in the solid that modulated the anisotropic g-tensor and hyperfine interaction. The spin-lattice relaxation rate was also observed to be anisotropic, as would be expected from such a mechanism. It is interesting to note that a direct measurement of the spin-lattice relaxation rate showed that it was the same in a deuterated and a protonated matrix, whereas progressive saturation estimates of the relaxation rate were sensitive to matrix deuteration. This illustrates the difficulties involved in determination of spin-lattice relaxation rates using continuous wave (cw) methods.

Brown and Sloop[120,121] have examined the spin-lattice relaxation of a number of organic free radicals in glasses. They found that the spin-lattice relaxation rate of orbitally degenerate free radicals is much larger than the rate of orbitally nondegenerate free radicals. In addition, the relaxation rate was found to be concentration dependent, possibly implying cross relaxation to rapidly relaxing radical clusters.

Michalik and Kevan[122] have examined the spin-lattice relaxation of CH_3 and CD_3 radicals in CH_3OH, CD_3OD, 3-methylhexane, and 2-methyltetrahydrofuran organic glasses from 5 to 100 K. The relaxation rate was linear with temperature at low temperatures and was fitted by a general process with an $\exp(\Delta E/kT)$ temperature dependence at higher temperatures. The process linear with temperature was attributed to an Orbach–Aminov process involving the tunneling of proton-containing groups in the matrix.[45] There was little difference in the rate of this process between CH_3 and CD_3 radicals in the same matrix, but deuteration of the matrix decreased the tunneling relaxation rate fourfold. The relaxation rate appears to be more rapid in those matrices that have the greater concentration of methyl groups (not radicals), suggesting that the tunneling group is a methyl group (but not the radical) in these glassy matrices. There also appears to be a correlation between the spin-lattice relaxation rate and the decay of methyl radicals in these matrices.

The spin-lattice relaxation rates from 20 to 120 K of anthroquinone anion radicals in ethanol glasses have been reported.[123] Relaxation is claimed to be due to several Orbach–Aminov processes in which the excited state involved is a vibrational mode of the radical hydrogen bonded to the ethanol solvent. A similar study of semiquinone radical anions in polymer resins found that relaxation was due to an Orbach–Aminov process that involved vibrational modes external to the radical.[124] It was also observed that rapid temperature changes of the sample were communicated more slowly to the vibrational modes responsible for electron spin-lattice relaxation.

The $CH_3C \cdot R_1R_2$ radical in irradiated acetyl-(d, l)-alanine was observed to have a spin-lattice relaxation rate between 2 and 300 K that was proportional to csch ($\Delta E/kT$).[47] The excited state in this Orbach–Aminov process is the tunneling rotation of the methyl group in the radical.

The CO_3^{3-} radical in calcite has been studied between 2.2 and 4.2 K.[125] From the temperature and magnetic field dependence of the electron spin-lattice relaxation rate, a direct process involving modulation of the electron nuclear dipolar interaction with surrounding nuclei is responsible for the relaxation.[125]

Dalton, Kwiram, and Cowen have examined the spin-lattice relaxation in a series of γ-irradiated hydro-, deutero-, and fluorocarbon single crystals.[126,127] The relaxation rate at low temperatures was often proportional to the square of the temperature, which they were not able to explain. They observed the direct, Raman, and Orbach–Aminov processes. In the fluorocarbon radicals the relaxation mechanism is modulation of the electron nuclear hyperfine interaction and the relaxation rate is quite anisotropic. In the other radicals the relaxation mechanism was reported to be the modulation of the electric field around the radical and the relaxation rate was isotropic. The Orbach–Aminov processes they observed involved the internal vibrational modes of the molecules or radicals.

5.3 Triplets

There is a great deal of interest in the intersystem crossing mechanisms in photoexcited triplet states. These mechanisms can often be studied using time domain magnetic resonance either at zero magnetic or finite magnetic fields. In analyzing the triplet kinetics to extract the intersystem crossing rates, the spin-lattice relaxation rates are often obtained. Unfortunately, the rates are often reported at a single field or single temperature that it is impossible to discern the relaxation mechanisms or processes involved. There have really been only a handful of experimental studies that examined the subject of triplet spin-lattice relaxation.

5.3.1 EXCHANGE-COUPLED PAIRS

There have been three reports of the spin-lattice relaxation of exchange-coupled radicals.[128–130] In all three cases the relaxation process involved was an Orbach–Aminov process that involved the singlet state of the pair that is located very near the triplet energy levels of the pair.

5.3.2 ZERO MAGNETIC FIELD

The spin-lattice relaxation rate between the three spin levels of the lowest excited triplet state of acridine in a biphenyl host indicates a direct process is involved.[131] There were three distinct rates corresponding to relaxation between the three pairs of levels. A direct process and a Raman process were observed for the relaxation of pyrazine-d_4 in a cyclohexane glass.[23] It was concluded that the relaxation mechanism involves the modulation of the zero-field-splitting tensor.

5.3.3 HIGH MAGNETIC FIELD

Wolfe[20] has reported a rather extensive investigation of the spin-lattice relaxation of a number of photoexcited triplets in durene and biphenyl hosts. The relaxation rates are anisotropic and indicate the presence of both direct and Raman processes. Since the relaxation rates for all the triplets are similar, as are the zero-field splittings, it would seem that modulation of these splittings is the relaxation mechanism. The spin-orbit coupling and hyperfine interaction of the different triplets studied were quite different.

Gille and Maruani[132] were able to explain qualitatively the anisotropy of the spin-lattice relaxation of pyrene with a model involving the modulation of both the magnitude and the direction of the zero-field-splitting tensor.

An examination of the relaxation in quinoxaline triplets in different hosts revealed the operation of both direct and Raman relaxation processes.[24] When the triplet energy of quinoxaline was near the triplet energy of the host molecules, the relaxation mechanism seemed to involve a modulation of the partial delocalization of the triplet state onto host molecules by an exchange interaction.

One study reports on relaxation of quinoxaline in a naphthalene crystal at 1.7 K at magnetic fields up to 180 kG.[133] The field dependence for the direct process observed is not incompatible with a mechanism involving modulation of the zero-field-splitting tensor.

An interesting report on the spin-lattice relaxation of the diphenyl-methylene triplet finds only a slight anisotropy in the relaxation rates.[134] However, although the multiple exponential decays expected in a triplet state were observed, only a single spin-lattice relaxation rate was reported.

6 CONCLUSIONS

At the present time it is difficult to draw any conclusions about the relative importance of the various relaxation mechanisms and processes for the electron spin-lattice relaxation of triplet state molecules. The few good observations that have been reported are not sufficient bases for good generalizations. In addition, many studies have been at rather high guest molecule concentrations, where clusters are easily formed and the relaxation may be influenced by the presence of triplet dimers or larger aggregates, as was pointed out by Rousslang and Kwiram.[135] It is to be hoped that in the future there will be a greater number of careful studies of the spin-lattice relaxation of triplet molecules as a function of temperature, particularly at zero field.

From the sparse data at present, it would appear that the spin-lattice relaxation of exchange-coupled radical pairs at low temperatures is largely due to an Orbach–Aminov process with the nearby singlet spin state of the pair.

Generalizations about spin-lattice relaxation in doublets are more easily made. The evidence indicates that the spin-lattice relaxation *mechanism* depends on the properties of the radical. When the radical has a large spin-orbit coupling, modulation of it is an effective mechanism. When there is large anisotropy, modulation of that, presumably by orientational modulations, is effective. When the hyperfine interaction is large and can be easily influenced by nuclear motions, that is an effective mechanism.

The relaxation *processes* important in spin-lattice relaxation in doublets are more a function of the matrix than of the radical. The obvious exception is when the radical contains low-lying excited states and an Orbach–Aminov process involving that state is possible. Otherwise it appears that direct and Raman processes are more commonly found in single crystals than in glasses. In glasses or disordered systems there seem to be many localized vibrational or tunneling modes that are very effective in producing spin-lattice relaxation. In most disordered systems, particularly those containing a large number of methyl groups, large amplitude nuclear motion modulating the electron nuclear hyperfine interaction between the radical and matrix nuclei seems to be a particularly effective mechanism for an Orbach–Aminov process.

When a very-low-energy vibration or tunneling splitting is involved as the excited state in an Orbach–Aminov process, it is not readily apparent how to distinguish it from a direct process, since both of them have the same temperature dependence in the temperature region readily accessible in spin-lattice relaxation experiments.

In the derivation of the Orbach–Aminov process involving tunneling,[45,98] the relaxation rate was shown to be inversely proportional to the square of the EPR frequency when hyperfine modulation was involved. The direct process, on the other hand, is proportional to the square or the fourth power of the EPR frequency. Unfortunately, the frequency dependence of the tunneling Orbach–Aminov process has not yet been measured.

Another method to distinguish direct and tunneling Orbach–Aminov processes is by the relative magnitude of the relaxation rate. The direct process is expected to be observed unless the Orbach–Aminov process is more effective. Table 1 lists the coefficient A of $T_1^{-1} = AT$ for the direct process and the high-temperature limit of the Orbach–Aminov process involving tunneling or low-energy vibrations for a number of radicals. Except for hydrogen atoms, there is a substantial gap between the direct process coefficients and those of the Orbach–Aminov process. This test, combined with matrix isotopic substitution experiments, can form a basis for distinguishing between these two processes.

In view of the evident importance of the Orbach–Aminov process in the spin-lattice relaxation of radicals and radical pairs in solids, it would seem to be a good method for the determination of the energy of various low-lying excited states. Although this has been done successfully in a few systems,[123–128,136] it certainly has not been utilized to the fullest possible extent. The splitting between the two lowest electronic states in a radical ion of a molecular dimer, zero-field splittings of transition metals in

Table 1 Comparison of Magnitudes for Electron Spin-Lattice Relaxation Rate Processes Linear in Temperature as Measured by the Coefficient A of $T^{-1} = AT$ for the Direct Process and the High-Temperature Limit of the Orbach–Aminov Process Involving Low-Energy Tunneling or Vibrational States

Radical	A (direct) s K	A (Orbach–Aminov) s K	Reference
H, D atoms	0.028		95
H atom	0.01–0.08		96
H, D atoms		0.2	97, 98
Ag atom		20–200	99
E'-centers	0.0011		75, 101
F-centers	0.00005		102, 103
Trapped electrons		1–40	97
CO_3^{3-}	0.00014		125
CH_3, CD_3		30–100	122
Irradiated carboxylic acids	0.005–0.04	1–300	126, 127

enzymes, and very-low-energy bending or rotational modes of radicals would all appear to be suitable subjects for study using spin-lattice relaxation as a probe. In the next few years the processes and mechanisms of spin-lattice relaxation of triplets and doublet radicals may become less of an object and more of a tool for research.

REFERENCES

1. A. Abragam and B. Bleaney, *Electron Paramagnetic Resonance of Transition Ions*, Clarendon, Oxford, 1970, Chap. 10.

2. L. Kevan and L. D. Kispert, *Electron Spin Double Resonance Spectroscopy*, Wiley-Interscience, New York, 1976.

3. (a) A. Carrington and A. D. McLachlan, *Introduction to Magnetic Resonance*, Harper & Row, New York, 1967, Chaps. 1, 2; (b) C. P. Schlichter, *Principles of Magnetic Resonance*, Harper & Row, New York, 1963, Chap. 3.

4. C. P. Poole and H. A. Farach, *Relaxation in Magnetic Resonance*, Academic, New York, 1971, Chaps. 1, 2, 8, 9.

5. A. M. Portis, *Phys. Rev.*, **91**, 1071 (1953).

6. T. G. Castner, *Phys. Rev.*, **115**, 1506 (1959).

7. E. L. Wolf, *Phys. Rev.*, **142**, 555 (1966).

8. N. A. Efremov and M. A. Kozhushner, *Teor. Eksp. Khim.*, **8**, 53 (1972).

9. M. K. Bowman, H. Hase, and L. Kevan, *J. Mag. Resonance*, **22**, 23 (1976).

10. O. P. Zhidkov, Ya. S. Lebedev, A. I. Mihailov, and B. N. Provotorov, *Theor. Exp. Chem.*, **3**, 135 (1967).

11. O. P. Zhidkov, V. I. Muromtsev, I. G. Akhvleiani, S. N. Safronov, and V. V. Kopylov, *Sov. Phys. Solid State*, **9**, 1095 (1967).

12. G. T. Trammell, H. Zeldes, and R. Livingston, *Phys. Rev.*, **110**, 630 (1958).

13. K. M. Salikhov and Yu. D. Tsvetkov, Chapter 7 of this book.

14. I. Brown, Chapter 6 of this book.

15. See reference 4, Chapter 4.

16. K. J. Standley and R. A. Vaughan, *Electron Spin Relaxation Phenomena in Solids*, Adam Hilger, London, 1969, Chaps. 7, 8, Appendix 1.

17. K. M. Salikhov, A. G. Semenov, and Yu. D. Tsvetkov, *Electron Spin Echo and Its Applications*, Nauka, Novosibirsk, 1976, Chap. 3 (in Russian).

18. M. Schwoerer and H. Sixl, *Z. Naturforsch.*, **24a**, 952 (1969).

19. H. Sixl and M. Schwoerer, *Z. Naturforsch.*, **25a**, 1383 (1970).

20. J. P. Wolfe, *Chem. Phys. Lett.*, **10**, 212 (1971).

21. J. Zuclich, J. U. von Schütz, and A. H. Maki, *Mol. Phys.*, **28**, 33 (1974).

22. H. Levanon and S. Vega, *J. Chem. Phys.*, **61**, 2265 (1974).

23. L. H. Hall and M. A. El-Sayed, *Chem. Phys.*, **8**, 272 (1975).

24. U. Konzelmann, D. Klipper, and M. Schwoerer, *Z. Naturforsch.*, **30a**, 754 (1975).

25. U. Eliav and H. Levanon, *Chem. Phys. Lett.*, **36**, 377 (1975).

26. C. C. Felix, S. S. Kim, and S. I. Weissman, *Chem. Phys. Lett.*, **48**, 29 (1977).

27. D. M. Daraseliya and A. A. Manenkov, *Zh. Eksp. Teor. Fiz. Pisma*, **11**, 337 (1970).

28. D. M. Daraseliya, A. S. Epifanov, and A. A. Manenkov, *Zh. Eksp. Teor. Fiz.*, **59**, 445 (1970).

29. J. H. Freed, *J. Phys. Chem.*, **78**, 1155 (1974).

30. A. D. Trifunac and M. C. Thurnauer, Chapter 4 of this book.

31. J. Schmidt and J. H. van der Waals, Chapter 9 of this book.

32. See reference 16, Chapter 5.

33. See reference 16, Chapter 7.

34. T. Förster, *Z. Naturforsch.*, **4a**, 321 (1949).

35. D. Tse and S. R. Hartmann, *Phys. Rev. Lett.*, **21**, 511 (1968).

36. A. D. Milov, K. M. Salikhov, and Yu. D. Tsvetkov, *Fiz. Tverd. Tela*, **14**, 2211 (1972).

37. A. D. Milov, K. M. Salikhov, and Yu. D. Tsvetkov, *Fiz. Tverd. Tela*, **14**, 2259 (1972).

38. W. S. Moore and T. Yalcin, *J. Mag. Resonance*, **11**, 50 (1973).

39. M. R. Smith and H. A. Buckmaster, *J. Mag. Resonance*, **17**, 29 (1975).

40. J. M. Ziman, *Electrons and Phonons*, Clarendon, Oxford, 1960, p. 30.

41. J. H. Van Vleck, *Phys. Rev.*, **57**, 426 (1940).

42. R. Orbach, *Proc. Roy. Soc. (Lond.)*, **A264**, 458 (1961).

43. L. K. Aminov, *Zh. Eksp. Teor. Fiz.*, **42**, 783 (1962).

44. J. Murphy, *Phys. Rev.*, **145**, 241 (1966).

45. M. K. Bowman and L. Kevan, *J. Phys. Chem.*, **81**, 456 (1977).

46. W. A. Phillips, *J. Low Temp. Phys.*, **7**, 351 (1972).

47. W. L. Gamble, I. Miyagawa, and R. L. Hartman, *Phys. Rev. Lett.*, **20**, 415 (1968).

48. A. A. Manenkov and R. Orbach, *Spin-Lattice Relaxation in Ionic Solids*, Harper & Row, New York, 1966.

49. I. Waller, *Z. Phys.*, **79**, 370 (1932).

50. R. de L. Kronig, *Physica*, **6**, 33 (1939).

51. C. B. P. Finn, R. Orbach, and W. P. Wolf, *Proc. Phys. Soc. (Lond.)*, **77**, 261 (1961).

52. R. Orbach, *Proc. Roy. Soc. (Lond.)*, **A264**, 485 (1961).

53. C. Y. Huang, *Phys. Rev.*, **154**, 215 (1967).

54. I. V. Aleksandrov and G. M. Zhidomirov, *Zh. Eksp. Teor. Fiz.*, **41**, 127 (1961).

55. I. V. Aleksandrov and A. V. Kessenikh, *Fiz. Tverd. Tela*, **6**, 1006 (1964).

56. I. V. Aleksandrov, *Zh. Eksp. Teor. Fiz.*, **48**, 869 (1965).

57. (a) I. V. Aleksandrov, *Teor. Eksp. Khim.*, **1**, 80 (1965); (b) I. V. Aleksandrov, *Teor. Eksp. Khim.*, **1**, 211 (1965).

58. I. V. Aleksandrov and A. V. Kessenikh, *Teor. Eksp. Khim.*, **1**, 221 (1965).

59. I. V. Aleksandrov, *Teor. Eksp. Khim.*, **2**, 67 (1966).

60. P. W. Atkins, *Mol. Phys.*, **12**, 201 (1967).

61. P. W. Atkins and M. T. Crofts, *Mol. Phys.*, **12**, 211 (1967).

62. P. W. Atkins and J. N. L. Connor, *Mol. Phys.*, **13**, 201 (1967).

63. V. L. Vinetskii and V. Ya. Kravchenko. *Fiz. Tverd. Tela.*, **7**, 319 (1965).

64. P. G. Klemens, *Phys. Rev.*, **125**, 1795 (1962).

65. D. W. Feldman, J. G. Castle, Jr., and J. Murphy, *Phys. Rev.*, **138**, A1208 (1965).

66. P. G. Klemens, *Phys. Rev.*, **138**, A1217 (1965).

67. I. V. Aleksandrov and V. P. Sakun, *Zh. Eksp. Teor. Fiz.*, **52**, 136 (1967).

68. V. P. Sakun, *Fiz. Tverd. Tela*, **9**, 2404 (1968).

69. R. Hernandez and M. B. Walker, *Phys. Rev.* **B4**, 3821 (1971).

70. V. Ya. Zevin and V. I. Konovalov, *Fiz. Tverd. Tela*, **14**, 866 (1972).

71. G. G. Sergeeva, *Fiz. Tverd. Tela*, **14**, 1511 (1972).

72. D. S. Tsitskishvili, *Fiz. Tverd. Tela*, **15**, 3422 (1973).

73. T. L. Reinecke and K. L. Ngai, *Phys. Rev.*, **B12**, 3476 (1976).

74. B. E. Vugmeister and V. I. Konovalov, *Fiz. Tverd. Tela*, **17**, 1219 (1975).

75. J. G. Castle, Jr., D. W. Feldman, P. G. Klemens, and R. A. Weeks, *Phys. Rev.*, **130**, 577 (1963).

76. J. Levy, *J. Phys.* **C, 2**, 1371 (1969).

77. V. Ya. Zevin, *Fiz. Tverd. Tela*, **3**, 599 (1961).

78. V. Ya. Kravchenko and V. L. Vinetskii, *Fiz. Tverd. Tela*, **7**, 3 (1965).

79. M. F. Deigen and I. I. Zheru, *Teor. Eksp. Khim.*, **2**, 366 (1966).

80. S. A. Al'tshuler, *Izv. Akad. Nauk SSSR*, **20**, 1207 (1956).

81. S. A. Al'tshuler, *Zh. Eksp. Teor. Fiz.*, **43**, 2318 (1962).

82. M. D. Glinchuk, V. G. Grachev, and M. F. Deigen, *Fiz. Tverd. Tela*, **8**, 3354 (1966).

83. L. L. Buishvili and M. D. Zviadadze, *Fiz. Tverd. Tela*, **9**, 1969 (1967).

84. V. G. Grachev, *Ukr. Fiz. Zh.*, **13**, 633 (1968).

85. V. I. Sugakov, *Fiz. Tverd. Tela*, **15**, 2042 (1973).

86. V. A. Andreev and V. I. Sugakov, *Fiz. Tverd. Tela*, **17**, 1963 (1975).

87. See reference 1.

88. See reference 4.

89. See reference 16.

90. S. A. Al'tshuler and B. M. Kozyrev, *Electron Paramagnetic Resonance*, Nauka, Moscow, 1972.

91. V. B. Kazanskii, G. B. Pariiskii, and A. I. Burshtein, *Opt. Spektr.*, **13**, 83 (1962).

92. S. A. Surin, G. M. Zhidomirov, B. N. Shemilov, and V. B. Kazanskii, *Teor. Eksp. Khim.*, **6**, 353 (1970).

93. J. C. Solem and G. A. Rebka, Jr., *Phys. Rev. Lett.*, **21**, 19 (1968).

94. A. M. Raitsimring and Yu. D. Tsvetkov, *Fiz. Tverd. Tela*, **11**, 1282 (1969).

95. D. W. Feldman, J. G. Castle, Jr., and J. Murphy, *Phys. Rev.*, **138**, A1208 (1965).

96. D. W. Feldman, J. G. Castle, Jr., and G. R. Wagner, *Phys. Rev.*, **145**, 237 (1966).

97. M. K. Bowman and L. Kevan, *Discuss. Faraday Soc.*, **63**, 7 (1977).

98. M. K. Bowman, Ph.D. Thesis, Wayne State University, 1975.

99. J. Michalik and L. Kevan, *J. Mag. Resonance*, **31**, 000 (1978).

100. B. L. Bales and L. Kevan, *J. Chem. Phys.*, **55**, 1327 (1971).

101. J. G. Castle, Jr., and D. W. Feldman, *Phys. Rev.*, **137**, A671 (1965).

102. D. W. Feldman, R. W. Warren, and J. G. Castle, Jr., *Phys. Rev.*, **135**, A470 (1964).

103. R. W. Warren, D. W. Feldman, and J. G. Castle, Jr., *Phys. Rev.*, **136**, A1347 (1964).

104. H. Panepucci and L. F. Mollenauer, *Phys. Rev.*, **178**, 589 (1969).

105. S. I. Weissman, G. Feher, and E. A. Gere, *J. Am. Chem. Soc.*, **79,** 5584 (1957).

106. T. S. Zhuravleva and A. V. Kessenikh, *Zh. Strukt. Khim.* **6,** 453 (1965).

107. A. M. Raitsimring and Yu. D. Tsvetkov, *Fiz. Tverd. Tela*, **11,** 1282 (1969).

108. G. Bürk and G. Schoffa, *Z. Naturforsch.*, **21a,** 296 (1966).

109. T. Gillbro and A. Lund, *Chem. Phys.*, **5,** 283 (1974).

110. M. Nechtschein and J. S. Hyde, *Phys. Rev. Lett.*, **24,** 672 (1970).

111. H. Shimizu, *J. Chem. Phys.*, **42,** 3603 (1965).

112. I. M. Brown, *J. Chem. Phys.*, **52,** 3836 (1970).

113. A. S. Fialkov, N. A. Mel'nikova, and B. G. Tarasov, *Zh. Fiz. Khim.*, **50,** 975 (1976).

114. I. V. Aleksandrov, L. Ya. Dzhavakhishvili, and G. D. Ketiladze, *Teor. Eksp. Khim.*, **2,** 97 (1966).

115. T. J. B. Swanenberg, N. J. Poulis, and G. W. J. Drewes, *Physica*, **29,** 713 (1963).

116. J. Turkevich, J. Soria, and M. Che, *J. Chem. Phys.*, **56,** 1463 (1972).

117. S. K. Rengan, M. P. Khakhar, B. S. Prabhananda, and B. Venkataraman, *J. Pure Appl. Chem.*, **32,** 287 (1972).

118. V. I. Muromtsev, N. A. Shteinshneider, S. N. Safronov, V. P. Golikov, A. I. Kuznetsov, and G. M. Zhidomirov, *Fiz. Tverd. Tela*, **17,** 813 (1975).

119. V. I. Muromtsev, V. V. Pomortsev, S. N. Safronov, V. P. Golikov, E. R. Klinshpont, V. K. Milinchuk, and G. M. Zhidomirov, *Fiz. Tverd. Tela*, **13,** 1062 (1971).

120. I. M. Brown, *J. Chem. Phys.*, **55,** 2377 (1971).

121. I. M. Brown and D. J. Sloop, *Chem. Phys. Lett.*, **1,** 579 (1968).

122. J. Michalik and L. Kevan, *J. Chem. Phys.*, **68,** 5325 (1978).

123. V. I. Muromtsev, G. A. Val'kova, K. K. Pukhov, D. N. Shigorin, and V. G. Cheredintsev, *Zh. Strukt. Khim.*, **16,** 909 (1975).

124. V. I. Volkov, V. I. Muromtsev, K. K. Pukhov, and V. G. Cheredintsev, *Fiz. Tverd. Tela*, **19,** 1230 (1977).

125. C. Y. Huang and S. A. Marshall, *Phys. Lett.*, **42A,** 49 (1972).

126. L. R. Dalton, A. L. Kwiram, and J. A. Cowen, *Chem. Phys. Lett.*, **14,** 77 (1972).

127. L. R. Dalton, A. L. Kwiram, and J. A. Cowen, *Chem. Phys. Lett.*, **17,** 495 (1972).

128. I. M. Brown and D. J. Sloop, *J. Chem. Phys.*, **47,** 2659 (1967).

129. E. A. Harris and K. S. Yngvesson, *J. Phys. C*, **1,** 1011 (1968).

130. L. Ya. Dzhabakhishbili, G. D. Ketiladze, and T. I. Sanadze, *Fiz. Tverd. Tela*, **10,** 2954 (1968).

131. D. A. Antheunis, B. J. Botter, J. Schmidt, P. J. F. Verbeek, and J. H. van der Waals, *Chem. Phys. Lett.*, **36,** 225 (1975).

132. X. Gille and J. Maruani, *18th Colloque Ampère*, 425 (1974).

133. K. F. Renk, H. Sixl and H. Wolfrum, *Chem. Phys. Lett.*, **52,** 98 (1977).

134. C. Cheng, T.-S. Lin, and D. J. Sloop, *Chem. Phys. Lett.*, **44,** 576 (1976).

135. K. W. Rousslang and A. L. Kwiram, *Chem. Phys. Lett.*, **39,** 226 (1976).

136. R. C. Herrick and H. J. Stapleton, *J. Chem. Phys.*, **65,** 4786 (1976).

4 TIME-RESOLVED ELECTRON SPIN RESONANCE OF TRANSIENT RADICALS IN LIQUIDS

Alexander D. Trifunac and Marion C. Thurnauer

Chemistry Division
Argonne National Laboratory
Argonne, Illinois

1   INTRODUCTION

Today there is increasing interest in the study of transient intermediates
in chemical reactions. Time-resolved electron spin resonance (ESR) spectroscopy is a tool that can provide many details of reaction dynamics of
transient radicals in solution. It is the purpose of this chapter to outline
and summarize the current achievements of time-resolved ESR spectroscopy in the microsecond and submicrosecond ($\sim$0.1–20 μs) domain in the
study of transient radicals in liquids. This work is developed along the
lines of chemical systems investigated. We discuss the work in areas of
radiation chemistry, photochemistry, and biophysics in the common basis
of all time-resolved ESR observations that is the phenomenon of chemically induced dynamic electron polarization (CIDEP).[1]

The study of transient free radicals has also been carried out using
steady-state ESR spectroscopy in radiation and photochemistry and by
various ESR-pulsed light techniques (flashlamp, light chopping, and light
modulation) with intermediate time resolution ($\sim$20–150 μs). These, so
far more common, approaches to the study of transient radicals in
solution have been covered in several reviews[2–5] and are not discussed
here.

1.1   History

The first studies of transient radicals in liquids were accomplished using
steady-state ESR with electron beam irradiation. The pioneering applications of ESR to radiation chemistry were reported by Fessenden and
Schuler.[6] In this study the first observation was made of what is now
called CIDEP, the phenomenon that is intimately related to all time-
resolved ESR studies of transient radicals in solution. The steady-state
ESR study of transient radicals in liquids by photolysis was extensively
applied by Livingston and Zeldes.[7,8]

The first microsecond time-resolved ESR spectrometer was described
by Smaller and co-workers at Argonne National Laboratory in 1968.[9]
Smaller's efforts were directed at the study of transient radicals in pulse
radiolysis using a Van de Graaff or linac accelerator as sources of the
electron pulse. Independently and at the same time at Oxford, Atkins and

McLauchlan have claimed development of a similar time-resolved ESR spectrometer.[1,10] The Oxford group used a pulsed nitrogen laser to study radicals produced by photolysis in liquids. The first time-resolved spectrometers achieved 2–5 μs time resolution using 2 MHz field modulation. More recently the use of direct (modulation-free) detection has allowed time-resolved ESR studies in the submicrosecond time domain.[11,12] Several groups have achieved time resolution of ~50–100 ns.[11,13] This is essentially the time resolution limit of an ESR spectrometer operating in X-band and using reasonably low-cavity Q ($Q \sim 1000$).

The transient radicals in liquids more often than not exhibit nonequilibrium electron spin population, CIDEP. Although first observed in 1963, it took almost 10 years to develop an adequate understanding of the CIDEP phenomena. The theoretical understanding of CIDEP has followed the development of the radical pair model of chemically induced dynamic nuclear polarization (CIDNP),[1] the analogous phenomenon involving observation of nonequilibrium population of nuclear spins in radical reaction products observed by nuclear magnetic resonance (NMR).

Indeed, as we illustrate, the study of radical reaction dynamics and CIDEP is the basis and purpose of most of the time-resolved ESR studies to date. The identification of transient radicals is usually only secondary, since these same radicals can often be identified by steady-state ESR methods or intermediate time resolution time-resolved ESR spectroscopy, both of which are experimentally easier techniques.

1.2 CIDEP Primer

Two mechanisms have been successfully used to explain the CIDEP observations to date: the radical pair model (RPM) and the triplet mechanism (TM) of CIDEP.[1] Both mechanisms can be operative separately and simultaneously in the same system, since triplet polarization is related to the process of radical creation while the radical pair polarization is created in the subsequent radical-radical interactions in solution. Triplet polarization results from the differing intersystem crossing population rates of the triplet state from the photoexcited singlet state and is found in the radicals produced from the triplet by electron or hydrogen transfer reactions. Also, regardless of their mode of creation, radicals diffuse in solution and the magnetic perturbations present in the radicals, as well as the Heisenberg exchange, may induce spin evolution that is ultimately observed as polarization (RPM).

The radical pair model of CIDEP was the first to be considered, since the radical pair model of CIDNP[14,15] had paved the way. Indeed, the first workers in the field of CIDNP speculated on the probable connection of CIDEP and CIDNP,[16] and the radical pair model of CIDEP was proposed by Kaptein and Oösterhoff at the same time along with the CIDNP model.[15] Further substantial development of RPM-CIDEP theory is due to Adrian.[17-19] The triplet model of CIDEP was developed by Atkins and McLauchlan[20,21] and independently by Wan, Wong, and Hutchison.[22] Recently, more sophisticated theoretical treatments have been advanced by several groups.[23-27] Notable is a series of papers by Pedersen and Freed[28-33] in which a theoretical treatment based on the stochastic Liouville equation is worked out for both radical pair and triplet CIDEP. Since it is not our purpose to dwell on the details of theoretical approaches, we present only a brief outline of ideas by which CIDEP can be understood.

1.2.1 RADICAL PAIR THEORY OF CIDEP

In the radical pair model nonequilibrium spin-state population is developed due to the radical-radical interactions between the radical pair components in solution. By the radical pair is meant that once two radicals encounter and separate in solution there is a finite chance that they will encounter again. Homolytic cleavage of a single bond in a molecule will give rise to a radical pair. In this case the two radicals will be initially correlated and one can consider the pair electron spin to be either a singlet (S) or triplet (T_{+1}, T_0, T_{-1}) state, depending on the multiplicity of the precursor. Radical pairs can also arise when two independently generated radicals collide. Such pairs initially have an equal distribution of all spin states.

The following sequence of events can give rise to nonequilibrium spin-state populations (polarization) in radicals. After radical creation, radical pairs with singlet character have greater probability of reaction on initial encounter, leaving in solution an excess of pairs with triplet character. The components of these pairs diffuse apart and subsequently undergo (nonreactive) reencounters. During the time interval between the initial and subsequent reencounters (several diffusive steps, each $\sim 10^{-12}$ s) the radical pair spin state evolves (singlet-triplet mixing) under the influence of magnetic interactions within the radicals and is further changed by the exchange interaction during reencounters. Although the magnetic interactions within the pair components cause the change of relative spin phasing, the exchange interaction translates this into nonequilibrium spin population in the radicals themselves.

The evolution of the radical pair wave function is given by the time-dependent Schrödinger equation. In units of $\hbar$,

$$[1 - J(2\hat{S}_1 \cdot \hat{S}_2 + \tfrac{1}{2}) + \mathcal{H}]\psi(t) = i\frac{\partial\psi}{\partial t}, \tag{1}$$

where $\mathcal{H}$ is the magnetic Hamiltonian.

$$\mathcal{H} = \beta(g_1\hat{S}_1 + g_2\hat{S}_2) \cdot \hat{H}_0 + \sum A_{1n}\hat{I}_{1n} \cdot \hat{S}_1 + \sum A_{2m}\hat{I}_{2m} \cdot \hat{S}_2, \tag{2}$$

where β is the Bohr magneton, $\hat{H}_0$ is the external magnetic field, g_1 and g_2 are the electron g-factors of radicals 1 and 2, $2J$ is $(E_S - E_T)$ the exchange splitting, $\hat{S}_1$ and $\hat{S}_2$ are the electron spin of radicals 1 and 2, respectively, and A_{1n} and $\hat{I}_{1n}$ are the isotropic hyperfine constant and nuclear spin of the nth nucleus of radical 1, respectively.

The off-diagonal matrix elements of $\mathcal{H}$ are illustrated:

$$\langle T_0, M_i, M_j \cdots | \mathcal{H} | S, M_i, M_j \cdots \rangle$$
$$= \tfrac{1}{2}[(g_1 - g_2)\beta H_0 + \sum A_i M_i - \sum A_k M_k], \tag{3}$$

$$\langle T_{\pm 1}, M_i \mp 1, M_j \cdots | \mathcal{H} | S, M_i, M_j \cdots \rangle$$
$$= \mp(8^{-1/2})[I_i(I_i + 1) - M_i(M_i \mp 1)]^{1/2}A_i, \tag{4}$$

where groups of magnetically equivalent nuclei give I_i, and M_i is the nuclear spin quantum number. For example, in the case of a radical pair $R_1H \cdots HR_2$ the $S-T_{-1}$ off-diagonal element connecting the indicated nuclear sublevels is

$$\langle T_{-1}\alpha_n\alpha_m | \mathcal{H} | S\beta_n\alpha_m \rangle = -8^{-1/2}A_1, \tag{4a}$$

where n and m indicate nuclear states of radicals R_1 and R_2, respectively.

Spins of the radicals experience different local magnetic fields, since their g-factors and hyperfine interactions differ, and thus they precess at different rates (about different directions). In the magnetic fields used in ESR experiments the S and T_0 levels are closest in energy, and $S-T_0$ mixing is thus dominant. As illustrated by the off-diagonal element of the magnetic Hamiltonian, Eq. (3), g-factor difference and the sums and differences of hyperfine coupling constants on radical pair components are responsible for $S-T_0$ mixing. In this $S-T_0$ spin sorting process the total number of α and β (electron and nuclear) spins is unchanged. Only the distribution of α and β electron spins in association with α_N and β_N nuclear spins is changed. Thus different hyperfine levels carry different polarization, or one component of the radical pair has more α spin while the other has more β spin if the g-factor difference of the two radical pair components is substantial.

If the mixing between S and $T_{\pm 1}$ levels is important, the overall number of α and β spins is changed. If the hyperfine mixing[34] is the dominant mixing mechanism, electron and nuclear spin flips occur. This is illustrated in the $S-T_{\pm 1}$ off-diagonal elements, Eq. (4).

Typically the observed CIDEP in transient radicals is as follows. The $S-T_0$ CIDEP yields an E/A pattern where the low field line is in emission (E) while the high field line is in enhanced absorption (A). The presence of a substantial g-factor difference can cause a shift of the whole spectrum to emission in the radical with higher g-factor and the complementary shift to enhanced absorption in the radical with the smaller g-factor.

Overall spectrum shift to emission in all radicals is observed if $S-T_{-1}$ ($J < 0$) mixing is contributing to radical polarization.

1.2.2 TRIPLET MECHANISM OF CIDEP

A whole class of photochemical reactions takes place from the lowest excited molecular triplet state. Typically, excitation takes place within the singlet manifold of states from which a spin-orbit-induced intersystem crossing occurs to the triplet state.

A molecule in the triplet state shows zero-field splitting due to spin-spin dipolar and spin-orbit interactions. Spin-orbit contributions to the zero-field splitting are usually small in the case of organic molecules. In zero magnetic field the spins are quantized along the molecular axes, whereas in high magnetic fields they are quantized along the field. The relation between the eigenstates in zero field T_u ($u = x, y, z$) and the eigenstates in a magnetic field T_i ($i = +1, 0, -1$) is

$$T_i(p, q, r) = \sum_u c_{iu}(p, q, r) T_u. \tag{5}$$

Here $c_{iu}(p, q, r)$ are the complex mixing coefficients that depend on the strength and direction of the magnetic field (H); p, q, and r are the direction cosines of the angles between H and x, y, and z, the molecular axes.[35] The rates of populating the three triplet sublevels via the spin-orbit-induced intersystem crossing from the lowest excited singlet state are not equal.[36] The population probabilities P_i ($i = +1, 0, -1$) are related to the values in zero field P_u ($u = x, y, z$) by the same coefficients c_{iu}:

$$P_i(p, q, r) = \sum_u |c_{iu}|^2 P_u. \tag{6}$$

It has often been assumed in studying the dynamics of intersystem crossing in triplet ESR experiments that the populations of the T_{+1} and T_{-1} spin sublevels are equal in the intermediate magnetic fields of an

X-band ESR experiment.[37,38] However, it is the small population difference due to mixing by the magnetic field together with the unequal zero-field population rates that give rise to polarization in the radicals created by further reaction of these triplet molecules before triplet spin relaxation destroys the spin polarization.

The first theoretical treatment of the TM considered a static random distribution of triplets.[20,22] It was concluded that the different relative rates of intersystem crossing (ISC) related to the molecular frame could indeed lead to polarization in the laboratory frame. Atkins[20,25] has given a nice discussion that shows that this is merely the consequence of the fact that the order of the triplet spin sublevels is independent of the molecular orientation, that is, for the case where at zero magnetic field the ordering of the eigenstates is $T_x > T_y > T_z$, in the high-field mixed state T_x and T_y will always contribute more to T_{+1} than T_z. Thus if population occurs primarily through T_x (in zero field), then in high-field T_{+1} will be more populated than T_{-1} and there will be more α than β spins.

The following expression can be deduced[22] for the electron spin polarization in the static case:

$$P_{+1/2} - P_{-1/2} \sim \left(\frac{4}{15}\right)\left(\frac{D}{H}\right)(P_x + P_y - 2P_z), \tag{7}$$

where $P_{+1/2}$ and $P_{-1/2}$ are the probabilities of finding a radical with spin $+\frac{1}{2}$ or $-\frac{1}{2}$, respectively. It can be seen from this that the polarization is very much dependent on the sign and magnitude of the zero-field-splitting parameter D relative to the magnetic field H and to the relative populating rates P_x, P_y, and P_z.

The theory has been generalized to include molecular rotation,[24,25] providing a more quantitative basis with which to test the theory and to set limits on the conditions necessary for the triplet mechanism to operate. The effects of using polarized light to excite the molecules are also included, following the original suggestion and analysis by Adrian.[39] Such effects result because the population of the excited singlet is proportional to $(\hat{\mu} \cdot \mathbf{E})^2$, where $\hat{\mu}$ is the electronic transition moment $(S_o \rightarrow S_1)$ and $\mathbf{E}$ is the electric field vector of the incident light. So by creating an initially anisotropic distribution of molecular orientations the electron spin polarization, which depends on the initial orientation of the triplet molecule with respect to the magnetic field, will be enhanced or reduced, depending on the orientation of $\mathbf{E}$ relative to the magnetic field.

The theoretical treatments provide basically three experimental tests of the triplet mechanism (TM). First, both radicals will show the same sign of polarization. Creation of a doublet radical via hydrogen atom abstraction from triplet molecules with excess population of the T_{+1} level (excess

α spins) will produce two doublet radicals with excess α spins. Second, there will be a dependence of the magnitude of polarization on concentration of the molecule with which the triplet reacts to produce two doublets. This changes the rate of the reaction competing with the triplet $^3T_1$ (electron spin relaxation) by which spin polarization of the triplet is reduced. Third, the extent of spin polarization will be influenced by using polarized light to excite the molecules as indicated above.

In the development of the triplet mechanism it has been assumed that the rates of reaction from the three triplet sublevels are equal. However, recent results[40,41] suggest it may be necessary to consider differing reaction rates as an additional polarization mechanism.

2 INSTRUMENTATION

The time-resolved ESR experiments are difficult because of sensitivity problems. The sensitivity is inversely proportional to the square root of the bandwidth. The effective bandwidth is limited by the overall time constant of the system, T_c (bandwidth $\propto 1/T_c$).[42] The conventional ESR experiment typically employs 100 kHz field modulation and synchronous detection, and the effective time constant is approximately 1 s. To provide time resolution in the microsecond range one must use a time constant of similar magnitude. Thus, compared to the conventional ESR experiment, time-resolved ESR is about three orders of magnitude less sensitive. It is possible to compensate for this loss by signal-averaging techniques and by using large radical concentrations ($10^{-4} - 10^{-5}M$ per pulse).

Several time-resolved ESR spectrometers using either 2 MHz field modulation[9,10,43,44] or direct detection[11-13] have been described. Smaller and co-workers have used 2.1 MHz modulation achieving time resolution of about 2 μs. Their spectrometer is schematically illustrated in Fig. 1. The 2 MHz modulation was introduced into the cavity by a hairpin loop. The cavity can also be modified using conventionally mounted modulation coils to provide a 7–10 G peak-to-peak modulation field.[45] Since the 2 MHz modulation approach has an effective time resolution of 1 μs and introduces 2 MHz sidebands on all narrow lines, the direct detection scheme is advantageous for some applications. The ultimate time resolution that can be achieved is then a function of the cavity response time that is given by the ratio of Q ($\sim$1000–3000) to the spectrometer frequency ($\sim$9.5 GHz). With direct detection time resolution of 50–100 ns has been achieved.[11,13]

Our direct detection ESR system is illustrated in Fig. 2. Some of the low-frequency noise is eliminated either through limited time observation

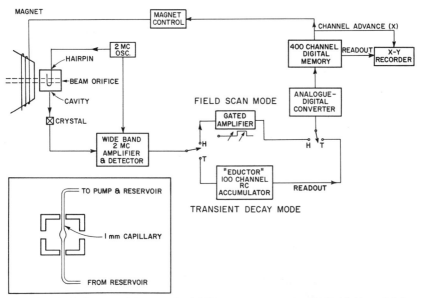

BLOCK DIAGRAM TRANSIENT EPR SPECTROMETER

Figure 1 Block diagram of time-resolved ESR spectrometer using 2 MHz field modulation. Microwave system is a conventional one and is not shown. (From Smaller, Avery, and Remko, reference 43).

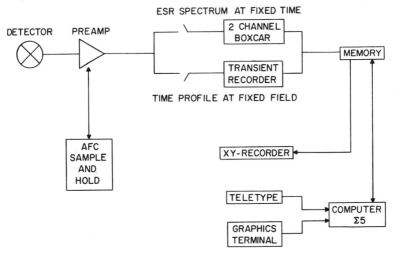

Figure 2 Block diagram of direct detection time-resolved ESR spectrometer in use at Argonne National Laboratory. Microwave system (not shown) is a conventional arrangement of klystron, circulator, and reflection cavity. AFC is a sample-and-hold system that updates klystron frequency before every electron pulse (~100–500 pps).

115

of signal decay (~100 μs) or by use of a two-channel boxcar averager in conjunction with electron or laser pulse repetition rates of 50–500 pps. Two modes of spectrometer operation are typically used. Spectra can be recorded at a fixed time delay after the pulse or the time profile at a fixed field can be obtained. Recording of the spectrum at fixed time delay is obtained by using a two-channel boxcar averager (PAR 162 with two 164 gated amplifiers). One gate is placed before the electron beam pulse, and the observing gate is placed after the electron pulse. We typically use an electron beam pulse width of 50–500 ns, and the observing gates of similar duration (100–500 ns gate width). The boxcar is operated in the A minus B mode so that any dc-level fluctuation that is slower than the time separation between the two gates (few μs) is eliminated. Furthermore, this mode of operation discriminates against steady-state ESR signals that may be present. The boxcar signal is stored in the Fabritek 1070 memory and is subsequently stored in the Sigma 5 computer, providing for further data manipulation. Monitoring the time decay of a signal at a fixed field is usually done using an appropriate multichannel analog to digital device. A PAR waveform educator that is limited to 1 μs per channel or a Biomation 8100 are used. Precise magnetic field–frequency control in such experiments becomes imperative when narrow ESR lines are studied.

Additional points are worth mentioning. When ESR spectra of transient radicals are recorded in the microsecond and submicrosecond domain, uncertainty broadening ($\Delta E \, \Delta t \simeq 1$) of spectral lines causes loss of intensity and spectral resolution.[46] When studies of signal decay at fixed field are carried out, one must be wary of artifacts produced by the laser or electron beam pulses.

3 APPLICATIONS—RADIATION CHEMISTRY

3.1 Identification and Study of Transient Free Radicals in Pulse Radiolysis

In their pioneering studies using time-resolved ESR spectroscopy Smaller and co-workers examined a variety of radicals produced in pulse radiolysis. Their first report[9] described the transient behavior of radicals produced by irradiation of linear and cyclic aliphatic hydrocarbons both in the absence and in the presence of radical scavangers (sulfur compounds and oxygen). The pertinent rate constants were determined by monitoring the decay of the appropriate radical. In Fig. 3 their observations on cyclopentyl radical are illustrated.

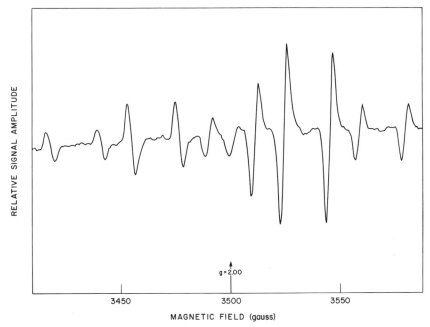

Figure 3 ESR spectrum of cyclopentyl radical (first derivative) taken in the 2–14 μs time window after the electron pulse. (From Smaller, Avery and Remko, reference 9.)

One of their more notable successes is the first ESR detection of the hydrated electron in liquid water. A basic aqueous solution containing methanol as OH scavenger was used in this experiment and is illustrated in Fig. 4.[48]

Although the hydrogen atom can be observed in steady-state radiolysis experiments, time-resolved ESR studies provide many interesting details of radical reactions that include the H radical. Smaller and co-workers were able to measure rate constants of a number of hydrogen atom reactions.[43]

In another interesting application the transient radicals of DNA bases were studied by pulse radiolysis–time-resolved ESR spectroscopy.[49] In particular, the effect of cysteine and cysteamine as radiation protectors was investigated.

Subsequently, many of these systems have been examined in greater detail, and the scope of time-resolved ESR studies in pulse radiolysis has been expanded. The behavior of the H atom signal and that of the hydrated electron have been analyzed in some detail.[12,47,50,51] Radicals

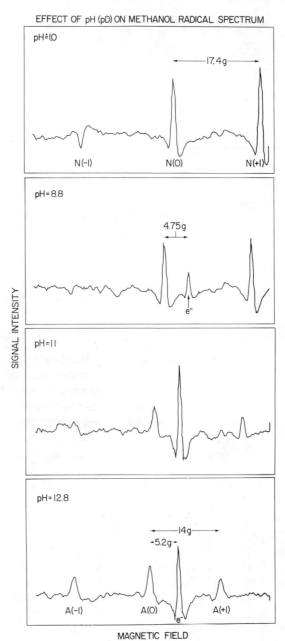

Figure 4 ESR spectra showing how the absorption line of e_{aq}^- can be seen in basic aqueous solutions containing methanol as an OH scavenger. Lines belonging to $\cdot CH_2OH$ and $\cdot CH_2O^-$ are labeled. (From Avery, Remko and Smaller, reference 48.)

118

from acetate ($\cdot CH_2COO^-$) and malonate ($\cdot CH(COO^-)_2$) and the hydroxy-cyclohexadienyl radical were studied.[52] Recently Fessenden has undertaken a detailed measurement and analysis of ESR time profiles.[12] Both the primary radicals of radiolysis (H and e_{aq}^-) and secondary radicals derived from their reactions have been studied.

In our work we have examined transient ESR spectra of aqueous solutions of simple organic acids and alcohols, micelles, and simple hydrocarbons.[53–60]

With the exception of the hydrated electron in solution many of the transient radicals mentioned previously were observed or can be observed using steady-state radiolysis methods. The time-resolved ESR observations are directed at the study of CIDEP and kinetics of radicals in solution. And it is for the study of these phenomena that the time-resolved ESR technique has been developed and is being applied.

3.2 CIDEP in Pulse Radiolysis

In the introduction we have provided a brief outline of the CIDEP phenomenon. This was discovered in the course of the study of steady-state spectra of the hydrogen atom.[6] Only recently the experimental verification of CIDEP theories has been undertaken. First it was necessary to determine whether there is an "initial" polarization occurring prior to or in the radical formation step or whether electron spin polarization is created by radical-radical reaction in solution. In 1973 Fessenden was able to show in the study of several simple radicals that polarization is developed primarily during radical reactions.[52] For example, it was observed that in $\cdot CH_2COO^-$ radical polarized signals last many tens of microseconds, whereas its relaxation time T_1 was measured to be 1.4 μs. This is illustrated in Fig. 5. Clearly no "initial" polarization could last for such a long time.

The radical pair model of CIDEP predicts that in the X-band ESR spectrometer the $S-T_0$ polarization pathway should be a dominant one. The $S-T_0$ polarization depends on the difference of g-factors between the radical pair components. Furthermore, there should be a hyperfine dependence of the relative polarization intensities in the radical lines. Trifunac et al. were able to observe the g-factor dependence and isotropic hyperfine dependence of CIDEP of radicals in pulse radiolysis substantiating the dominance of the $S-T_0$ radical pair polarization mechanism.[53–55]

The $S-T_0$ radical pair model of CIDEP predicts that the polarization enhancement for the pair of lines corresponding to the same hyperfine coupling will be proportional to the magnitude of that coupling

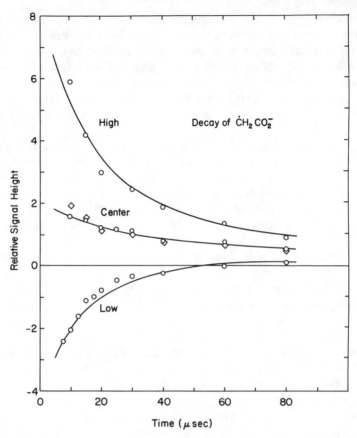

Figure 5 Time dependence of the amplitudes of the three lines of the CH_2COO^- radical following the irradiation pulse. (From Fessenden, reference 52.)

$(\sum A_i M_i)$.[54,55] That is, we expect to see that lines away from the center of the spectrum will be more intensely polarized. So rather than seeing the "normal" relative ESR line intensities as given by the binomial coefficients, one observes emission and enhanced absorption lines with intensities proportional to the hyperfine term $(\sum A_i M_i)$.

The spectrum of ethylene glycol radical itself (Fig. 6a), as well as the spectrum of isopropanol, ethanol, and n-propanol radicals (Fig. 7), illustrate how signal enhancements are larger in the lines further away from the center of spectra. The outside lines carry more polarization, the center line is not polarized and the lines close to the center of the spectrum are weakly polarized. In Table 1 the proportionality of the observed enhancement and the calculated enhancement using Adrian's

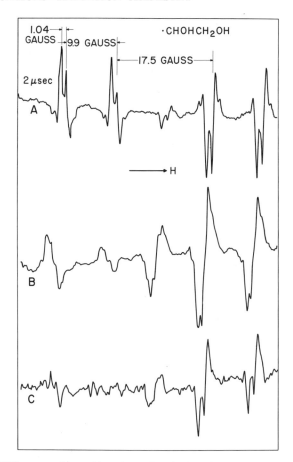

Figure 6 ESR spectra observed in pulse radiolysis of 10% by volume ethylene glycol in (*a*) water, (*b*) plus ~ 2% $CHCl_3$, or (*c*) plus ~ 1% $CHBr_3$. Spectra were obtained using 2 MHz modulation. Positive signals mean emission, and negative signals mean absorption. (From Trifunac and Thurnauer, reference 55.)

CIDEP model[17] is illustrated. Figure 8 illustrates the $\sum A_i M_i$ dependence.[55]

The g-factor experiments are best illustrated by the observations involving radicals from ethylene glycol (Fig. 6).[55] The $S–T_0$ radical pair theory of CIDEP predicts that the polarization of a radical will be in emission if this radical encounters another radical with a smaller g-factor and that it will be in absorption if the opposite is true. Since most organic radicals do not have very large g-factor differences, spectra in complete emission or enhanced absorption are not observed. In order to maximize

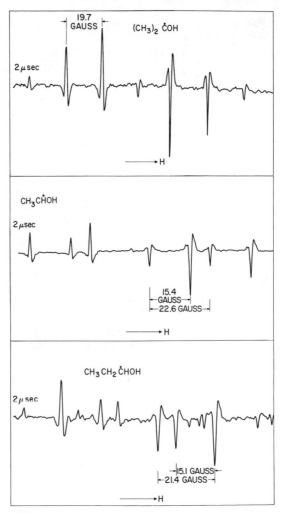

Figure 7. Time-resolved ESR spectra of several simple alcohols in aqueous solutions. Spectra were obtained using 2 MHz modulation. (Adapted from Trifunac and Thurnauer, reference 55.)

the g-factor difference between radical pair components we have employed mixed radical systems containing $\cdot R$ and $\cdot RX$ where $\cdot RX$ is a halogen-containing radical. Since in most cases the ESR spectra of the halogen containing radicals are poorly resolved, we studied $\cdot R$ (i.e., the radical with the smaller g-factor that will then be shifted to enhanced absorption). The presence of the halogen-containing radicals increased relaxation of all radicals in solution, so it was necessary to minimize the concentration of $\cdot RX$ in order to observe the g-factor effects. CIDEP effects in the ESR spectra of the radical from the aqueous solution of ethylene glycol ($\cdot CHOHCH_2OH$) are illustrated in Fig. 6; alone (Fig. 6a) and in the presence of radicals containing chlorine and bromine (most likely $\cdot CHCl_2$ and $\cdot CHBr_2$) in Fig. 6b and Fig. 6c, respectively. The change of relative line intensities when radicals with larger g-factors are present shows how there is a shift to absorption in the high-field lines and the low-field lines. The change of relative intensities indicates that inner (toward the center of spectrum) lines are more sensitive to the contribution from g-factor polarization since they have smaller polarization due to the hyperfine term $\sum A_i M_i$.

Recently in our laboratory we have made several interesting observations relating to the S-$T_{\pm 1}$ polarization mechanism of CIDEP.[57–60] As described previously, in X-band ESR fields S-T_0 interaction is the dominant polarization pathway. However, conditions exist when the contribution from the S-$T_{\pm 1}$ polarization can be seen superimposed on the S-T_0 polarization effect. Examination of the off-diagonal elements of the spin Hamiltonian [Eq. (3) and (4)] illustrates the difference between S-T_0 and S-$T_{\pm 1}$ mixing. Whereas the S-T_0 pathway is determined by the g-factor and sums and differences of hyperfine terms of radical pair components, the S-$T_{\pm 1}$ pathway depends on a single hyperfine term only. Furthermore, in contrast to the spin selection in S-T_0 mixing, where there is simply a change in association between electron and nuclear spins, in

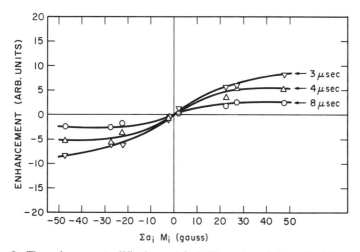

Figure 8 The enhancements (V) observed in ESR spectra of the radical from ethanol $CH_3\dot{C}HOH$ are plotted as a function of $\Sigma A_i M_i$. (From Trifunac and Thurnauer, reference 55.)

Table 1   Observed and Calculated Signal Enhancements

$|\Sigma A_i M_i|$	V observed[a]	$\dfrac{\rho}{\rho_e}$ calc.	Proportionality Factor
$|CH_3|_2\dot{C}OH$	$|a_H| = 19.7\ G$		
1. 59.10	−12.0	−57.36	0.478
2. 39.4	−10.5	−43.93	0.418
3. 19.7	−5.0	−24.59	0.492
4. 0	−	0.0	
5. 19.7	5.0	24.59	
6. 39.4	10.5	43.93	
7. 59.10	12.0	57.36	
		mean	0.46 ± 0.04
$CH_3\dot{C}HOH$	$|A_H^{\alpha}| = 15.4\ G$	$|A_H^{\beta}| = 22.2\ G$	
1. 41.0	−8.67	−11.71	0.74
2. 25.6	−6.67	−8.45	0.79
3. 18.8	−3.89	−6.19	0.63
4. 3.4	−1.11	−1.69	0.65
5. 3.4	1.11	1.69	
6. 18.8	3.89	6.19	
7. 25.6	6.67	8.45	
8. 41.0	8.67	11.71	
		mean	0.70 ± 0.08
$CH_3CH_2\dot{C}HOH$	$|A_H^{\alpha}| = 15.1\ G$	$|A_H^{\beta}| = 21.4\ G$	
1. −28.8	−15.67	−6.68	2.35
2. −13.85	−9.67	−3.76	2.57
3. −7.55	−4.33	−1.91	2.27
4. 7.55	4.33	1.91	
5. 13.85	9.67	3.76	
6. 28.95	15.67	6.68	
		mean	2.40 ± 0.16

Table 1 Contd.

CH$_2$OHCHOH	$\|A_H^\alpha\| = 9.9\,G$	$\|A_H^\beta\| = 17.5\,G$	
1. −22.45	−12.78	−5.87	2.18
2. −12.5	−6.18	−3.67	1.68
3. −4.95	−^{c}	−1.48	
4. 4.95	−	1.48	
5. 12.5	6.18	3.67	
6. 22.45	12.78	5.87	
		mean	1.93 ± 0.35

^{a} V is the observed enhancement ($V = (H_H - H_L)/(H_H + H_L)$, where H_H and H_L refer to the peak heights of the high and low field lines belonging to the same $\|M_i\|$.) Enhancements shown were obtained 3 μsec after the electron pulse. All spectra were obtained with 6 mA peak current at the same microwave settings.
^{b} Calculated intrinsic enhancement, $\rho/\rho_e = V(\tau_c/T_1)$.
^{c} Enhancement of the middle lines of the radical from ethylene glycol could not be measured.

$S\text{-}T_{\pm 1}$ mixing nuclear and electron spin flips occur. Whereas, $S\text{-}T_0$ mixing occurs in the region of small exchange, $S\text{-}T_{-1}$ mixing occurs in the region where exchange matches the Zeeman splitting of the triplet levels. Thus the observations of $S\text{-}T_{-1}$ polarization might be possible when large hyperfine splitting makes for efficient $S\text{-}T_{-1}$ mixing during the short time period when there is near degeneracy of $S\text{-}T_{-1}$ states, or when the reduction of radical diffusion might also extend the time interval for this $S\text{-}T_{-1}$ mixing. Our qualitative observations of CIDEP in several systems bear this out.

An example of the hyperfine effect can be found in CIDEP of reactions involving hydrogen radicals. In the hydrogen radical itself the low-field emission was found to be more intense than the high-field absorption.[12,61] However, it is difficult to monitor simultaneously two lines ~500 G apart, and the first study of hydrogen radicals did not note this 30% difference in intensity.[47] We chose to study CIDEP in ·CH$_2$COOH (produced by OH abstraction from acetate) in the presence of hydrogen radicals.[57] These results are illustrated in Fig. 9—top three spectra. In slightly acidic solution ·CH$_2$COOH exhibits the $S\text{-}T_0$ CIDEP. The low-field line is in emission, the center line is not polarized, and the high field line is in enhanced absorption. [The intensity of the low-field emission is either equal to or slightly less than (relaxation) the intensity of the high-field enhanced absorption.] When the pH of the solution was lowered (e_{aq}^- converted to hydrogen radicals), we observed that ·CH$_2$COOH radical

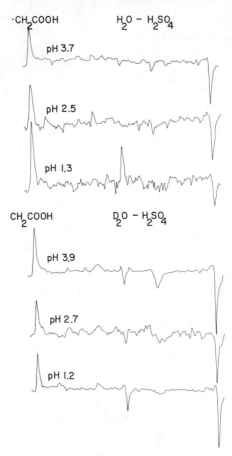

Figure 9 ESR spectra of the acetate radical $\cdot CH_2COO^-$ ($a_H = 21.2$ G) in solutions of varying acidity (sodium acetate $0.33M$) in H_2O and D_2O. All spectra were obtained in the 2–2.5 μs time window as measured from the beginning of the electron pulse (500 ns pulse width). Spectra were obtained using direct detection. Positive signals mean emission, and negative signals mean absorption. (Adapted from Trifunac and Nelson, reference 57.)

CIDEP was significantly shifted to emission. The low-field line is more intensely in emission than the high-field enhanced absorption, and the central line now also is in emission (third spectrum in Fig. 9). In (neutral) and slightly acidic solution the S-T_0 CIDEP observed in the acetate radical originates in radical pairs involving only this radical. At low pH there should be almost equal amounts of H and $\cdot CH_2COOH$ radicals. Under these conditions the dominant radical pair species is $H \cdot CH_2COOH$, but the acetate radical pairs are also present. Since the spectra are not completely in emission S-T_0 polarization is still the dominant polarization pathway. Obviously although S-T_{-1} processes occur in radical pairs involving hydrogen radicals (because of the large hyperfine of hydrogen), these same pairs give rise to S-T_0 polarization along with the acetate pairs.

A simple check of this hypothesis is obtained when D is substituted for H (the identical experiment is carried out in D_2O). Figure 9 (bottom) illustrates that with D there is no observable shift to emission as the pH is lowered. Thus a change from H hyperfine (506 G) to that of D (75 G) completely eliminates significant S-T_{-1} contribution to polarization. We have observed these effects in several other systems (e.g., citrate and malonate). It appears that H radical can produce S-T_{-1} CIDEP with substrates that react relatively slowly with it (rates $10^{-5} - 10^{-6} M^{-1} s^{-1}$); otherwise hydrogen is consumed by reacting with substrate before appreciable S-T_{-1} polarization can develop.

Examples of the effect of radical diffusion on CIDEP are found in viscous solutions[59,60] and in radicals from micelles.[58] In viscous solutions containing $\cdot CH_2COO^-$ radical we have been able to observe a shift to emission by changing viscosity with temperature. A nice example of this is illustrated in the experiment in 65 volume % quadrol in water where quadrol is $R_2NCH_2CH_2NR_2$ where $R = CH_2CHOHCH_3$.[60] (See Fig. 10.) In a solution of lower viscosity (higher temperature) the usual S-T_0 CIDEP pattern is observed. When viscosity is increased by cooling the solution the contribution from S-T_{-1} CIDEP becomes visible by a shift of

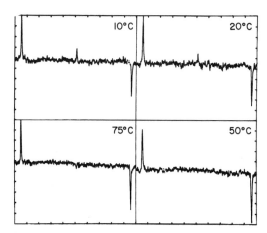

Figure 10 ESR spectra of acetate radical (from $ClCH_2COO\,Na$) in aqueous solution containing 65 volume % Quadrol at various temperatures (viscosities). Positive signals mean emission and negative signals mean absorption. Spectra were obtained using direct detection. (Quadrol = $R_2NCH_2CH_2NR_2$, where $R = CH_2CH(OH)CH_3$). (From Trifunac et al. reference 60.)

the ESR spectrum to emission. Again the low-field line becomes more intensely in emission and the high-field line enhanced absorption is smaller (by ~20%) than the low-field emission and the center line becomes emissive. The same effect can also be observed when the solution viscosity is varied by changing the amount of glycerol in water at constant temperature.[59]

The last example of S-T_{-1} mixing can be found in pulse radiolysis of micelles.[58,60] In this case the bulk viscosity of the solution is not changed; however, radical diffusion is limited by aggregation. When one obtains CIDEP-ESR in solutions when surfactant molecules are aggregated [above the critical micelle concentration (CMC)], [62] significant excess emission is observed (Fig. 11). When the surfactant is not aggregated (at or below the CMC) the "normal" S-T_0 equal emission–absorption is seen. At concentrations exceeding the CMC, aggregation restricts radical diffusion and affords S-T_{-1} contributions to the polarization. While the micellar aggregation is a dynamic phenomenon occurring in ~10^{-4} s; on the microsecond time scale the radicals are either predominantly in aggregates (with critical micelle concentration of surfactant monomer always present) or predominantly in a monomeric situation. Thus S-T_0 and S-T_{-1} effects are superimposed. These effects were observed in all micellar systems that we have examined and is quite independent of the anionic, cationic, or neutral nature of the surfactant head group.

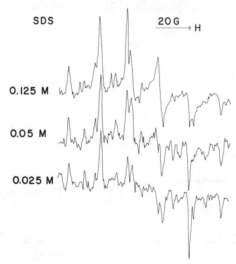

Figure 11 ESR spectrum obtained in pulse radiolysis of sodium dodecyl sulfate (SDS) aqueous solutions at 2 μs after the pulse. Surfactant concentrations are indicated. CMC of SDS is $3.3 \times 10^{-2}M$. Spectra were obtained using direct detection, positive signals mean emission and negative signals mean absorption. (From Trifunac and Nelson, reference 58.)

In conclusion, in the study of CIDEP in these varied radical systems a consistent picture was obtained of the involvement of the "minor" polarization pathway $(S\text{-}T_{-1})$. This is especially relevant to all CIDEP studies carried out when there is restriction to radical diffusion. Biological systems fall into this category.

Although for the most part these studies of $S\text{-}T_0$ and $S\text{-}T_{\pm 1}$ CIDEP by time-resolved ESR are qualitative or semiquantitative, we will also discuss how in some simple chemical systems a more quantitative approach has been applied.

3.3 Submicrosecond Studies

The development of the direct detection ESR spectrometer system has allowed us to study CIDEP in radicals as soon as ~100 ns after the electron beam pulse.[11] Such studies can be carried out only with intensely polarized radical systems such as the $\cdot CH_2COO^-$ radical. Line broadening is severe (uncertainty broadening) and signal intensities are poor, since the spin system has not had a chance to develop fully.[63] However, several interesting observations can be made. In the study of the submicrosecond domain we were able to observe initial spin population in radicals that was transferred to them by their precursors.

The $\cdot CH_2COO^-$ radical was studied.[11] It was produced by OH abstraction from acetate or by dissociative electron capture by chloro or iodoacetate. ESR spectra were recorded in the 100–200 ns time window after the end of a 100 ns electron pulse (Fig. 12). When the $\cdot CH_2COO^-$ radical was obtained from an OH reaction the spectra showed essentially a $1:2:1$ intensity pattern (slightly distorted by the growing CIDEP), whereas the radical produced from an e^-_{aq} reaction showed $1:0:1$ relative intensities from the growing CIDEP. This difference may be explained if it is assumed that the OH radical relaxes very fast and in reacting it transfers an essentially Boltzmann spin population to the newly formed radical. In contrast, the e^-_{aq} relaxes slowly, and it transfers its equal population of α and β electron spins to the new radical. These explanations assume that spin is conserved during reaction; for example,

$$OH(\alpha) + CH_3COO^- \rightarrow \cdot CH_2COO^-(\alpha)$$

$$OH(\beta) + CH_3COO^- \rightarrow \cdot CH_2COO^-(\beta); \qquad \frac{N_\alpha}{N_\beta} \sim \exp\left(-\frac{g\beta H}{kT}\right),$$

and

$$e^-_{aq}(\alpha) + ClCH_2COO^- \rightarrow \cdot CH_2COO^-(\alpha)$$

$$e^-_{aq}(\beta) + ClCH_2COO^- \rightarrow \cdot CH_2COO\ (\beta); \qquad N_\alpha \sim N_\beta.$$

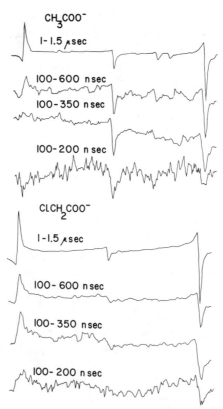

Figure 12 Submicrosecond ESR spectra of the ·CH$_2$COO$^-$ radical obtained by OH reaction and with acetate by e_{aq}^- reaction with chloroacetate. Spectra were obtained using direct detection; time windows are as indicated and were measured from the end of the electron pulse. For the 100 ns experiments a 100 ns electron pulse was used.

As we see later, Fessenden and Verma have reached identical conclusions in their analysis of these same systems.[12]

3.4 Quantitative Studies

The time dependence of the transient signal intensity can be analyzed using modified Bloch equations. These must take into account that the equilibrium value of the magnetization changes as the radicals are eliminated by reaction, since the fast reactions are similar to relaxation in that they cause loss of magnetization. Also, CIDEP effects on the magnetization must be included. We will outline Fessenden's treatment, since he is the only one to date to make a detailed study of ESR time profiles.[12] The modified Bloch equations in the rotating frame are

$$\dot{M}_x = \Delta\omega M_y - \left(T_2^{-1} + \frac{\dot{R}}{R}\right)M_x \tag{8a}$$

$$\dot{M}_y = \Delta\omega M_x - \left(T_2^{-1} + \frac{\dot{R}}{R}\right)M_y - \omega_1 M_z \tag{8b}$$

$$\dot{M}_z = \omega_1 M_y - \left(T_1^{-1} + \frac{\dot{R}}{R}\right)M_z + M_0 T_1^{-1}F(t) + PM_0(F(t))^2, \tag{8c}$$

where T_1 and T_2 are the electron relaxation times, $\omega_1 = \gamma H_1$ (H_1 is the rotating component of the microwave magnetic field and γ is the electron magnetogyric ratio), $\Delta\omega = \omega - \omega_0$ is the offset of the observing field from the center of resonance in angular frequency. In addition to the usual relaxation terms (i.e., M_x/T_2), there are terms $M_x(\dot{R}/R)$ that represent the relaxation brought about by loss of radical concentration by chemical reaction. It is assumed that the loss of magnetization follows the loss of radical concentration. The form of $\dot{R}/R$ is a function of the radical reaction order (i.e., $\dot{R}/R = -k$ for first-order decay where k is the rate constant). The term $M_0 T^{-1}F(t)$, where $F(t) = R/R_0$, takes into account the loss of equilibrium magnetization as radicals decay. Thus the form of $F(t)$ is $\exp(-kt)$ for first-order decay and $(1 + t/\tau_c)^{-1}$ for second-order radical decay where τ_c is the half-life. When there is no radical decay for stable radicals, $F(t) = 1$. CIDEP produced by radical pair polarization processes is represented by a production term $PM_0(F(t))^2$ in the equation for M_z. The factor P has different values at different initial concentrations. $P = V/T_1$ for slow radical decay where V is the polarization enhancement equal to $(S - S_0)/S_0$, where S is the signal intensity.

For exponential radical decay an analytical solution exists. From this it is possible, in the limit $T_1^{-1} \gg k$, to replace the first-order terms $\exp(-kt)$ by the second-order decay function to get the solution for second-order decay:

$$S(t) = \frac{M_0\tau_c}{t + \tau_c}\left(1 + \frac{V\tau_c}{t + \tau_c}\right)\left(\frac{\omega_1 T_1}{1 + (\delta^2 + 1)(\omega_1 T_1)^2}\right)$$

$$\times \left[1 - \exp\left(-\frac{t}{T_1}\right)\cos(\delta^2 + 1)^{1/2}\omega_1 t - \frac{\exp(-t/T_1)}{\omega_1 T_1(\delta^2 + 1)^{1/2}}\sin(\delta^2 + 1)^{1/2}\omega_1 t\right]$$

$$+ f\left(\frac{M_0\tau_c}{t + \tau_c}\right)\frac{\exp(-t/T_1)}{(\delta^2 + 1)^{1/2}}\sin(\delta^2 + 1)^{1/2}\omega_1 t, \tag{9}$$

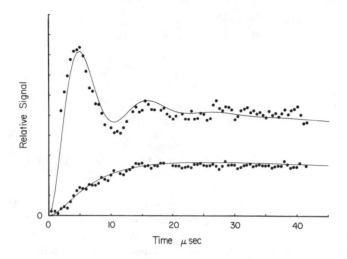

Figure 13 Time dependence of the high-field line of $^-O_2CCH = \dot{C}CO_2^-$ in a solution of 2 mM bromomaleate at pH 9. The upper curve was taken at a power level of -5 dB and the lower was at -25 dB. The calculated curves took field inhomogeneity into account. (From Verma and Fessenden, reference 12.)

where $\delta = \Delta\omega/\omega_1$. The initial value of $M_z = fM_0$, where f is the fraction of the Boltzmann value. T_1 is assumed to be equal to T_2, which is verified for $\cdot CH_2COO^-$. The transient terms involving $\exp(-t/T_1)$ are damped out over several relaxation periods. If H_1 is small then the terms with $\sin \omega_1 t$ are zero; in contrast, if H_1 is large, the sine and cosine terms lead to an oscillatory behavior. Nice examples of this oscillatory behavior of the spin system can be seen in the maleate radical (Fig. 13).

The actual analysis of the ESR decay curves is carried out by fitting a curve described by Eq. (9). The parameters are T_1, τ_c, H_1, f, V, and a scaling factor. Some of these parameters can be measured independently (e.g., T_1 and H_1). H_1 was determined by use of Torrey oscillations[66] or from the measurement of the cavity Q and power levels. Curve fitting at several different power levels further restricts the range of some parameters. The authors claim that different portions of the curves are affected differently by various parameters so that a unique fit may be feasible.

Field inhomogeneity is a dominant factor influencing linewidth and it was taken into account by using a Gaussian inhomogeneity line shape factor. Both primary radicals in aqueous radiolysis (H, e_{aq}^-) and several secondary radicals were examined.

As we have mentioned before, in the time-resolved ESR spectrum of the hydrogen radical, the amplitude of the low field line was found to be about 30% greater than that of the high field line. For a suggested explanation, see Section 3.2.

In the study of e_{aq}^{-} several interesting observations were made. It was found that the electron signal was in emission in solutions containing carbonate, hydroquinone, and phenol counterradicals.[50] Since the electron has the lowest g-factor of all radicals in these systems, the radical pair model of CIDEP would predict the opposite polarization (enhanced absorption) of the e_{aq}^{-} signal. It was suggested that in these systems, e_{aq}^{-} radical reaction leads to the triplet state of the product. Although this view is still highly speculative, recent CIDNP results in photochemical systems indicate plausibility of that interpretation.[67]

In another observation on the e_{aq}^{-} transient signal in these systems it was found that the amplitude of the emission signal was proportional to radical concentration that depends on dose rather than the expected square dependence on concentration. It was suggested that increased relaxation of e_{aq}^{-} electron spin at higher concentration might be due to Heisenberg spin exchange. However, studies of the dependence of relaxation on radical concentration do not appear to support such views.[68]

In the study of secondary radicals,[12] notably $\cdot CH_2COO^{-}$ produced both using e_{aq}^{-} reaction with chloroacetate or OH abstraction with acetate, it was found that these reactions occur with spin conservation and that the population difference in the primary radical (e_{aq}^{-} and OH) is transferred to the resulting acetate radical. This initial magnetization in the radicals was found to be zero in radicals produced from e_{aq}^{-} (Fig. 14). Radicals

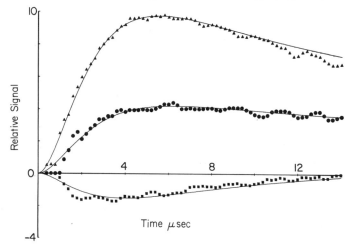

Figure 14 Time dependence of all three lines of $\cdot CH_2COO^{-}$ formed from $ClCH_2COO^{-}$. Curves were calculated using $H_1 = 13$ mG, $T_1 = 1.4$ μs, and $f = 0$. The values of V were -1.6, 0, and 1.6 for the low-field, center, and high-field lines, respectively. (From Verma and Fessenden, reference 12.)

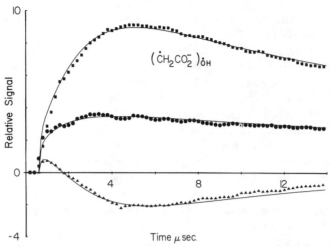

Figure 15 Data for $\cdot CH_2COO^-$ formed from OH abstraction from CH_3COO^-. Calculated curves use the same parameters as in Fig. 14 except $f = 1$ and $V = -2$, 0, and 2. Note the initial positive signal for the low-field line at short times. (From Verma and Fessenden, reference 12.)

produced from OH indicate that the initial magnetization of 1 (Boltzmann units, $f = 1$) was inherited by them from the OH precursor (Fig. 15), implying that the OH radical relaxes to the Boltzmann distribution before reaction and transfers this population distribution to the product radicals. Analyses of several reaction systems indicate that the OH relaxation time is less than 1 ns. In contrast, e_{aq}^- has a relaxation time of several microseconds and its initial magnetization is zero ($f = 0$). (Similar conclusions on the acetate system by Trifunac et al.[11] were discussed in Section 3.3.) In a particularly nice series of experiments these authors[12] were able to prepare other than an equilibrium population of e_{aq}^- by using CO_3^- to develop an inverted population (see earlier) and were able to observe transfer of this inverted population to the product radicals of e_{aq}^- reacting with fumarate.

It was found that, as expected, the polarization enhancement (V) was proportional to the $\cdot CH_2COO^-$ radical concentration (half-life τ_c) at least up to V of 3.6 and $\tau_c \sim 15$ μs.

Examples of transfer of nonequilibrium spin population were also found in radicals produced from the hydrogen addition reaction with acetylene dicarboxylate ($^-O_2CCH = \dot{C}CO_2^-$). It was observed that the transfer of a population difference from hydrogen to the resulting radical was dependent on the acetylene dicarboxylate concentration. This was taken to suggest that the population difference is built up in the hydrogen radical when the addition reaction occurs over a longer period of time as is the case for lower concentrations of acetylene dicarboxylate. However, lack of proportionality to the radical concentration led to suggestions that processes may exist for producing a population difference in the hydrogen radical that are not dependent on homogeneous radical-radical reactions. Thus it was speculated that this population difference arises when hydrogen radicals are produced or by reactions within the spur.

It should be pointed out that in studies of CIDNP it has been observed in numerous instances that nuclear population transfer, also called the *memory effect*,[69] is occurring. We have observed this in many studies of radiolysis by the CIDNP–nuclear magnetic resonance (NMR) technique.[70]

The effect of the ionic strength on CIDEP was studied.[56] The effect on the polarization of radicals from acetate and malonate is moderate and is more pronounced at lower ionic strength. These findings are in qualitative agreement with theoretical predictions.[31] For the acetate radical, it was found that in the acetate concentration range from $0.05–1M$ there is approximately 30% increase in the polarization, whereas the change in the rate constant was about 10%. In the course of this work a surprising observation was made. The polarization enhancement of radicals from acetate, methanol, and malonate, examined at high radical concentrations ($\sim 2–10$ μs half-life), was independent of concentration. It was speculated that either polarization quenching, possibly via Heisenberg exchange, or the initial nonuniform spatial distribution of radicals in spurs may be responsible. At lower radical concentrations polarization enhancement was found to depend on radical concentrations.[12]

3.5 Summary

In summary, the time-resolved ESR studies in radiation chemistry have provided observations of transient radicals down to 100 ns after their creation by an electron pulse. The radical pair model of CIDEP was substantiated by observations of the S-T_0 polarization pathway in pulse radiolysis. In addition, observations were made of the less usual S-T_{-1} CIDEP. This and the quantitative studies of the time behavior of ESR spectra have allowed the development of a better understanding of the CIDEP phenomenon and thus the underlying chemistry of the transient radical species produced in pulse radiolysis.

4 APPLICATIONS—PHOTOCHEMISTRY

4.1 In Quest of a CIDEP Mechanism

The fast time-resolved photo-ESR work has generally been carried out with 2 MHz detection systems coupled with laser flash photolysis. The time resolution has been in the range of 1–5 μs. Much of the work in this area has been directed toward understanding the mechanism of the CIDEP in photochemical systems,[71–73] and the development of these ideas has included studies where CIDEP has been observed using 100 kHz phase sensitive detection with various pulsed light sources and thus a somewhat limited time resolution. A discussion of the latter experiments is included when relevant to the development of photo-CIDEP mechanisms.

The reactions studied to date have primarily involved photoreaction of carbonyl compounds. In such systems the carbonyl is excited to its first excited singlet state, and intersystem crossing takes place to the first excited triplet followed by either electron or hydrogen atom transfer from an appropriate donor (DH).

$$A \xrightarrow{\ h\nu\ } {}^{1*}A \xrightarrow{\ \text{ISC}\ } {}^{3*}A + DH \Big\langle \begin{array}{l} A^{\cdot -} + DH^{\cdot +} \\[1em] AH^{\cdot} + D^{\cdot} \end{array}$$

Atkins et al. reported the time dependence of the amplitude of the ESR spectrum of the radicals formed from irradiation of several carbonyl compounds in liquid paraffin or as neat solutions at room temperature.[20,74,75] The radicals were produced in the ESR cavity by excitation of the carbonyl with a pulse of unpolarized 337.1 nm laser light with subsequent hydrogen atom abstraction from the solvent. The solvent radicals (counterradicals) were not observed. Observation was either with 100 kHz phase sensitive detection (~100 μs time resolution)[75] or with 2 MHz modulation (~5 μs time resolution).[74] It was noted that in several of the cases studied the entire spectrum appeared in emission when observed at short times after radical creation. This included irradiation of benzophenone, benzaldehyde, 2- and 4-chlorobenzaldehyde, acetophenone, dibenzil, and anthraquinone. In certain instances the phase of the early signals was dependent on the experimental conditions. The ketyl radical from benzaldehyde in low-viscosity paraffin was in absorption at short times after the flash, but in high-viscosity paraffin it was in emission.[20] In general, the observed emission decayed on the order of 10–50 μs into absorption that decayed on a much slower time scale (ms) as the radicals reacted.

To explain the results the authors rejected the TM believing that the excited triplet molecule would come to spin equilibrium before hydrogen atom extraction. Instead they proposed a radical pair mechanism based on S-T_{-1} mixing induced by a hyperfine or spin rotation perturbation.[74]

The spin rotation mixing mechanism was favored because of the apparent independence of the intensity of emission on the nuclear magnetic quantum number corresponding to a particular line. However, a closer examination of the field-dependent emission spectrum taken at a given time after the laser pulse actually showed the low field lines to be more intensely in emission than the high field lines.[20,75] In cases where this effect was more obvious, the spectra did, however, show net emission.[20]

In a study of the dependence of polarization on viscosity in the photoreduction of perdeuterobenzophenone, Atkins et al.[76] favored an S-T_0 radical pair mechanism for producing polarization, since they observed the functional dependence on viscosity that was predicted from Adrian's diffusion model.[19] In a later paper[77] it was pointed out that S-T_0 polarization should give marked hyperfine distortion, and it was shown that the observed dependence of polarization on viscosity in heavy paraffin solvents could also be predicted via a triplet mechanism (TM).

In another study in systems in which electron transfer was believed to take place via exiplex formation, total emission was observed. It was suggested that the retarded diffusion in this case would allow S-T_{-1} mixing to take place.[78] It was later shown that the quantitative data on these systems could be accounted for by the TM.[77,79]

In 1972 Wong and Wan[80] reported emission from all hyperfine components in the photochemically generated 1,4-naphthosemiquinone radicals in liquid 2-propanol. They used 100 kHz field modulation to detect radicals generated by irradiating 1,4-naphthoquinone in 2-propanol with a 200 W superpressure mercury lamp coupled to a rotating sector. At constant magnetic field the signal was in emission immediately after the light pulse. The emission reverted to absorption as the light pulse decayed with a decay constant of $\sim 300\,\mu s$. The absorption then decayed further with second-order kinetics. It was suggested that the net emission was due to the TM, since the rate of hydrogen atom abstraction from the excited triplet of 1,4-naphthoquinone could compete with the triplet spin-lattice relaxation time to preserve polarization in the subsequent radicals.

Until this time the observation of the counterradical had eluded the two groups. This was important since for the TM one expects the two radicals to have the same sign of polarization. In the radical pair theory for S-T_0 mixing there is expected to be no "net" effect. So for the RPM to be operative in the cases that had been observed, the counterradical would have to be in total absorption (mixing by Δg effect). For S-$T_{\pm1}$ mixing one expects a net emission (absorption), in both radicals; however, in most of the cases studied, the mixing perturbation would probably be via a hyperfine interaction, and a noticeable intensity dependence on nuclear quantum number would be expected.

Wong et al.[81] made the first observation of the counterradical in the system of duroquinone in acetic acid with 2,6-di-*t*-butylphenol as hydrogen donor. All hyperfine components of both radicals were emissive. Furthermore, they observed that the "initial" (within 150 μs) emission of the phenoxy radicals was enhanced by increasing concentration. Thus if the spin depolarization rate is relatively constant, this would increase the hydrogen abstraction rate and hence the polarization.

The results of a study of the photolysis of 1,4-naphthoquinone in the presence of various alcohols were analyzed in terms of the RPM and TM.[82] The RPM via S-T_0 mixing could readily be ruled out, since total emission was observed in photolysis of 1,4-naphthoquinone in the presence of 1,4-dihydroxynaphthalene as hydrogen donor. For S-T_0 mixing no net polarization is expected for two identical radicals in the radical pair. It was shown in this system that the intensity of the "initial" emission was dependent on 1,4-dihydroxynaphthalene concentration even though the absorption part of the curves showed that the concentrations of the 1,4-naphthosemiquinone radicals were about equal in all solutions (see Fig. 16). Thus it was concluded that, although an S-T_{-1} RPM could explain the total emissive nature and same sign of polarization for both

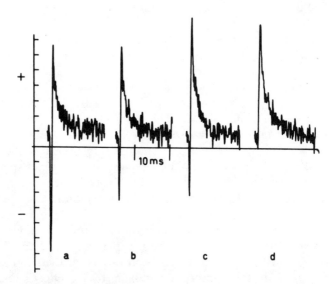

Figure 16 Polarization of 1,4-naphthosemiquinone radical produced in the photolysis of 1,4-naphthoquinone in acetic acid with varying concentrations of 1,4-dihydroxynaphthalene: (*a*) 0.167*M*; (*b*) 0.084*M*; (*c*) 0.056*M*; (*d*) 0.028*M*. Quinone concentrations are 0.024*M* in (*a*) and (*c*) and 0.012*M* in (*b*) and (*d*). (From Wong et al., reference 82.)

radicals in the systems studied, only the TM could explain the dependence on concentration of the hydrogen donor. It was pointed out that the most likely rationale involves a combination of TM and RPM mechanisms.

Atkins, Dobbs, and McLauchlan[83] also observed the counterradical during photolysis of duroquinone and anthraquinone in the presence of 2,6-di-t-butlyphenol. Both the resulting semiquinone and phenoxy radicals were observed in emission with no dependence of intensity on hyperfine components up to 60 μs after the flash observed using 100 kHz detection. They pointed out that an S-T_{-1} radical pair mechanism could explain that both radicals were in emission with no hyperfine distortion, if the mixing is by an anisotropic g-factor with slow rotation or by a hyperfine mechanism if the nuclear relaxation is sufficiently rapid to remove hyperfine distortion. It was noted that both the TM and RPM are sensitive to the time intervals involved; namely, the triplet relaxation rate in the TM and the T-S mixing time and nuclear relaxation rate in the RPM. Within the time scale of the experiment both mechanisms could be operative. The need to be able to observe time-resolved spectra at very short times was noted. Subsequently, Dobbs reported that the emission observed in radicals produced in these systems is hyperfine independent as early as 5 μs after the photolytic flash.[77]

It was suggested by Adrian[39] that a possible test of the TM would be to monitor the effect on the spin polarization of using polarized light to generate the radicals. Since the transition moment $\hat{\mu}$ for the initial $S_0 \rightarrow S_1$ excitation is polarized with respect to the molecular axes, there will be a selection of orientations of excited molecules with respect to the magnetic field when the system is excited with linearly polarized light within the ESR cavity. If spin polarization is produced via the triplet mechanism that depends on the orientation of the molecules relative to the magnetic field, the degree of spin polarization will vary with the orientation of the electric vector of the light with respect to the field. Adrian predicted that the spin polarization should vary as $(3 \cos^2 \psi - 1)$, where ψ is the angle between the electric vector of the light and the magnetic field direction. For static, randomly oriented, axially symmetric molecules this gives a 20% increase in the spin polarization as ψ goes from 0 to 90° if the transition dipole moment lies along the principal molecular axis and a 10% decrease in the polarization as ψ goes from 0 to 90° if it is perpendicular to it.

Dobbs and McLuachlan[84] studied the radical produced when irradiating duroquinone in isopropanol in the presence of triethylamine with a linearly polarized light source. They observed very small effects in the direction supporting the TM.

Using 2 MHz modulation with a pulsed laser source, Adeleke, Choo, and Wan[85] were also able to test for the role of the TM in photochemical CIDEP. They studied the effects of polarized light on the reactions of a series of 1,4-quinones with 2,6-di-t-butylphenol. The systems were such that the singlet-singlet transition of the quinone was polarized either parallel or perpendicular to the principal molecular axis. The results agree fairly closely with the quantitative predictions of the theory. In one case, however, the variation between parallel and perpendicular was larger than expected from theory. The many problems inherent in performing these experiments suggest that one should consider only the qualitative trends that support the TM.

Thus by 1975 it was generally accepted that the TM could, and in most cases did, explain the net emission observed in the photochemically generated CIDEP. In the studies described previously it was always recognized that the RPM could operate simultaneously in the same systems. Close examination of the photo CIDEP combined with CIDNP results has shown this to be true. Some of these experiments will be described below. Perhaps the controversy over mechanisms has emphasized that the observation of net emission does not necessarily signify one mechanism or another until each case is examined in detail with the fastest time resolution possible.

4.2 Reaction Mechanism and Relaxation Studies

The utility of time-resolved photo-ESR together with the usual accompanying CIDEP has been demonstrated in many instances.[71–73] For instance, if the dominance of the TM in cases where a "net" effect is observed can be established, then a qualitative assessment of the relative intersystem crossing rates can be made in some of these systems when the sign of the zero field splitting parameter D is known.[86] Conversely, since optically detected magnetic resonance techniques usually provide the molecular intersystem crossing dynamics, the study of CIDEP in these systems can provide the sign of D.

The photo-ESR techniques provide a direct approach to the study of reaction mechanisms along with the time scales involved with various events in the reaction.[73] Unfortunately, in some instances it has been necessary to sacrifice time resolution in order to gain sensitivity. This was true in a study of the photolysis of pivalophenone ($PhCOCMe_3$).[87] In this molecule irradiation generates the benzoyl radical ($\cdot PhCO$) and the t-butyl radical ($\cdot CMe_3$) via the triplet state in inert solvents, and a reduction product in the presence of hydrogen-donor solvents. Most of the observations under varying conditions could be readily explained by operation of either the TM or the RPM; however, it should be emphasized that possible explanations of certain observations could only be tested with faster time resolution.

It has been shown in several studies that a combined photo-ESR CIDEP and photo-NMR CIDNP study will give mechanistic information impossible to obtain from either study alone. Thus ESR provides a positive identification of the transient radicals, whereas the actual reaction pathways can be established by studying the diamagnetic products by NMR–CIDNP. Atkins et al.[88] studied the photolysis of benzaldehyde in this fashion. They have demonstrated that data from the two techniques are complementary so that a more complete picture is attained.

In a CIDNP-CIDEP study of the photolysis of pyruvic acid in various hydrogen-donor solvents, Choo and Wan demonstrated the simultaneous operation of the RPM and the TM.[89] In these systems the RPM is dominant and the spectrum of the pyruvic acid radical exhibits a multiplet effect with a small amount of enhanced absorption due to the TM. (See Fig. 17.) It was suggested that the primary photochemical reaction of the excited triplet molecules contributes to polarization via the TM, whereas

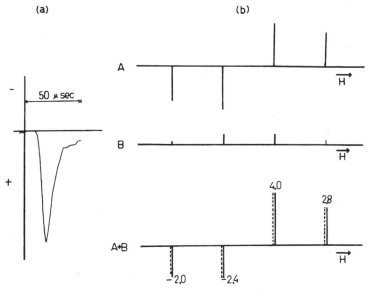

Figure 17 (a) Typical time dependence polarization of $CH_3\dot{C}(OH)COOH$ radicals after a laser pulse; photolysis of pyruvic acid in lactic acid. (b) Reconstruction of the CIDEP spectrum. A. Pure multiplet effect from RPM (low-field emission, high-field absorption). B. Net enhanced absorption from TM [from Eq. (7) either $D > 0$ and $P_z > P_x, P_y$ or $D < 0$ and $P_z < P_x, P_y$ to give enhanced absorption]; $A + B$ is the total overall spectrum as compared with experimental results (dotted lines). (From Choo and Wan, reference 89.)

the secondary reactions of the primary radical pair contribute via the RPM. The hyperfine couplings in pyruvic acid are relatively large, and therefore the dominance of the RPM can be observed. The contribution from each mechanism will vary with the conditions. In the photoreduction of various quinones by 2,6-di-t-butylphenol it was demonstrated that the RPM plays a minor role compared to the TM.[90] It is perhaps fortunate that these were the systems originally studied by photo-CIDEP, since the coexistence of the TM in systems that exhibit a distinct EA pattern could only be demonstrated by a detailed study of the polarization ratios of each individual line.[89]

The combined ESR-NMR studies have stimulated interest in observing the CIDNP resulting from transfer of polarization produced through the TM via an Overhauser cross-relaxation mechanism.[90–92] This was observed in the CIDNP of photoreduction of tetrafluoro-1,4-benzoquinone with 1,4-tetrafluorohydroquinone, including a study of the effect of plane polarized light on the nuclear spin polarization.[91] Apparently the nuclear polarization results are more complicated, however, in view of recent results of Roth et al.[93]

The decay of the spin-polarized signals from a non-Boltzmann situation to Boltzmann populations is related to the spin-lattice relaxation time T_1. Thus a method has been developed to determine T_1 (and also T_2) in systems that exhibit CIDEP.[94] The technique depends on the ability to induce spin polarization. This process must be faster than T_1. Further, it requires that the chemical decay processes are much longer than T_1 so that they may be ignored. The decay of emission to absorption was monitored and the effective decay time extrapolated to zero microwave power. For the ketyl radical derived from irradiating perdeuterobenzophenone in liquid paraffin, the values for T_1 and T_2 obtained were on the order of 10^{-4} and 10^{-8} s, respectively.

A further application of T_1 determinations was suggested by McLauchlan and Sealy.[95] They showed that electron polarization could be transferred in secondary reactions; that is, after the initial polarized radicals are created by the TM, one of the polarized radicals may react with another ground state molecule and will transfer its polarization. This is called the *memory effect*. (See examples in Section 3.4.) They caution that this can undoubtedly confuse the issue in determining polarization ratios, but it can also provide a means to obtain polarized radicals from species that are difficult to obtain by direct techniques, so that one can follow the decay of polarization to determine T_1's. In a related study[96] they showed that one can obtain information on primary polarization and subsequent physical and chemical processes from a study of secondary, polarized radicals.

Once it can be determined whether or not the TM is dominant in producing the resultant polarization in a particular system, Atkins et al.[79,97] suggested a method for determining $^3T_1$, the triplet spin-lattice relaxation time in solution. They studied the reaction

$$DQ + Et_3N \xrightarrow{h\nu} DQ\cdot^- + Et_3N\cdot^+,$$

where DQ is duroquinone and ET_3N is triethylamine. The polarization was studied as a function of triethylamine concentration. They obtained a $^3T_1$ on the order of 3–17 ns, depending on solvent. They found that the doublet relaxation time changed in approximately the same way as $^3T_1$ with solvent viscosity.

4.3 Submicrosecond Studies

Recently, direct detection techniques have been applied to time-resolved photo-ESR in liquids. Trifunac, Thurnauer, and Norris at Argonne[98] have employed a Molectron N_2 1 MW laser for an excitation source, and using an unmodified Varian E-line microwave bridge together with a Par two-channel boxcar have obtained spectra as soon as 100 ns after the laser pulse. By observing the complete spectrum at various time windows after the laser pulse, one can readily ascertain which polarization mechanism is important in a particular case. This is illustrated in Fig. 18, which shows spectra obtained in the system duroquinone in isopropanol. The spectrum of the monoprotonated durosemiquinone radical is observed at several time windows after the laser pulse. At very early time windows the spectrum clearly shows a superposition of the triplet and radical pair polarization pathways with a dominance of the triplet mechanism as evidenced by the overall shift to emission. In the spectrum taken at 60 μs after the laser pulse one can observe essentially the equilibrium spin populations.

As has been pointed out before[89,90] the contribution of each mechanism will vary with the nature of the system. In the monoprotonated p-benzosemiquinone radical formed from photolysis of p-benzoquinone in isopropanol it was apparent that the radical pair polarization made a much smaller contribution to the polarization than in the monoprotonated durosemiquinone radical. This agrees with the conclusions of Pedersen et al.[99] that in the p-benzoquinone system in ethylene glycol, the RPM contributes between 7 and 23% toward the observed polarization.

The expectation that the overall contribution of the radical pair produced CIDEP increases as the hyperfine coupling of the radical increases[89]

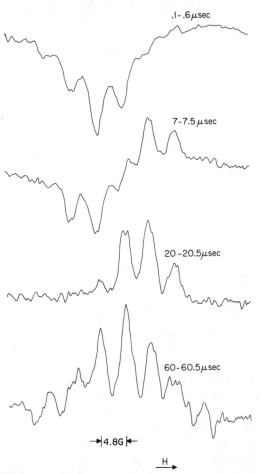

Figure 18 Monoprotonated durosemiquinone radical formed from photolysis of duro-quinone in isopropanol (concentration ~0.5M using a flat quartz EPR cell with a flow system) taken at the indicated times after the laser pulse. The spectra are all normalized to about the same intensity and emission is down and absorption up. (From Trifunac et al., reference 98.)

was evident in the observation of the radical formed from isopropanol ($(CH_3)_2CHOH$, dominant hyperfine 19.7 G) observed in the system benzophenone in isopropanol. The radical shows polarization predominantly from a radical pair mechanism even at the shortest times observed (0.1–0.6 μs). Nevertheless, triplet CIDEP can be observed at short times as the overall emission superimposed on the E/A radical pair polarization pattern. The ketyl radical from benzophenone can also be seen in the center of the spectrum, and its emission is more pronounced.

These experiments illustrate the ease and the advantages of direct detection ESR with submicrosecond time resolution and should prove to be of some help both in reexamination of some chemical systems and in the detection of hitherto unseen transient radicals in laser photolysis.

4.4 Biophysics

Since the primary events of photosynthesis involve electron transfer reactions with the consequent creation of radical ions, ESR spectroscopy has been a major tool in identifying the early reactants.[100] So it is perhaps only natural that one of the exciting new applications of time-resolved ESR spectroscopy has been the study of photosynthesis. Actually CIDEP was first observed in these systems in a steady-state ESR experiment involving the reaction of bacteriochlorophyll with a quinone.[101]

The first time-resolved ESR experiments in photosynthesis were reported by Blankenship et al.[102] They used 1 MHz field modulation together with xenon lamp flashes 10 μs long or 1 μs laser flashes. The spectrometer response time was reported to be 2 μs. They studied the time-resolved ESR after light excitation of spinach chloroplasts flowing through the spectrometer cavity at room temperature. A signal at $g = 2.0037$ was observed to be in emission. The rise and decay times of the signal were instrument limited. It was proposed that the signal was due to the primary electron acceptor of photosystem I and that the emission was due to the TM.

Recently, Dismukes et al.[103] in the same laboratory have obtained new data on several photosynthetic systems that compel them to retract their original TM explanation of the observed CIDEP in favor of the RPM. Furthermore, they now believe that the observed polarized signal is due to P_{700}^{+}, the primary electron donor, and its unusual properties relative to the steady-state P_{700}^{+} signal reflect interaction with a counterradical with ESR properties resembling a reduced iron-sulfur protein. In these experiments they have employed a Q-switched ruby laser with 50 ns pulse width with the 1 MHz ESR spectrometer. They observe a large transient change in intensity during the laser pulse followed by a slowly decaying signal (Fig. 19). They find that for some field values the fast signal can invert in an oriented flowing system compared to a nonoriented nonflowing system. Examples of the dependence of magnetic field strength on the fast and slow signals for flow and nonflow cases are shown in Fig. 20. They have explained these results (emission in nonoriented and mixed emission and absorption in the oriented case) within the framework of S-T_0 RPM, where rapid electron transfer takes the place of radical diffusion. Using an average of the known g-tensors for two possible

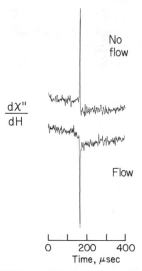

Figure 19 Kinetic traces of ESR signal intensity observed following laser pulse excitation of broken spinach chloroplasts showing the large transient change in intensity occurring during the laser pulse followed by a slower decaying signal. The effect of sample flow (flow rate 0.5 mL/min) on the polarity of the initial transient signal for some field positions is shown here. (From Dismukes et al., reference 103.)

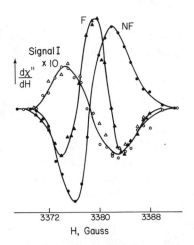

Figure 20 First derivative of the ESR absorption as a function of magnetic field for slow (signal I) and fast (polarized) kinetic components of Fig. 18. Signal I: △, flowing sample and ○, nonflowing sample. Polarized signal: ▲, flowing sample (F) showing emission and ●, nonflowing sample (NF) showing mixed emission-absorption. (From Dismukes et al., reference 103.)

146

candidates for the counter radical, they speculate that the Δg term [first term in Eq. (3)] will dominate in the nonoriented case and give emission in the observed radical. In the oriented case, assuming that only two components of the g-tensor are oriented in the plane of the membrane, they show how the Δg term will decrease and the hyperfine term [second term in Eq. (3)] will dominate so the mixed low-field emission–high-field absorption is observed. In photosynthetic systems where a flow gradient is not expected to orient the particles, only an emission signal is observed.

Other workers have studied photosynthetic systems using modified 100 kHz detection systems. McIntosh and Bolton[104] have observed initial signals on a 50 μs time scale in emission in green algal systems and spinach chloroplasts. They believe that these signals arise from photosystem II and attribute the polarization to the TM primarily on the basis that there is restricted diffusion in these systems. Our discussions previously have shown that this latter conclusion is not necessarily warranted.

Similar studies have been carried out by Hoff et al.[105] on iron-depleted bacterial reaction centers. Using 100 kHz modulation with a 20–50 μs time resolution, these workers observed a flash-induced ESR signal that displayed a combination of enhanced absorption and emission in kinetic traces taken at different field values. They believe that the observed signal is a composite of the primary electron donor P^{+}_{865} and the primary electron acceptor X^{-}.[106] They claim to be able to simulate the observed field-dependent spectrum by assuming a RPM to produce the polarization. It was also alluded that a somewhat better fit to the data was obtained by admixture of polarization that varied nonlinearly with the hyperfine coupling perhaps indicative of some S-T_{-1} mixing. In view of the results of Trifunac et al.[60] (Section 3.2) on the role of S-T_{-1} mixing in diffusion-restricted systems, these processes may also be important in these systems.

It is interesting that no fast-decaying signal was observed in bacterial systems that were not depleted in iron. This was also true when studied in a system with 10 μs time response.[103] Such experiments may prove useful in understanding the nature of the magnetic interactions of the iron in reaction centers.

5 SUMMARY AND OUTLOOK

We hope to have summarized and illustrated the major successes of the time-resolved ESR study in liquids. In these last few years we have seen experiments that go a long way toward explaining the CIDEP observed. Whether the RPM or TM or both apply, the study of CIDEP has contributed to our knowledge and understanding of radical reaction mechanisms in solution.

Unfortunately as most of the work we discussed was carried out by less than half a dozen research groups this illustrates that very fast time-resolved ESR studies are not widespread. Perhaps with the advent of suitable dye lasers we will see some growth in the application of the time-resolved ESR. Although we feel that substantial inroads have been made in the study of photo- and radiation-produced transient free radicals, there are many areas, especially in the photo-ESR field, where faster time-resolved work is needed. Also we hope to see increasing application of the quantitative methods of analysis that so far have been limited to but a few radiolytic studies. There are many more subtle effects that need further study and analysis. Finally, it has been demonstrated that the time-resolved ESR technique can be applied to the study of biological systems, and it will certainly provide us with an invaluable tool in these areas.

ACKNOWLEDGMENT

The Argonne work described in this chapter was performed under the auspices of the Division of Basic Energy Sciences of the Department of Energy.

We wish to thank members of our respective research groups, some for their work, others for many useful discussions.

We acknowledge the essential support we have received from the Argonne Chemistry Division Accelerator Section in the maintenance, development, and running of the Van de Graaff accelerator.

Above all we wish to acknowledge Kenneth W. Johnson for his design of our time-resolved ESR spectrometers, Van de Graaff pulse systems, and accessory equipment. Without his inspired and dedicated support many advances in application of magnetic resonance to radiation and photochemistry would not have been possible.

REFERENCES

1. A. R. Lepley and G. L. Closs, Eds., *Chemically Induced Magnetic Polarization*, Wiley-Interscience, New York, 1973.

2. P. B. Ayscough, T. E. English, and D. A. Tong, *Fast Processes in Radiation Chemistry and Biology*, Vol. 2, 1973, pp. 76–81; G. E. Adams, E. M. Fielden, and B. D. Michael, Eds., Institute of Physics and Wiley, London, 1975.

3. J. T. Warden and J. R. Bolton, in W. R. Ware, Ed., *Creation and Detection of the Excited States*, Marcel Dekker, New York, 1974.

4. H. H. Günthard, *Ber. Bunsenges. Phys. Chem.*, **78**, 1110 (1974).

5. H. Levanon, in M. Dorio and J. H. Freed, Eds., *Multiple Electronic Resonance*, Chap. 13, Plenum, New York, 1977.

6. R. W. Fessenden and R. H. Schuler, *J. Chem. Phys.*, **39**, 2147 (1963).

7. R. Livingston and H. Zeldes, *J. Chem. Phys.*, **44**, 1245 (1966).

8. R. Livingston and H. Zeldes, *J. Chem. Phys.*, **59**, 4891 (1973).

9. B. Smaller, J. R. Remko, and E. C. Avery, *J. Chem. Phys.*, **48**, 5174 (1968).

10. P. W. Atkins, R. C. Gurd, and A. F. Simpson, *J. Phys.* **E, 3,** 547 (1970).

11. A. D. Trifunac, K. W. Johnson, B. E. Clifft, and R. H. Lowers, *Chem. Phys. Lett.*, **35,** 566 (1975).

12. N. C. Verma and R. W. Fessenden, *J. Chem. Phys.*, **65**, 2139 (1976).

13. S. S. Kim and S. I. Weissman, *J. Mag. Resonance*, **24**, 167 (1976).

14. (a) G. L. Closs, *J. Am. Chem. Soc.*, **91**, 4552 (1969); (b) G. L. Closs and A. D. Trifunac, *J. Am. Chem. Soc.*, **92**, 2183 (1970).

15. (a) R. Kaptein and L. J. Oösterhoff, *Chem. Phys. Lett.*, **4**, 195 (1969); (b) R. Kaptein and L. J. Oösterhoff, *Chem. Phys. Lett.*, **4**, 214 (1969).

16. H. Fischer and J. Bargon, *Acc. Chem. Res.*, **2**, 110 (1969).

17. F. J. Adrian, *J. Chem. Phys.*, **54**, 3918 (1971).

18. F. J. Adrian, *Chem. Phys. Lett.*, **10**, 70 (1971).

19. F. J. Adrian, *J. Chem. Phys.*, **57**, 5107 (1972).

20. P. W. Atkins and K. A. McLauchlan, in A. R. Lepley and G. L. Closs, Eds., *Chemically Induced Magnetic Polarization*, Wiley-Interscience, New York, 1973, Chap. 2.

21. P. W. Atkins and G. T. Evans, *Chem. Phys. Lett.*, **25**, 108 (1974).

22. S. K. Wong, D. A. Hutchinson, and J. K. S. Wan, *J. Chem. Phys.*, **58,** 985 (1973).

23. G. T. Evans, P. D. Fleming, and R. G. Lawler, *J. Chem. Phys.*, **58**, 2071 (1973).

24. P. W. Atkins and G. T. Evans, *Mol. Phys.*, **27**, 1633 (1974).

25. P. W. Atkins and G. T. Evans in I. Prigogine and S. A. Rice, Eds., *Advances in Chemical Physics*, Vol. 35, Wiley, New York, 1976, pp. 1–29.

26. G. T. Evans, *Mol. Phys.*, **31,** 777 (1976).

27. G. T. Evans, P. D. Fleming, and R. G. Lawler, *Mol. Phys.*, **31,** 1337 (1976).

28. J. B. Pedersen and J. H. Freed, *J. Chem. Phys.*, **57,** 1004 (1972).

29. J. B. Pedersen and J. H. Freed, *J. Chem. Phys.*, **58,** 2746 (1972).

30. J. B. Pedersen and J. H. Freed, *J. Chem. Phys.*, **59,** 2869 (1973).

31. J. B. Pedersen and J. H. Freed, *J. Chem. Phys.*, **62,** 1706 (1975).

32. J. B. Pedersen and J. H. Freed, *J. Chem. Phys.*, **62,** 1790 (1975).

33. J. H. Freed and J. B. Pedersen, *Adv. Mag. Resonance*, **8,** 1 (1975).

34. Other possible but less likely perturbations that can cause S-$T_{\pm 1}$ mixing are anisotropic g-factor and spin rotation interaction. (See reference 24.)

35. For a review of triplet state properties and ESR see *Molecular Spectroscopy of the Triplet State*, S. P. McGlynn, T. Azumi, and M. Kinoshita, Prentice-Hall, Englewood Cliffs, N.J., 1969.

36. For a discussion of intersystem crossing see reference 20.

37. M. Schwoerer and H. Sixl, *Z. Naturforsch.*, **24a,** 952 (1969).

38. H. Levanon and S. Vega, *J. Chem. Phys.*, **61,** 2265 (1974).

39. F. J. Adrian, *J. Chem. Phys.*, **61,** 4875 (1974).

40. M. Leung and M. A. El-Sayed, *J. Am. Chem. Soc.*, **97,** 669 (1975).

41. B. Dellinger, R. M. Hochstrasser, A. B. Smith, III, *J. Am. Chem. Soc.*, **99,** 5843 (1977).

42. C. P. Poole, Jr., *Electron Spin Resonance: A Comprehensive Treatise on Experimental Techniques,* Wiley, New York, 1967, Chap. 14.

43. B. Smaller, E. C. Avery, and J. R. Remko, *J. Chem. Phys.*, **55,** 2414 (1971).

44. G. E. Smith, R. E. Blankenship, and M. P. Klein, *Rev. Sci. Instrum.*, **48,** 282 (1977).

45. K. W. Johnson and B. E. Clifft, unpublished results.

46. There are at least two ways of looking at this uncertainty broadening. One way is to consider fast radical creation as turning on the microwave power. At short times the frequency of this source cannot be well defined. This frequency spread results in a line broadening. Another way involves analysis of the Bloch equations [see Eq. (9)] and has to do with the nonzero value of δ, the reduced offset. At short times sine terms involving δ contribute over a wider field around the resonance value. (See reference 47.)

47. N. C. Verma and R. W. Fessenden, *J. Chem. Phys.*, **58,** 2501 (1973).

48. E. C. Avery, J. R. Remko, and B. Smaller, *J. Chem. Phys.*, **49,** 951 (1968).

49. G. Nucifora, B. Smaller, R. Remko, and E. C. Avery, *Radiation Research,* **46,** 96 (1972).

50. R. W. Fessenden and N. C. Verma, *J. Am. Chem. Soc.*, **98,** 243 (1976).

51. F. P. Sargent and E. M. Gardy, *Chem. Phys. Lett.*, **39,** 188 (1976).

52. R. W. Fessenden, *J. Chem. Phys.*, **58,** 2489 (1973).

53. A. D. Trifunac and E. C. Avery, *Chem. Phys. Lett.*, **28,** 141 (1974).

54. A. D. Trifunac and E. C. Avery, *Chem. Phys. Lett.*, **28,** 294 (1974).

55. A. D. Trifunac and M. C. Thurnauer, *J. Chem. Phys.*, **62,** 4889 (1975).

56. A. D. Trifunac, *J. Am. Chem. Soc.*, **98,** 5202 (1976).

57. A. D. Trifunac and D. J. Nelson, *J. Am. Chem. Soc.*, **99,** 289 (1977).

58. A. D. Trifunac and D. J. Nelson, *Chem. Phys. Lett.*, **46,** 346 (1977).

59. A. D. Trifunac, *Chem. Phys. Lett.*, **49,** 457 (1977).

60. A. D. Trifunac, D. J. Nelson, and C. Mottley, *J. Mag. Resonance,* **30,** 263 (1978).

61. H. Shiraishi, H. Kodoi, Y. Katsumura, Y. Tabata, and K. Oshima, *J. Phys. Chem.*, **80,** 2400 (1976).

62. (a) J. F. Fendler and E. J. Fendler, *Catalysis in Micellar and Macromolecular Systems,* Academic, New York, 1975; (b) K. Shinoda, T. Nakagawa, B. Tamamashi, and T. Isemura, *Colloidal Surfactants,* Academic, New York, 1963.

63. At low H_1 [ω_1 in Eq. (8)] the signal will grow as $1-\exp(-t/T_1)$; so it takes approximately a T_1 time period to fully develop the ESR signal. However, experimental conditions sometimes require use of larger H_1 fields, and oscillatory behavior can sometimes be observed. See references 12, 47, 64, and 65. Also see Section 3.4.

64. P. W. Atkins, A. J. Dobbs, and K. A. McLauchlan, *Chem. Phys. Lett.*, **25,** 105 (1974).

65. S. S. Kim and S. I. Weissman, *Chem. Phys.*, **27,** 21 (1978).

66. H. C. Torrey, *Phys. Rev.*, **76**, 1059 (1949).

67. G. L. Closs and M. S. Czeropski, *J. Am. Chem. Soc.*, **99**, 6127 (1977).

68. (a) G. J. Krüger, *Adv. Mol. Relaxation Processes*, **3**, 235 (1972); (b) S. K. Rengan, M. P. Khalchar, B. S. Brabhananda, and B. Venkataraman, *J. Pure Appl. Chem.*, **32**, 287 (1972).

69. (a) J. A. Den Hollander, Ph.D. Thesis, Leiden, 1976; (b) J. A. Den Hollander, *Chem. Phys.*, **10**, 167 (1975).

70. A. D. Trifunac and D. J. Nelson, *J. Am. Chem. Soc.*, **99**, 1745 (1977).

71. P. W. Atkins, *Organic Magnetic Resonance*, **5**, 239 (1973).

72. J. K. S. Wan, S. K. Wong, and D. A. Hutchinson, *Acc. Chem. Res.*, **7**, 58 (1974).

73. J. K. S. Wan and A. J. Elliot, *Acc. Chem. Res.*, **10**, 161 (1977).

74. P. W. Atkins, J. C. Buchanan, R. C. Gurd, K. A. McLauchlan and A. F. Simpson, *J. Chem. Soc. D, Chem. Commun.*, **513** (1970).

75. P. W. Atkins, R. C. Gurd, K. A. McLauchlan, and A. F. Simpson, *Chem. Phys. Lett.*, **8**, 55 (1971).

76. P. W. Atkins, J. K. Duggan, K. A. McLauchlan, and P. W. Percival, *Chem. Phys. Lett.*, **24**, 565 (1974).

77. A. J. Dobbs, *Mol. Phys.*, **30**, 1073 (1975).

78. P. W. Atkins, K. A. McLauchlan, and P. W. Percival, *J. Chem. Soc. D, Chem. Commun.*, **121** (1973).

79. P. W. Atkins, A. J. Dobbs, G. T. Evans, K. A. McLauchlan, and P. W. Percival, *Mol. Phys.*, **27**, 769 (1974).

80. S. K. Wong and J. K. S. Wan, *J. Am. Chem. Soc.*, **94**, 7197 (1972).

81. S. K. Wong, D. A. Hutchinson, and J. K. S. Wan, *J. Am. Chem. Soc.*, **95**, 622 (1973).

82. S. K. Wong, D. A. Hutchinson, and J. K. S. Wan, *Can. J. Chem.*, **52**, 251 (1974).

83. P. W. Atkins, A. J. Dobbs, and K. A. McLauchlan, *Chem. Phys. Lett.*, **22**, 209 (1973).

84. A. J. Dobbs and K. A. McLauchlan, *Chem. Phys. Lett.*, **30**, 257 (1975).

85. B. B. Adeleke, K. Y. Choo, and J. K. S. Wan, *J. Chem. Phys.*, **62**, 3822 (1975).

86. S. K. Wong and J. K. S. Wan, *J. Chem. Phys.*, **59**, 3859 (1973).

87. P. W. Atkins, A. J. Dobbs, and K. A. McLauchlan, *J. Chem. Soc. Faraday I*, **71**, 1269 (1975).

88. P. W. Atkins, J. M. Frimston, P. G. Frith, R. C. Gurd, and K. A. McLauchlan, *J. Chem. Soc. Faraday II*, **69**, 1542 (1973).

89. K. Y. Choo and J. K. S. Wan, *J. Am. Chem. Soc.*, **97**, 7127 (1975).

90. B. B. Adeleke and J. K. S. Wan, *J. Chem. Soc. Faraday Trans. I*, **72**, 1799 (1976).

91. H. M. Vyas and J. K. S. Wan, *Chem. Phys. Lett.*, **34**, 470 (1975).

92. H. M. Vyas and J. K. S. Wan, *Can. J. Chem.*, **54**, 979 (1976).

93. H. D. Roth, R. S. Hutton, and M. L. Manion Schilling, in O. P. Strausz, Ed., *Reviews of Chemical Intermediates*, to be published (1979).

94. P. W. Atkins, K. A. McLauchlan, and P. W. Percival, *Mol. Phys.*, **25**, 281 (1973).

95. K. A. McLauchlan and R. C. Sealy, *Chem. Phys. Lett.*, **39**, 310 (1976).

96. K. A. McLauchlan, R. C. Sealy, and J. M. Wittmann, *J. Chem. Soc. Faraday Trans. II*, **73**, 926 (1977).

97. P. W. Atkins, *Chem. Phys. Lett.*, **29**, 616 (1974).

98. A. D. Trifunac, M. C. Thurnauer, and J. R. Norris, *Chem. Phys. Lett.*, **57,** 471 (1978).

99. J. B. Pedersen, C. E. M. Hensen, H. Farbo, and L. T. Muus, *J. Chem. Phys.*, **63,** 2398 (1975).

100. Reviews of ESR applied to the study of photosynthesis include: (a) D. H. Kohl, in J. R. Bolton and D. C. Borg, Eds., *Biological Applications of Electron Spin Resonance*, Wiley-Interscience, New York, 1972, pp. 213–264; (b) J. T. Warden and J. R. Bolton, *Acc. Chem. Res.*, **7,** 189 (1974).

101. J. R. Harbour and G. Tollin, *Photochem. Photobiol.*, **19,** 163 (1974).

102. R. Blankenship, A. McGuire, and K. Sauer, *Proc. Nat. Acad. Sci. USA*, **72,** 4943 (1975).

103. C. G. Dismukes, A. McGuire, R. Blankship, and K. Sauer, *Biophys. J.*, **21,** 239 (1978).

104. A. R. McIntosh and J. R. Bolton, *Nature*, **263,** 443 (1976).

105. A. J. Hoff, P. Gast, and J. C. Romijn, *FEBS Lett.*, **73,** 185 (1977).

106. X is believed to be a ubiquinone-iron complex, and gives a very anisotropic ESR signal with peaks at $g = 2.00$, 1.82, and 1.68. With the iron depleted the ubiquinone signal is near the P_{865}^{+} signal ($g = 2.0026$) at $g = 2.0046$.

5 THEORY OF ELECTRON SPIN ECHOES IN NONVISCOUS AND VISCOUS LIQUIDS

Arthur E. Stillman and Robert N. Schwartz

Department of Chemistry
University of Illinois at Chicago Circle
Chicago, Illinois

1   INTRODUCTION

The line shapes of magnetic resonance spectra contain information concerning the dynamic magnetic environment surrounding the spins.[1,2] Thus from spin-relaxation studies the diffusional processes of the spin-containing molecules may be elucidated. The analysis, however, is often complicated because frequently the lines consist of a superposition of spin packets arising from static local magnetic field differences.[3,4] These may typically be caused by Zeeman field inhomogeneities or unresolved hyperfine interactions. The problem lies in that the relaxation theories are only valid for the individual spin packets and not the whole inhomogeneously broadened line. One must in some way deconvolute the signal into the individual spin packets, the widths of which are related to the various dynamic processes affecting the spins.

One way of accomplishing this is to assume a particular model for the source of inhomogeneous line broadening and then to calculate the line shapes for a series of spin-packet widths.[5,6] Interpolation of a plot of the envelope width versus the spin-packet width yields the desired result. The difficulty with this approach is that the obtained intrinsic line width is *strongly* correlated to the model used for the source of inhomogeneous line broadening.[6] Thus if the major source of inhomogeneity is due to unresolved hyperfine interactions and if the corresponding hyperfine model is not accurately known, the spin-packet width so obtained will be in error.

Another method for obtaining the intrinsic line width of inhomogeneously broadened lines is to perform spin-echo studies.[3,4,7] Pulse techniques have proved to be invaluable in nuclear magnetic resonance (NMR) spectroscopy, and it is expected that they will prove to be likewise important in electron paramagnetic resonance (EPR) spectroscopy. The major difficulty encountered in time domain EPR experiments has been in the technology in producing the high-power short microwave pulses and fast data acquisition necessary because of the comparatively shorter time scale of the EPR experiment. Also, the sensitivity of time domain EPR spectrometers has been a source of difficulty. There have recently been great strides made in microwave technology and digital electronics, however, and commercial instruments are expected to become available within the next few years.

It is clear that if electron spin-echo spectroscopy is to become a viable research tool, the theory of electron spin-echo phase memory decay must be well understood. Considerable progress over the past 10 years has been made in this direction for electron spin echoes in solids.[4,7] However, very little attention has been paid to free radicals dissolved in liquids. This is no doubt a direct result of the greater difficulty in performing this experiment because of the relatively shorter relaxation times often encountered in liquid solutions. With the improving technology, however, one may expect that liquid-phase electron spin-echo spectroscopy will become an easier experiment to perform and possibly of greater general utility than its solid-phase counterpart.

Several experimental studies of electron spin echoes in liquids have been reported.[8–12] Whereas most of the early studies related the obtained phase memory times to the relaxation times found in the time-dependent Bloch equations, Brown studied several free radicals over a large temperature range and attempted to analyze the phase memory times in terms of radical and matrix motions.[11] This is clearly a step in the right direction, since a more detailed picture of the dynamics of the system is explicitly taken into account. However, Brown's approach suffers in that it is still a phenomenological theory and that it incorrectly predicts that the phase memory times are the same for each hyperfine manifold of a radical containing magnetically coupled nuclei. The latter is a direct result of assuming a dipolar Hamiltonian with no g-tensor anisotropy; Brown's theory does not appear to lend itself to cases for which more complicated spin Hamiltonians are appropriate. Moreover, Brown's theory is only valid in limiting cases of radical–matrix molecular motions rather than the whole range of correlation times of interest in molecular dynamics studies.

In this chapter we are concerned with other more general approaches to the study of electron spin-echo phase memory in liquids and a density matrix formalism proves to be useful for this purpose. The electron spin-echo phase memory decay is determined from the evolution in time of the spin-density matrix and may be considered to arise from either a relaxation operator[13] or, equivalently, an explicit time dependence of the spin Hamiltonian.[14] The former approach is more useful from a computational viewpoint and is discussed in Section 3. This method permits the consideration of effects resulting from relaxation during the microwave pulses as well as the study of more complicated pulse experiments such as ENDOR spin-echo spectroscopy.[15a] The relation between the electron spin-echo phase memory decay and the time dependence of the spin

Hamiltonian is revealed through a cumulant expansion of the Liouville equation and is discussed in Section 4. This method is seen to provide considerable physical insight for both motionally narrowed as well as slow-tumbling cases and allows one to study short time relaxation behavior.

2 SEMICLASSICAL DESCRIPTION OF ELECTRON SPIN ECHOES

A phenomenological description of an electron spin echo is obtainable from the vector dipole model for a two-pulse electron spin echo as shown in Fig. 1.[3,4] In a static magnetic field defining the z-direction, a system of spin packets precesses about the magnetic field and hence has a net magnetic moment $\mathbf{M}$ along the z-axis. A microwave pulse along the x-axis of sufficient duration and amplitude nutates $\mathbf{M}$ 90° from the z-direction into the y-axis. As soon as the microwave power is removed, the spin packets precess apart in the $x–y$ plane because there is a distribution of resonance frequencies due to local field differences[16] and thereby the projection of the magnetization along the y-axis is decreased. Since the EPR signal intensity is proportional to the projection of magnetization along the y-direction, there is a corresponding decrease in the signal intensity. This phenomenon is called the free-induction decay (FID). The spin packets precess apart for a time τ and are then subjected to a 180° microwave pulse that nutates each spin packet by 180° in the $x–y$ plane. The spin packets then rephase together along the $-y$-axis τ seconds following the second pulse, and an echo or increase of signal amplitude is observed.

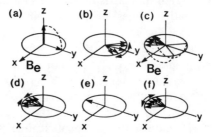

Figure 1 Classical vector dipole model for a two-pulse echo. (a) The spin packets initially precess about the static Zeeman field B_0 and hence have a net magnetic moment along the z-axis. The rf pulse nutates this magnetization along the y-axis. (b) The spin packets dephase for τ seconds following the application of the first pulse. This is the free-induction decay. (c) The second microwave pulse nutates the spin packets to the -y-axis. (d) The spin packets begin to rephase together. (e) Maximum rephasing occurs τ seconds following the second pulse and the echo is observed. (f) The spin packets continue to dephase.

Experimentally one studies the two-pulse spin-echo amplitude as a function of the time τ between the microwave pulses. Because of randomly fluctuating magnetic fields, the echo amplitude decreases irreversibly as τ is increased. In liquids these randomly fluctuating fields result from the motional modulation of the various magnetic interactions of the spin system. We thus distinguish the dephasing of the spin packets as arising from two different sources. The reversible component is due to the distribution of resonance frequencies that constitute the inhomogeneous line broadening. It is reversible in the sense that the second microwave pulse causes the spin packets to rephase. Those dynamic processes that result in spin packets that do not rephase after the second pulse are due to irreversible dephasing processes and are of three types: secular, pseudosecular, and nonsecular. Secular modulation of the energy levels renders the resonance frequencies time dependent. Any change in the resonance frequency following the application of the first pulse will cause the corresponding spin packet to be at some position other than along the $-y$-axis after a time τ seconds following the second pulse. Hence the spin packet will not contribute to the echo. Pseudosecular transitions result in a different magnetic environment for the unpaired electron and thus change the resonance frequency of the corresponding spin packet. It follows that the pseudosecular transitions are also a source of irreversible dephasing. Nonsecular transitions take the spin packets out of the x–y plane altogether with a corresponding increase in magnetization along the z-axis. This likewise gives rise to an irreversible decay of the echo amplitude.

One point should be made clear about the free-induction decay (FID) signal. The FID is isomorphic with the inhomogeneously broadened continuous wave (cw) line through its Fourier transform, whereas the irreversible dephasing completely determines the homogeneous spin-packet linewidth and is a consequence of dynamic field fluctuations at the electron. It is thus the irreversible component that is of interest for measuring rotational and translational correlation times in spin-relaxation studies.

Although the vector dipole model for electron spin echoes serves as a useful descriptive picture, the actual physics of the echo process is better represented from a quantum-mechanical viewpoint. In this case the echo is considered to result from a spontaneous emission from a superradiant state created by the phase coherence of the microwave radiation source.[17] This phase coherence produces off-diagonal elements in the spin-density matrix that decay in time as the phase memory of the spin system is lost. The spontaneous emission rate is, however, proportional to these off-diagonal density matrix elements so that the echo intensity decreases with the loss of phase memory. Thus the evolution in time of the spin-density matrix completely describes the electron spin-echo envelope decay. In the following discussion we consider the echo to result from the time dependence of the absorption signal. However, it is important to keep in mind that, although this is consistent with common convention, it is actually somewhat of a misnomer.[18]

3 DENSITY MATRIX–RELAXATION OPERATOR FORMALISM FOR ELECTRON SPIN ECHOES IN LIQUIDS

3.1 Motionally Narrowed Region

3.1.1 GENERAL METHODS

The time evolution of the spin-density matrix $\boldsymbol{\sigma}(t)$ is given by the generalized Liouville equation[1,19]

$$\dot{\boldsymbol{\sigma}}(t) = -i(\mathcal{H}_0^x + \boldsymbol{\varepsilon}(t)^x)\boldsymbol{\sigma}(t) - \boldsymbol{\Gamma}(\boldsymbol{\sigma} - \boldsymbol{\sigma}_0), \tag{1}$$

where $\mathcal{H}_0$ is the static zero-order contribution to the spin Hamiltonian

$$\hbar\mathcal{H}(t) = \hbar[\mathcal{H}_0 + \mathcal{H}_1(t) + \boldsymbol{\varepsilon}(t)], \tag{2}$$

$\mathcal{H}_1(t)$ is a random time-dependent perturbation, $\boldsymbol{\Gamma}$ is the stochastic relaxation operator, $\boldsymbol{\sigma}_0$ is the spin-density matrix at thermal equilibrium, and, where for arbitrary operators $\mathbf{A}$ and $\mathbf{B}$,

$$\mathbf{A}^x\mathbf{B} \equiv \mathbf{AB} - \mathbf{BA}. \tag{3}$$

Here $\boldsymbol{\varepsilon}(t)$ is the perturbation due to the radiation field and is given by

$$\boldsymbol{\varepsilon}(t) = \tfrac{1}{2}\gamma_e B_e[\mathbf{S}_+ \exp(-i\omega t) + \mathbf{S}_- \exp(+i\omega t)], \tag{4}$$

where B_e is the magnitude of the microwave field in the rotating frame with frequency ω and γ_e is the gyromagnetic ratio. It is convenient to define a reduced-density matrix by

$$\boldsymbol{\chi}(t) = \boldsymbol{\sigma}(t) - \boldsymbol{\sigma}_0. \tag{5}$$

The matrix elements of $\boldsymbol{\chi}$ may then be expressed as a generalized Fourier expansion:[20]

$$\chi_{\alpha,\alpha'}(t) = \sum_{n=-\infty}^{\infty} Z_{\alpha,\alpha'}^{(n)}(t) \exp(in\omega t). \tag{6}$$

The nth harmonic of the EPR absorption due to the $\alpha \to \alpha'$ transition is proportional to the imaginary part of $Z_{\alpha,\alpha'}^{(n)}(t)$ whereas the dispersion is proportional to its real part. Normally it is the first harmonic that is detected by the pulse spectrometer, so that only the $n = 1$ term of Eq. (6) need be retained for the off-diagonal spin states.

The condition of motional narrowing is defined by $|\mathscr{H}_1(t)\tau_c|^2 \ll 1$ where τ_c is a correlation time for $\mathscr{H}_1(t)$.[1] We may assume in this case that the steady-state EPR linewidth is inhomogeneously broadened by unresolved intramolecular isotropic hyperfine interactions; that is, each line is actually a superposition of homogeneously broadened resonances. Other sources of inhomogeneity are generally small enough to be neglected. The NMR analogous case is the echo arising from a multiplet structure.[1] The observed echo signal is therefore given by the sum of the $\text{Im}\{Z_{\lambda_j}^{(1)}\}$ over all the λ_j transitions for each of the constituent lines. In this notation j serves as an index for the degenerate transitions and λ_j corresponds to a $\langle\lambda_j^-||\lambda_j^+\rangle$ matrix element with $\pm$ referring to the respective M_s quantum state.

Substituting Eq. (6) into Eq. (1) one obtains after some algebraic manipulation the following supermatrix equation for $\tau \gg \tau_c$ in the high-temperature limit:[20]

$$
\begin{bmatrix}
\frac{1}{\sqrt{2}}\dot{\mathbf{Z}}(t) \\[2mm]
\frac{1}{\sqrt{2}}\dot{\mathbf{Z}}^*(t) \\[2mm]
\frac{1}{2}\dot{\hat{\mathbf{X}}}(t) \\[2mm]
\frac{1}{2}\dot{\tilde{\mathbf{X}}}(t)
\end{bmatrix}
=
\begin{bmatrix}
\mathbf{R}-i\mathbf{K} & 0 & \sqrt{2}\,i\hat{\mathbf{d}} & \sqrt{2}\,i\hat{\mathbf{d}} \\[2mm]
0 & \mathbf{R}+i\mathbf{K} & -\sqrt{2}\,i\hat{\mathbf{d}} & -\sqrt{2}\,i\hat{\mathbf{d}} \\[2mm]
\sqrt{2}\,i\hat{\mathbf{d}}^T & -\sqrt{2}\,i\hat{\mathbf{d}}^T & -\hat{\mathbf{W}} & -\mathscr{W} \\[2mm]
\sqrt{2}\,i\tilde{\mathbf{d}}^T & -\sqrt{2}\,i\tilde{\mathbf{d}}^T & -\mathscr{W}^T & -\tilde{\mathbf{W}}
\end{bmatrix}
\times
\begin{bmatrix}
\frac{1}{\sqrt{2}}\mathbf{Z}(t) \\[2mm]
\frac{1}{\sqrt{2}}\mathbf{Z}^*(t) \\[2mm]
\frac{1}{2}\hat{\mathbf{X}}(t) \\[2mm]
\frac{1}{2}\tilde{\mathbf{X}}(t)
\end{bmatrix}
+
\begin{bmatrix}
\frac{i\mathbf{Q}}{\sqrt{2}} \\[2mm]
\frac{-i\mathbf{Q}}{\sqrt{2}} \\[2mm]
0 \\[2mm]
0
\end{bmatrix},
$$

$$\tag{7}$$

where $\hat{\mathbf{X}}(t)$ and $\tilde{\mathbf{X}}(t)$ are defined by

$$
\hat{X}_{\lambda_j} \equiv \chi_{\lambda_j^+,\lambda_j^+} - \chi_{\lambda_j^-,\lambda_j^-}, \tag{8a}
$$

and

$$
\tilde{X}_{\lambda_j} \equiv \chi_{\lambda_j^+,\lambda_j^+} + \chi_{\lambda_j^-,\lambda_j^-}. \tag{8b}
$$

In this expression the relaxation matrix $\mathbf{R}$ and the transition probability matrices $\hat{\mathbf{W}}$, $\tilde{\mathbf{W}}$, and $\mathscr{W}$ are defined by the various matrix elements of $\mathbf{\Gamma}$, the matrices $\hat{\mathbf{d}}$ and $\tilde{\mathbf{d}}$ contain elements proportional to the various transition moments defined as[19]

$$
d_{\lambda_j} \equiv \tfrac{1}{2}\gamma_e B_e S_{-\lambda_j}, \tag{9}
$$

and the vector $\mathbf{Q}$ has elements given by

$$Q_{\lambda_j} = \omega_{\lambda_j} d_{\lambda_j}\left(\frac{\hbar}{NkT}\right), \tag{10}$$

where ω_{λ_j} is the resonance frequency of the λ_j transition, k is Boltzmann's constant, T is the absolute temperature, and N is the number of eigenstates of $\mathcal{H}_0$. $\mathbf{K}$ is the coherence matrix that is found to be diagonal in transition space for the present example. Its elements are given by

$$K_{\lambda_j,\lambda_j} = \omega - \omega_{\lambda_j}. \tag{11}$$

If the transition probabilities satisfy

$$W_{\alpha\pm,\beta\mp} = W_{\alpha\mp,\beta\pm}, \tag{12a}$$

and

$$W_{\alpha\pm,\beta\pm} = W_{\alpha\mp,\beta\mp}, \tag{12b}$$

then the matrix $\mathcal{W} = \mathbf{0}$ (a null matrix). This is a common occurrence.[21] If it is further assumed that the radiation field excites only EPR transitions, then $\tilde{\mathbf{d}} = \mathbf{0}$ and the $\tilde{\mathbf{X}}(t)$ solution is decoupled from the remaining three. The relevant equation in this case is

$$\begin{bmatrix} \frac{1}{\sqrt{2}}\dot{\mathbf{Z}}(t) \\[2mm] \frac{1}{\sqrt{2}}\dot{\mathbf{Z}}^*(t) \\[2mm] \frac{1}{2}\dot{\hat{\mathbf{X}}}(t) \end{bmatrix} = \begin{bmatrix} \mathbf{R}-i\mathbf{K} & \mathbf{0} & \sqrt{2}\,i\hat{\mathbf{d}} \\[2mm] \mathbf{0} & \mathbf{R}+i\mathbf{K} & -\sqrt{2}\,i\hat{\mathbf{d}} \\[2mm] \sqrt{2}\,i\hat{\mathbf{d}}^T & -\sqrt{2}\,i\hat{\mathbf{d}}^T & -\hat{\mathbf{W}} \end{bmatrix} \times \begin{bmatrix} \frac{1}{\sqrt{2}}\mathbf{Z}(t) \\[2mm] \frac{1}{\sqrt{2}}\mathbf{Z}^*(t) \\[2mm] \frac{1}{2}\hat{\mathbf{X}}(t) \end{bmatrix} + \begin{bmatrix} \frac{i\mathbf{Q}}{\sqrt{2}} \\[2mm] \frac{-i\mathbf{Q}}{\sqrt{2}} \\[2mm] \mathbf{0} \end{bmatrix}. \tag{13}$$

If there are degeneracies, the line shape may often be averaged so that[21]

$$Z_\lambda = \sum_j Z_{\lambda_j}, \tag{14a}$$

$$\hat{X}_\lambda = \sum_j \hat{X}_{\lambda_j}, \tag{14b}$$

$$R_{\lambda,\eta} = R_{\lambda_j,\eta_k}, \tag{14c}$$

$$K_{\lambda_j,\eta_k} = K_{\lambda,\eta}\,\delta_{jk}\,\delta_{\lambda\eta}, \tag{14d}$$

$$\hat{W}_{\lambda,\lambda} = \hat{W}_{\lambda_j,\lambda_j}, \tag{14e}$$

$$\hat{W}(\mathrm{ex})_{\lambda,\eta} = \hat{W}(\mathrm{ex})_{\lambda_j,\eta_k}, \qquad \lambda_j \neq \eta_k \tag{14f}$$

[for Heisenberg exchange],

$$\hat{W}(\mathrm{END})_{\lambda,\eta} = \hat{W}(\mathrm{END})_{\lambda_j,\eta_k}, \qquad \lambda \neq \eta \tag{14g}$$

[for electron-nuclear dipolar (END) relaxation],

$$\hat{d}_{\lambda_j, \eta_k} = \hat{d}_{\lambda, \eta} \, \delta_{jk} \, \delta_{\lambda\eta}, \tag{14h}$$

and

$$D_\lambda Q_\lambda = \sum_j Q_{\lambda_j}, \tag{14i}$$

where D_λ is the degeneracy of the λth transition. One then has for this subspace that

$$
\begin{bmatrix}
\dfrac{1}{\sqrt{2}}\dot{\mathbf{Z}}(t) \\[2mm]
\dfrac{1}{\sqrt{2}}\dot{\mathbf{Z}}^*(t) \\[2mm]
\tfrac{1}{2}\dot{\hat{\mathbf{X}}}(t)
\end{bmatrix}
=
\begin{bmatrix}
\mathbf{R} - i\mathbf{K} & \mathbf{0} & \sqrt{2}\,i\hat{\mathbf{d}} \\[2mm]
\mathbf{0} & \mathbf{R} + i\mathbf{K} & -\sqrt{2}\,i\hat{\mathbf{d}} \\[2mm]
\sqrt{2}\,i\hat{\mathbf{d}}^T & -\sqrt{2}\,i\hat{\mathbf{d}}^T & -\hat{\mathbf{W}}
\end{bmatrix}
\times
\begin{bmatrix}
\dfrac{1}{\sqrt{2}}\mathbf{Z}(t) \\[2mm]
\dfrac{1}{\sqrt{2}}\mathbf{Z}^*(t) \\[2mm]
\tfrac{1}{2}\hat{\mathbf{X}}(t)
\end{bmatrix}
+
\begin{bmatrix}
\dfrac{i\mathbf{DQ}}{\sqrt{2}} \\[2mm]
\dfrac{-i\mathbf{DQ}}{\sqrt{2}} \\[2mm]
0
\end{bmatrix},
\tag{15}
$$

where $\mathbf{Z}$, $\hat{\mathbf{X}}$, $\mathbf{Q}$, $\mathbf{R}$, $\mathbf{K}$, and $\hat{\mathbf{d}}$ are all redefined in this subspace and $\mathbf{D}$ is a diagonal matrix of the degeneracies.

This supermatrix equation bears a formal similarity to the time-dependent Bloch equations.[1] We may therefore obtain the analytic solution to Eq. (15) by generalization of the technique employed by Atkins et al.[22] for the solution of this simpler problem. For compactness of notation we rewrite Eq. (15) as

$$\dot{\mathbf{b}} = \mathbf{Ab} + \mathbf{c}. \tag{16}$$

Taking the Laplace transform of both sides of this equation one has that[13]

$$\mathscr{L}\{\dot{\mathbf{b}}\} = \mathscr{L}\{\mathbf{Ab}\} + \mathscr{L}\{\mathbf{c}\}$$
$$= \rho\mathbf{B}(\rho) - \mathbf{b}(0), \tag{17}$$

where

$$\mathscr{L}\{\mathbf{b}(t)\} \equiv \int_0^\infty e^{-\rho t}\mathbf{b}(t)\, dt = \mathbf{B}(\rho). \tag{18}$$

If $\mathbf{A}$ is *independent* of time, Eq. (17) may be solved for $\mathbf{B}(\rho)$ in terms of $\mathbf{b}$:

$$\mathbf{B}(\rho) = (-\mathbf{A} + \rho\mathbf{I})^{-1}\mathbf{b}(0) + \frac{(-\mathbf{A} + \rho\mathbf{I})^{-1}\mathbf{c}}{\rho}. \tag{19}$$

By making use of the identities

$$\mathscr{L}^{-1}\{(-\mathbf{A} + \rho\mathbf{I})^{-1}\} \equiv \exp(\mathbf{A}t), \tag{20}$$

and

$$\mathscr{L}^{-1}\left\{\frac{(-\mathbf{A}+\rho\mathbf{I})^{-1}\mathbf{c}}{\rho}\right\} \equiv \int_0^t \exp\left[\mathbf{A}(t-u)\right]\mathbf{c}\,du, \tag{21}$$

one obtains, after taking the inverse Laplace transform of both sides of Eq. (19),

$$\mathbf{b}(t) = \exp\left(\mathbf{A}t\right)\mathbf{b}(0) - \mathbf{A}^{-1}[\mathbf{I} - \exp\left(\mathbf{A}t\right)]\mathbf{c}. \tag{22}$$

A convenient form of this equation is obtained by using the diagonalizing transformation $\mathbf{S}^{-1}\mathbf{A}\mathbf{S} = \mathbf{\Lambda}$, where $\mathbf{\Lambda}$ is the eigenvalue matrix

$$\mathbf{b}(t) = \mathbf{S}\exp\left(\mathbf{\Lambda}t\right)\mathbf{S}^{-1}\mathbf{b}(0) - \mathbf{S}\mathbf{\Lambda}^{-1}[\mathbf{I} - \exp\left(\mathbf{\Lambda}t\right)]\mathbf{S}^{-1}\mathbf{c}. \tag{23}$$

The real part of Λ_{ii} is always found to be negative. It follows that $\mathbf{b}(t)$ is a superposition of oscillatory decays plus constant "driving" functions.

Consider the typical two-pulse scheme diagrammed in Fig. 2. During the pulses of microwave power in regions I and III, $\hat{\mathbf{d}}$ is diagonal with elements given by $-\gamma_e B_e/2$.[21] In regions II and IV, $\hat{\mathbf{d}}$ is a matrix of zeros since $B_e = 0$. Implicit in the earlier derivation is the assumption that $\mathbf{A}$ is a constant function of time. Clearly it is not in the preceding example. However, $\mathbf{A}$ is constant in each region. In fact, $\mathbf{A}_{II} = \mathbf{A}_{IV}$. Generally the microwave power is the same during each pulse so one also has that $\mathbf{A}_I = \mathbf{A}_{III}$.

We seek solutions of the form of Eq. (23) requiring continuity at the boundaries of each region. The initial conditions of region I are given by $\mathbf{Z}(0) = \mathbf{Z}^*(0) = \hat{\mathbf{X}}(0) = \mathbf{0}$ so that in region I

$$\mathbf{b}_I(t) = -\mathbf{S}_I\mathbf{\Lambda}_I^{-1}[\mathbf{I} - \exp\left(\mathbf{\Lambda}_I t\right)]\mathbf{S}_I^{-1}\mathbf{c}. \tag{24}$$

In regions II and IV the vector $\mathbf{Q}$ and hence $\mathbf{c}$ are zero, since the elements of $\mathbf{Q}$ are proportional to B_e. Thus for times $t_1 \leq t \leq t_2$ one has that

$$\mathbf{b}_{II}(t) = \mathbf{S}_{II}\exp\left[\mathbf{\Lambda}_{II}(t-t_1)\right]\mathbf{S}_{II}^{-1}\mathbf{b}_{II}(t_1). \tag{25}$$

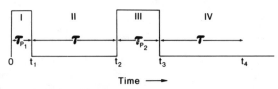

Figure 2 Two-pulse spin-echo pulse sequence. The echo occurs with maximum amplitude at $t = t_4$. [Reprinted with permission from A. E. Stillman and R. N. Schwartz, *Mol. Phys.*, **32**, 1045 (1976). Copyright by Taylor & Francis Ltd.]

Since the solutions are assumed to be continuous at the region boundaries, it follows that

$$\mathbf{b}_\mathrm{I}(t_1) = \mathbf{b}_\mathrm{II}(t_1). \tag{26}$$

The initial boundary conditions at $t = 0$ are known. This guarantees a unique solution to the differential equation in region I. Once this solution is found, $\mathbf{b}_\mathrm{I}(t_1)$ may be calculated. The requirement that the solutions between region boundaries are continuous defines the initial conditions for region II. Hence the solution for region II may be determined. The solution for other regions and pulse schemes is straightforward.

For computational purposes it is preferable to perform a numerical integration of Eq. (15) over each region using the appropriate boundary conditions and requiring continuity of the solutions as discussed earlier.[13] The reason for this is that there are problems with numerical instabilities in the computation of matrix exponentials such as appear in the analytic solution.[23] It is found that a real form of Eq. (15) is useful for numerical purposes. This may be obtained by transformation with the unitary matrix $\mathbf{U}$ where

$$\mathbf{U} = \begin{bmatrix} \dfrac{1}{\sqrt{2}}\mathbf{I} & \dfrac{1}{\sqrt{2}}\mathbf{I} & \mathbf{0} \\[2ex] -\dfrac{i}{\sqrt{2}}\mathbf{I} & \dfrac{i}{\sqrt{2}}\mathbf{I} & \mathbf{0} \\[2ex] \mathbf{0} & \mathbf{0} & \mathbf{I} \end{bmatrix}. \tag{27}$$

It should be understood that the identity matrix $\mathbf{I}$ in this expression varies in dimensionality so as to be consistent with the partitioned matrices in Eq. (15). One obtains

$$\begin{bmatrix} \dot{\mathbf{Z}}'(t) \\ \dot{\mathbf{Z}}''(t) \\ \tfrac{1}{2}\dot{\hat{\mathbf{X}}}(t) \end{bmatrix} = \begin{bmatrix} \mathbf{R} & \mathbf{K} & \mathbf{0} \\ -\mathbf{K} & \mathbf{R} & 2\hat{\mathbf{d}} \\ \mathbf{0} & -2\hat{\mathbf{d}} & -\hat{\mathbf{W}} \end{bmatrix} \times \begin{bmatrix} \mathbf{Z}'(t) \\ \mathbf{Z}''(t) \\ \tfrac{1}{2}\hat{\mathbf{X}}(t) \end{bmatrix} + \begin{bmatrix} \mathbf{0} \\ \mathbf{DQ} \\ \mathbf{0} \end{bmatrix}, \tag{28}$$

where $\mathbf{Z}(t) = \mathbf{Z}'(t) + i\mathbf{Z}''(t)$.

These equations may be solved separately:

$$\mathbf{Z}(t) = \mathbf{S}_1 \exp\left[\mathbf{\Lambda}_1(t - t_3)\right]\mathbf{S}_1^{-1}\mathbf{Z}(t_3), \tag{30a}$$

$$\mathbf{Z}^*(t) = \mathbf{S}_2 \exp\left[\mathbf{\Lambda}_2(t - t_3)\right]\mathbf{S}_2^{-1}\mathbf{Z}^*(t_3), \tag{30b}$$

and

$$\hat{\mathbf{X}}(t) = \mathbf{S}_3 \exp\left[\mathbf{\Lambda}_3(t - t_3)\right]\mathbf{S}_3^{-1}\hat{\mathbf{X}}(t_3), \tag{30c}$$

where

$$\mathbf{S}_1^{-1}(\mathbf{R} - i\mathbf{K})\mathbf{S}_1 = \mathbf{\Lambda}_1, \tag{31a}$$

$$\mathbf{S}_2^{-1}(\mathbf{R} + i\mathbf{K})\mathbf{S}_2 = \mathbf{\Lambda}_2, \tag{31b}$$

and

$$\mathbf{S}_3^{-1}(-\hat{\mathbf{W}})\mathbf{S}_3 = \mathbf{\Lambda}_3. \tag{31c}$$

It may be assumed that the relaxation times for each of the hyperfine lines that give rise to the inhomogeneous broadening are the same. This is a common assumption for spin labels.[26] Although this assumption is often a good one, each of these relaxation times must in fact be different.[27] If $|\gamma_e B_e| \tau_c \ll 1$ so that the relaxation matrix is not significantly affected by the radiation field[19] and if Heisenberg exchange and END effects are neglected, one obtains for completely equivalent nuclei by making use of the preceding approximation:

$$\mathrm{Re}\,\{\mathbf{\Lambda}_1\} = \mathrm{Re}\,\{\mathbf{\Lambda}_2\} = -T_2^{-1}\mathbf{I}, \tag{32a}$$

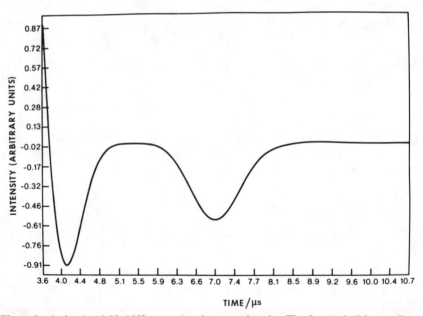

Figure 3 A simulated 90–180° two-pulse electron spin echo. The first "echo" is actually a superposition of free-induction decay signals and may be ignored. The simulations were performed using 2,2,6,6-Tetramethyl-4-piperdinol-l-oxyl proton coupling constants (see reference 26) as the source of inhomogeneous broadening with $T_1^{-1}/|\gamma_e| = T_2^{-1}/|\gamma_e| = 0.0$ G and $B_e = 10$ G. The time is as measured from the beginning of the first pulse. (From Stillman and Schwartz, reference 13.)

and

$$\mathrm{Re}\{\mathbf{\Lambda}_3\} = -T_1^{-1}\mathbf{I},\tag{32b}$$

where $\mathbf{I}$ is the identity matrix. One also has in this case that the matrices $\mathbf{R} \pm i\mathbf{K}$ and $\hat{\mathbf{W}}$ are diagonal[19] so that

$$\mathbf{S}_1 = \mathbf{S}_2 = \mathbf{S}_3 = \mathbf{I}.\tag{33}$$

Equation (28) may then be numerically integrated by standard means.[24,25]

The complexity of this formulation is such that it is difficult to predict that an echo should be created from a two-pulse sequence as diagrammed in Fig. 2. For this we defer to the phenomenological theory from which it is known that an echo occurs with maximum amplitude at $t = \tau_{p_1} + \tau_{p_2} + 2\tau$.[4] However, stepwise numerical integration of Eq. (28) does indeed yield an echo and for $T_1^{-1} = T_2^{-1} = 0$, the maximum occurs at the same time as predicted by the phenomenological theory.[13] Figures 3 and 4 show simulated 90–180° electron spin echoes generated by these calculations. As seen in Fig. 4, the echo will not in general have a maximum amplitude at $t = \tau_{p_1} + \tau_{p_2} + 2\tau$ when relaxation effects are included in the calculations.

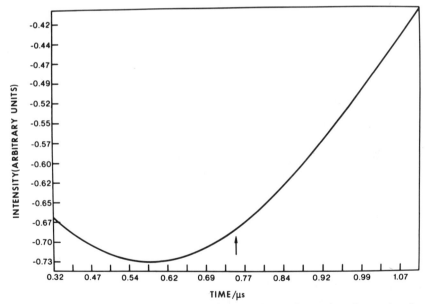

Figure 4 An electron spin echo demonstrating an asymmetric envelope from relaxation effects. The arrow indicates the time at which the isochromats rephase ($t = t_4$). $T_1^{-1}/|\gamma_e| = T_2^{-1}/|\gamma_e| = 0.27\,\mathrm{G}$. All other parameters are as in Fig. 3. (From Stillman and Schwartz, reference 13).

3.1.2   ELECTRON SPIN-ECHO PHASE MEMORY DECAY

The analytical solution given by Eq (23) may be used to obtain some qualitative understanding of electron spin-echo phase memory decay.[13] The electron spin-echo appears in region IV of Fig. 2. Since there is no microwave power in this region, the matrix $\mathbf{A}$ is block diagonal and the solutions of Eq. (15) are decoupled. One then has that for $t \geq t_3$

$$\dot{\mathbf{Z}}(t) = (\mathbf{R} - i\mathbf{K})\mathbf{Z}(t), \tag{29a}$$

$$\dot{\mathbf{Z}}^*(t) = (\mathbf{R} + i\mathbf{K})\mathbf{Z}^*(t), \tag{29b}$$

and

$$\dot{\hat{\mathbf{X}}}(t) = -\hat{\mathbf{W}}\hat{\mathbf{X}}(t). \tag{29c}$$

With these assumptions the echo amplitude at t_4 is given by

$$I(t_4) = \operatorname{Im}\{\mathbf{1}^T \mathbf{Z}(t_4)\}$$

$$= \operatorname{Im}\left\{\mathbf{1}^T \exp\left[-\frac{(t_4 - t_3)}{T_2}\right] \exp\left[-i\mathbf{K}(t_4 - t_3)\right]\right.$$

$$\left. \times \exp\left(i\mathbf{\theta}t_3\right)\mathbf{Z}(t_3)|\right\}, \tag{34}$$

where $\mathbf{Z}(t_3) = \exp(i\mathbf{\theta}t_3)|\mathbf{Z}(t_3)|$ and $\mathbf{1}$ is a vector of ones. Since the isochromats rephase at $t = t_4$, the phase factor in Eq. (34) must be of order $-i\mathbf{I}$ for a turning angle approximately equal to π and hence

$$I(t_4) = -\exp\left[-\frac{(t_4 - t_3)}{T_2}\right]\mathbf{1}^T |\mathbf{Z}(t_3)|. \tag{35}$$

$|\mathbf{Z}(t_3)|$ is a function of τ, τ_{p_1}, τ_{p_2}, B_e, ω, T_1, and T_2. The phase memory decay rate in between pulses is T_2^{-1}. In nonviscous liquids it must also be approximately equal to T_2^{-1} during the pulses. This can be seen by analogy to Torrey's solution of the time-dependent Bloch equations.[28] For $|\gamma_e B_e| \gg T_2^{-1}$ the decay constant is given by

$$T_p^{-1} = T_2^{-1} - (T_2^{-1} - T_1^{-1})\left(\frac{\gamma_e B_e}{\Omega}\right)^2, \tag{36}$$

where $\Omega = [\gamma_e^2 B_e^2 + (\omega_\gamma - \omega)^2]^{1/2}$. In nonviscous liquids $T_1 \approx T_2$, hence, $T_p \approx T_2$. In addition, the pulse times are generally less than the interpulse time so that the phase memory decays essentially as T_2.

Utilizing the preceding assumptions, one has that

$$|\mathbf{Z}(t_3)| \approx \mathbf{\Phi} \exp\left(-\frac{t_3}{T_2}\right), \tag{37}$$

where $\mathbf{\Phi}$ is a function of the pulse parameters, ω, B_e, τ_{p_1}, and τ_{p_2}. The echo decay is therefore given by

$$I(t_4) \approx -\exp\left(-\frac{t_4}{T_2}\right)\mathbf{1}^T\mathbf{\Phi}. \tag{38}$$

It is seen that the phase memory time is approximately T_2 and that it is *independent* of the hyperfine model that gives rise to the inhomogeneous broadening. This result is corroborated by the exact computer calculations[13] and also the cumulant theory presented in Section 4.1.

It should be pointed out that for small B_e fields nonexponential behavior may be observed. Figure 5 shows deviation from exponential behavior for $B_e = 1$ G. The pulse times are on the same order of magnitude as τ for this case and the spin packets are incompletely nutated. Even though the preceding assumptions are not rigorously met in this example, least squares regression yields a phase memory time equal to T_2 within two significant figures. For B_e larger than about 5 G, essentially pure exponential behavior is observed.

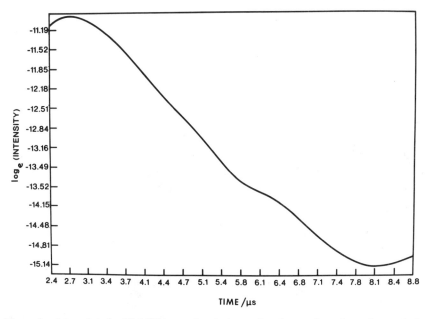

Figure 5   A log plot of a 90–180° two-pulse electron spin echo envelope decay demonstrating nonexponential behavior from incomplete nutation and pulse durations on the order of τ. In this simulation $B_e = 1$ G and $T_1^{-1}/|\gamma_e| = T_2^{-1}/|\gamma_e| = 0.27$ G. All other parameters are as in Fig. 3. Essentially linear plots are found for $B_e \gtrsim 5$ G. (From Stillman and Schwartz, reference 13.)

3.1.3 HEISENBERG SPIN EXCHANGE

One may also use the formalism presented earlier to examine the effect of Heisenberg exchange on electron spin-echo phase memory times.[13] In this case neither the relaxation matrix $\mathbf{R}$ nor the transition probability matrix $\hat{\mathbf{W}}$ are diagonal.[6] Following the preceding reasoning, one obtains

$$I(t_4) = \mathrm{Im}\,\{\mathbf{1}^T \mathbf{S}_1 \exp[(t_4 - t_3)\mathbf{\Lambda}_1]\mathbf{S}_1^{-1}\mathbf{Z}(t_3)\}$$
$$= -\mathbf{1}^T \boldsymbol{\nu}_1 \exp[(t_4 - t_3)\,\mathrm{Re}\,(\mathbf{\Lambda}_1)]\,|\boldsymbol{\nu}_1^{-1}\mathbf{Z}(t_3)|\,, \qquad (39)$$

where $\mathbf{S}_1 = \boldsymbol{\nu}_1 \exp(i\boldsymbol{\theta}_1)$. Thus even if the phase memory time in the absence of exchange, $T_2(0)$, is assumed to be the same for each iso-chromat, the real parts of the eigenvalues of $\mathbf{R} - i\mathbf{K}$ are in general different, and the echo decay is actually a superposition of decaying solutions. However, one would not expect that the solutions would be decaying at very different rates. It is possible therefore to assume that

$$\mathrm{Re}\,(\mathbf{\Lambda}_1) \approx -T_2^{-1}\,(\mathrm{ex})\mathbf{I}, \qquad (40)$$

so that

$$I(t_4) \approx -\exp\left[-\frac{t_4}{T_2(\mathrm{ex})}\right]\mathbf{1}^T \boldsymbol{\nu}_1 \mathbf{\Phi}, \qquad (41)$$

where

$$|\boldsymbol{\nu}_1^{-1}\mathbf{Z}(t_3)| = \mathbf{\Phi}\exp\left[-\frac{t_3}{T_2(\mathrm{ex})}\right]. \qquad (42)$$

An estimate for $T_2(\mathrm{ex})$ may be found by noting the structure of the relaxation matrix in the presence of exchange. For example, a radical with completely equivalent spin $I = \frac{1}{2}$ nuclei has diagonal elements of $\mathbf{R}$ given by[6]

$$R_{ii} = -T_2^{-1}(0) + \frac{\omega_{ex}D_{ii}}{\sum\limits_k D_{kk}} - \omega_{ex}. \qquad (42)$$

An "average" phase memory decay rate may be defined by

$$-\frac{1}{N}\,\mathrm{Tr}\,\{\mathbf{R}\} = T_2^{-1}(0) - \frac{\omega_{ex}}{N} + \omega_{ex}, \qquad (43)$$

so that

$$T_2^{-1}(\mathrm{ex}) \approx T_2^{-1}(0) + \omega_{ex} \qquad (44)$$

for large N, where N is the order of the matrix $\mathbf{R}$. This result compares well with that found from the exact computer calculations and is corroborated by the experiments and simple phenomenological theory of Milov et al.[9a] It may be readily demonstrated that this result is also valid for more complex spin systems such as nitroxides.[13]

3.2 Slow-Motional Region

3.2.1 GENERAL METHOD

The results of Section 3.1 are strictly valid only if the condition $|\mathcal{H}_1(t)\tau_c|^2 \ll 1$ is fulfilled. It is often found in spin-label studies, however, that $|\mathcal{H}_1(t)\tau_c| \gtrsim 1$.[29] It is thus of interest to examine electron spin echoes for radicals undergoing slow-rotational reorientation.

A generalized time domain slow-motional theory may be obtained from the stochastic Liouville equation:[20,29]

$$\dot{\boldsymbol{\sigma}}(\Omega, t) = -i[\mathcal{H}_0^x + \boldsymbol{\varepsilon}(t)^x + \mathcal{H}_1(\Omega)^x + i\mathbf{R}' - i\Gamma_\Omega]$$
$$\times [\boldsymbol{\sigma}(\Omega, t) - \boldsymbol{\sigma}_0(\Omega)], \qquad (45)$$

where $\boldsymbol{\sigma}(\Omega, t)$ is the orientation-dependent spin-density matrix, $\mathbf{R}'$ is the orientation-independent relaxation matrix, Γ_Ω is the Markov operator for rotational tumbling, and $\boldsymbol{\sigma}_0(\Omega)$ is the orientation-dependent density matrix at thermal equilibrium. Equation (45) is solved by expanding $[\boldsymbol{\sigma}(\Omega, t) - \boldsymbol{\sigma}_0(\Omega)]$ in the basis that diagonalizes Γ_Ω, that is,

$$\boldsymbol{\sigma}(\Omega, t) - \boldsymbol{\sigma}_0(\Omega) = \sum_n \mathbf{C}_n(t)\mathbf{G}_n(\Omega). \qquad (46)$$

Equation (45) may then be written in the form

$$\frac{1}{\sqrt{2}}\dot{\mathbf{C}}_m(t) = \sum_n (\mathbf{R}_{m,n} - i\mathbf{K}_{m,n}) \frac{1}{\sqrt{2}}\mathbf{C}_n(t) + i\sqrt{2}\,\hat{\mathbf{d}}_{m,\hat{m}}(\tfrac{1}{2}\mathbf{b}_{\hat{m}}(t)) + i\frac{1}{\sqrt{2}}\mathbf{Q}_m$$
$$(47)$$

as well as the complex-conjugate of Eq. (47) for the off-diagonal spin states, and

$$\tfrac{1}{2}\dot{\mathbf{b}}_{\hat{m}}(t) = i\sqrt{2}\,\hat{\mathbf{d}}_{\hat{m},m}^{\mathrm{T}}\left[\frac{1}{\sqrt{2}}\mathbf{C}_m(t) - \frac{1}{\sqrt{2}}\mathbf{C}_m^*(t)\right] - \sum \hat{\mathbf{W}}_{\hat{m},\hat{n}}(\tfrac{1}{2}\mathbf{b}_{\hat{n}}(t)) \qquad (48)$$

for the diagonal spin states of Eq. (45).[20] The absorption due to the $\alpha \to \alpha'$ transition is then proportional to $\mathrm{Im}\{C_0(t)_{\alpha\alpha'}\}$.[29] For Brownian or jump rotational diffusion in isotropic liquids, $\mathbf{G}_n(\Omega)$ is given by the normalized Wigner rotation matrices

$$\mathbf{G}_n(\Omega) \rightarrow \left[\frac{(2L+1)}{8\pi^2}\right]^{1/2} \mathscr{D}^L_{KM}(\Omega), \tag{49}$$

and $R_{L,L'}$ is given by

$$R_{L,L'} = -[T_2^{-1} + B_L L(L+1)\mathscr{R}]\,\delta_{L,L'}, \tag{50}$$

where B_L is the model parameter and $\mathscr{R}$ is the rotational diffusion coefficient.[29]

Equation (47), its complex-conjugate, and Eq. (48) form an infinite set of coupled first-order ordinary differential equations of the form given by Eq. (16). This system of equations may be solved using the methods described in Section 3.1. The basis set is truncated at some value of n such that little change in the phase memory decay is noted if further terms in the expansion are retained. The final result provides the electron spin-echo intensity due to single isochromat. This is sufficient to describe the electron spin-echo envelope decay for a single line such as the secular g-tensor case with $|\gamma_e B_e| \gg |\hbar^{-1}\beta_e B_0(g_\parallel - g_\perp)|$. More generally, however, one must take the inhomogeneous line broadening into account. The reason for this is that lines far off of resonance are incompletely nutated so that they are not rephased by the microwave pulses. One then obtains an incorrect oscillatory behavior for the echo envelope decay. This problem may in principle be removed by averaging $\mathbf{b}(t)$ of Eq. (16) over a distribution of resonance frequencies

$$\langle \mathbf{b}(t;\omega) \rangle = \sum_i g(\omega - \omega_i)\mathbf{b}(t; \omega - \omega_i). \tag{51}$$

This procedure is, however, prohibitively costly in computer time so that more efficient computational methods need to be developed properly to take the inhomogeneous broadening into account.

In practice the analytic solution given by Eq. (23) proves to be more useful than numerical integration techniques. The reason for this is that for slow-motional problems Eq. (16) may be stiff;[24,30] that is, the real parts of the eigenvalues of $\mathbf{A}$ may differ greatly so that some solutions decay very quickly while others decay slowly. The step size in the numerical integration must be small for the rapidly decaying solutions. However, it is computationally inefficient to use the same small step size for the slowly decaying solutions. Although numerical algorithms exist for stiff problems, they are inherently slow. Thus rather than numerically integrate Eq. (16), we have chosen to use the analytic solution given by Eq. (23). However, in view of the difficulties that may be encountered in the computation of matrix exponentials,[23] one must take care that the obtained results are physically consistent and occasionally compare them to the solutions obtained by numerical integration methods.

3.2.2 SECULAR g-TENSOR EXAMPLE

The simple case for which the phase memory decay results from the motional modulation of an electron spin $S = \frac{1}{2}$ radical with an axially symmetric secular g-tensor spin Hamiltonian may be readily evaluated using the formalism presented earlier.[14] In this example $\mathbf{K}$ is given by

$$K_{L,L'} = (\omega - g_s\beta_e\hbar^{-1}B_0)\,\delta_{L,L'} - N(L,L')\begin{pmatrix} L & 2 & L' \\ 0 & 0 & 0 \end{pmatrix}^2 \mathscr{F}, \tag{52}$$

and

$$W_{L,L'} = [2W_e + B_L L(L+1)\mathscr{R}]\,\delta_{L,L'}, \tag{53}$$

where $\begin{pmatrix} L & 2 & L' \\ 0 & 0 & 0 \end{pmatrix}$ is a 3-j symbol,

$$g_s = \tfrac{1}{3}\mathrm{Tr}\,\{\mathbf{g}\}, \tag{54}$$

$$N(L, L') = [(2L+1)(2L'+1)]^{1/2}, \tag{55}$$

$$\mathscr{F} = \tfrac{2}{3}\hbar^{-1}\beta_e B_0(g_\parallel - g_\perp), \tag{56}$$

and W_e is the electron spin-flip rate. One then obtains the following coupled sets of equations:

$$\frac{\dot{C}_{00}^{L(1)}(t)}{\sqrt{2}} = [-i(\omega - \omega_\lambda) - (T_2^{-1} + B_L L(L+1))]\left(\frac{C_{00}^{L(1)}(t)}{\sqrt{2}}\right)$$

$$+ i\mathscr{F}\sum_{L'}N(L, L')\begin{pmatrix} L & 2 & L' \\ 0 & 0 & 0 \end{pmatrix}^2\left(\frac{C_{00}^{L'(1)}(t)}{\sqrt{2}}\right)$$

$$- i\sqrt{2}\,d_\lambda\left(\frac{\hat{b}_{00}^{L(0)}(t)}{2}\right) + \frac{iq\omega_\lambda d_\lambda\,\delta(L, 0)}{\sqrt{2}}, \tag{57a}$$

as well as the complex-conjugate of Eq. (57a) and

$$\frac{\hat{b}_{00}^{L(0)}(t)}{2} = -[2W_e + B_L L(L+1)\mathscr{R}]\left(\frac{\hat{b}_{00}^{L(0)}(t)}{2}\right)$$

$$- i\sqrt{2}\,d_\lambda\left[\left(\frac{C_{00}^{L(1)}(t)}{\sqrt{2}}\right) - \left(\frac{C_{00}^{L(1)}(t)^*}{\sqrt{2}}\right)\right], \tag{57b}$$

where

$$\omega_\lambda = g_s\beta_e\hbar^{-1}B_0, \tag{58}$$

and

$$\hat{b}_{00}^{L(0)}(t) = [C_{00}^{L(0)}(t)]_{\lambda^+} - [C_{00}^{L(0)}(t)]_{\lambda^-}. \tag{59}$$

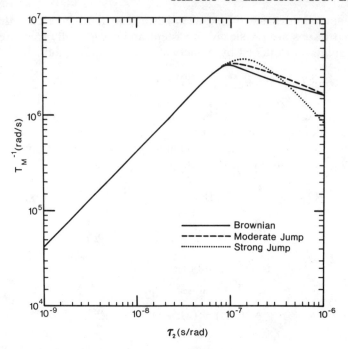

Figure 6. Phase memory decay rates T_M^{-1} calculated from time domain stochastic Liouville theory for an axially symmetric secular g-tensor hamiltonian with $g_\parallel = 2.00235$, $g_\perp = 2.00310$, $B_e = 5\,G$, $B_0 = 3300\,G$ for Brownian, moderate-jump (free), and strong-jump rotational diffusion models (see reference 29). [Reprinted with permission from A. E. Stillman and R. N. Schwartz, *J. Chem. Phys.*, **69**, 3532 (1978). Copyright by the American Institute of Physics.]

The $C_{KM}^{L(n)}(t)$ notation results from the association given by Eq. (49) and the superscript n refers to the appropriate harmonic.

Piecewise integration of Eq. (57a,b) over each region of Fig. 2 as a function of the interpulse time τ yields the slow-motional phase memory decay behavior for the axially symmetric secular g-tensor example. Continuous wave EPR line shapes in the slow-tumbling region are known to be sensitive to both the rotational correlation time for an interaction described by a tensor of second rank, τ_2, and the assumed stochastic diffusional model.[29,31] As shown in Fig. 6, phase memory times are also sensitive to the diffusional model and display a functional dependence on τ_2. Note that the phase memory time is predicted to have a minimum value. This is consistent with the observations of Brown for radicals with an assumed dipolar spin Hamiltonian.[11] Since τ_2 is known to have a nearly linear dependence on the ratio of the coefficient of viscosity to the absolute temperature η/T,[31] this suggests that if τ_2 is calculated from the phase memory times using various rotational diffusion models and is compared with that calculated by extrapolation of the motionally narrowed results, it may be possible to determine which model is operative in experimental systems. This idea is similar in concept to that proposed by Freed for correlation times estimated from cw EPR spectra by a semiempirical technique.[29b] The electron spin-echo method does offer the advantage, however, that it is not complicated by inhomogeneous line broadening as cw EPR spectra often are.[5,6,31]

3.3 ENDOR Spin-Echo Spectroscopy

3.3.1 GENERAL METHOD

As a final application of the time domain formalism discussed earlier, we consider the electron nuclear double resonance (ENDOR) spin-echo experiment.[15a] This is a double resonance technique wherein at some particular time of an electron spin-echo pulse sequence, a radiofrequency (rf) pulse is applied that is resonant with the hyperfine-coupled nuclei. A typical pulse scheme is diagrammed in Fig. 7 for a three-pulse stimulated electron spin echo.[4] Several experimental studies have been reported for paramagnetic centers in various crystals.[32-35] The ENDOR effect is manifested in the reduction of the electron spin-echo intensity when the rf field is on nuclear resonance. Thus a plot of echo intensity versus rf frequency yields the ENDOR spectrum.

This technique offers several distinct advantages over conventional ENDOR spectroscopy.[36] Since there is no microwave power during the

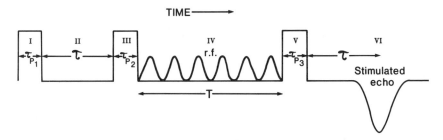

Figure 7   A three-pulse stimulated electron spin-echo pulse sequence. The resonant rf field is applied between the second and third microwave pulses to induce the ENDOR transitions. [Reprinted with permission from A. E. Stillman and R. N. Schwartz, *Mol. Phys.*, **35**, 301 (1978). Copyright by Taylor & Francis ltd.]

observation of the electron spin echo, klystron noise is largely eliminated.[37] Moreover, there is an important additional advantage in that, unlike conventional ENDOR spectroscopy, the detection of ENDOR spin echoes does not depend on a critical balance of the rf and microwave powers and the various relaxation rates. In fact, Mims claims that an ENDOR spin echo should be observable providing only that the electron spin echo is detectable and that sufficient rf power is supplied so as to induce enough nuclear transitions.[32]

Previous theoretical investigations of ENDOR spin echoes have been confined to solids. The most detailed study is due to Liao and Hartmann.[34] They arrived at an analytic expression for the three-pulse stimulated electron spin-echo intensity utilizing a density matrix formalism with the assumptions of neglect of relaxation rates, negligibly short microwave pulse durations, and a rf field that is on nuclear resonance. The predicted ENDOR spin-echo spectrum thus resembles a "stick" diagram. They also assumed that the transverse relaxation times are short enough compared to the duration of the rf pulse that the off-diagonal elements of the spin-density matrix could be neglected. Thus the effects of the rf on only the M_z component of the magnetization were considered. Since coherence effects are manifested in the off-diagonal elements of the density matrix, it follows that the coherence effects of the microwave and rf fields were also neglected. These effects have been shown to be important for conventional ENDOR spectroscopy of free radicals dissolved in liquid solution when large microwave and/or rf powers are used.[38] As the experimental conditions require such high power levels, it would be desirable to ascertain what role, if any, coherence effects play in ENDOR spin-echo spectroscopy. In addition, it is known that the line shapes of conventional ENDOR spectra possess a wealth of information. This is true for free radicals dissolved in both solid[39] and liquid[40] solutions. It would clearly be desirable to assess what information content is contained in ENDOR spin-echo line shapes.

Although no liquid-phase ENDOR spin echoes have yet been reported, if Mims' assertions are valid, such spectra should be experimentally realizable. The previously discussed methods may be applied to this problem and the particularly simple example of an electron spin $S = \frac{1}{2}$ radical coupled to two completely equivalent spin $I = \frac{1}{2}$ nuclei undergoing rapid rotational reorientation in the high field limit is considered below. While this theory is rigorously valid only for motionally narrowed ENDOR spin-echo spectra, it is expected to account roughly for the relaxation and coherence effects found in solid-state ENDOR spin echoes.

For an ENDOR experiment the perturbation due to the radiation field is given by[19]

$$
\begin{aligned}
\hbar\varepsilon(t) = &\tfrac{1}{2}\hbar\gamma_e B_e[\mathbf{S}_+ \exp(-i\omega_e t)+\mathbf{S}_- \exp(+i\omega_e t)]\\
&+\tfrac{1}{2}\hbar\gamma_r B_n[\mathbf{J}_{r+} \exp(-i\omega_n t)+\mathbf{J}_{r-} \exp(+i\omega_n t)]\\
&+\tfrac{1}{2}\hbar\gamma_e B_n[\mathbf{S}_+ \exp(-i\omega_n t)+\mathbf{S}_- \exp(+i\omega_n t)]\\
&+\tfrac{1}{2}\hbar\gamma_r B_e[\mathbf{J}_{r+} \exp(-i\omega_e t)+\mathbf{J}_{r-} \exp(+i\omega_e t)],
\end{aligned}
\tag{60}
$$

where $\mathbf{J}_r$ is the coupled nuclear spin angular momentum operator of the rth group of completely equivalent nuclei, that is,

$$
\mathbf{J}_r = \sum_{i\in r}\mathbf{I}_i.
\tag{61}
$$

In Eq. (60) the microwave field B_e is at frequency ω_e and the rf field B_n is at frequency ω_n. Since the last term in Eq. (60) is too far off of nuclear resonance to induce transitions and since the third term gives rise to negligible contributions for the small hyperfine couplings that give rise to the inhomogeneous broadening,[19] these terms are omitted from the following discussion.

Equation (60) may be solved for the ENDOR spin-echo example by expanding $\chi(t)$ in a generalized Fourier expansion:

$$
\chi_{\alpha,\alpha'}(t) = \sum_{m,n=-\infty}^{\infty} \exp[i(m\omega_e + n\omega_n)t]Z_{\alpha,\alpha'}^{(m,n)}(t).
\tag{62}
$$

As only the $m=1$, $n=0$ harmonic is detected by the EPR spectrometer, only those (mixed) harmonics that couple to $Z_{\alpha,\alpha'}^{(1,0)}(t)$ need be considered. It should be pointed out that the $|m|+|n|>1$ harmonics correspond to multiple quantum transitions and that the absorption signal of the $\alpha \to \alpha'$ transition is proportional to the imaginary part of $Z_{\alpha,\alpha'}^{(1,0)}(t)$, whereas the dispersion signal is proportional to its real component.[19] From substitution of Eq. (62) into Eq. (60) one obtains the following supermatrix expression in the high-temperature limit:[15a,20]

$$
\begin{bmatrix}
\dfrac{1}{\sqrt{2}}\dot{\mathbf{Z}}(t)\\[2ex]
\dfrac{1}{\sqrt{2}}\dot{\mathbf{Z}}^*(t)\\[2ex]
\tfrac{1}{2}\dot{\hat{\mathbf{X}}}(t)\\[2ex]
\tfrac{1}{2}\dot{\tilde{\mathbf{X}}}(t)
\end{bmatrix}
=
\begin{bmatrix}
\mathbf{R}-i\mathbf{K} & i\mathbf{K}' & \sqrt{2}\,i\hat{\mathbf{d}} & \sqrt{2}\,i\tilde{\mathbf{d}}\\[2ex]
-i\mathbf{K}' & \mathbf{R}+i\mathbf{K} & -\sqrt{2}\,i\hat{\mathbf{d}} & -\sqrt{2}\,i\tilde{\mathbf{d}}\\[2ex]
\sqrt{2}\,i\hat{\mathbf{d}}^T & -\sqrt{2}\,i\hat{\mathbf{d}}^T & -\hat{\mathbf{W}} & -\mathcal{W}\\[2ex]
\sqrt{2}\,i\tilde{\mathbf{d}}^T & -\sqrt{2}\,i\tilde{\mathbf{d}}^T & -\mathcal{W}^T & -\tilde{\mathbf{W}}
\end{bmatrix}
\times
\begin{bmatrix}
\dfrac{1}{\sqrt{2}}\mathbf{Z}(t)\\[2ex]
\dfrac{1}{\sqrt{2}}\mathbf{Z}^*(t)\\[2ex]
\tfrac{1}{2}\hat{\mathbf{X}}(t)\\[2ex]
\tfrac{1}{2}\tilde{\mathbf{X}}(t)
\end{bmatrix}
+
\begin{bmatrix}
\dfrac{i\mathbf{Q}}{\sqrt{2}}\\[2ex]
-\dfrac{i\mathbf{Q}}{\sqrt{2}}\\[2ex]
\mathbf{0}\\[2ex]
\mathbf{0}
\end{bmatrix},
\tag{63}
$$

where the diagonal elements of the coherence matrix $\mathbf{K}$ are given by

$$K_{\lambda_j,\lambda_j}^{(m,n)} = m\omega_e + n\omega_n - \omega_{\lambda_j}. \tag{64}$$

Both the off-diagonal matrix elements of $\mathbf{K}$ and the $\mathbf{K}'$ matrix are proportional to the various transition moments given by Eq. (9) and

$$d_{\eta_k} \equiv \tfrac{1}{2}\lambda_r B_n J_{r-\eta_k}. \tag{65}$$

$\mathbf{K}'$ is a coherence-like matrix not found in previous ENDOR studies.[38] It arises primarily from inclusion of terms that are far off of resonance. They are retained for generality because at some point in the frequency-swept experiment they may become important.

As in Section 3.1, we consider only average transitions and assume that the transition probabilities obey Eq. (12a,b). One then has that

$$
\begin{bmatrix}
\dfrac{1}{\sqrt{2}}\dot{\mathbf{Z}}(t) \\[2mm]
\dfrac{1}{\sqrt{2}}\dot{\mathbf{Z}}^*(t) \\[2mm]
\tfrac{1}{2}\dot{\hat{\mathbf{X}}} \\[2mm]
\tfrac{1}{2}\dot{\tilde{\mathbf{X}}}
\end{bmatrix}
=
\begin{bmatrix}
\mathbf{R}-i\mathbf{K} & i\mathbf{K}' & \sqrt{2}\,i\hat{\mathbf{d}} & \sqrt{2}\,i\tilde{\mathbf{d}} \\[2mm]
-i\mathbf{K}' & \mathbf{R}+i\mathbf{K} & -\sqrt{2}i\hat{\mathbf{d}} & -\sqrt{2}\,i\tilde{\mathbf{d}} \\[2mm]
\sqrt{2}\,i\hat{\mathbf{d}}^T & -\sqrt{2}\,i\hat{\mathbf{d}}^T & -\hat{\mathbf{W}} & 0 \\[2mm]
\sqrt{2}\,i\tilde{\mathbf{d}}^T & -\sqrt{2}\,i\tilde{\mathbf{d}}^T & 0 & -\tilde{\mathbf{W}}
\end{bmatrix}
\times
\begin{bmatrix}
\dfrac{1}{\sqrt{2}}\mathbf{Z}(t) \\[2mm]
\dfrac{1}{\sqrt{2}}\mathbf{Z}^*(t) \\[2mm]
\tfrac{1}{2}\hat{\mathbf{X}}(t) \\[2mm]
\tfrac{1}{2}\tilde{\mathbf{X}}(t)
\end{bmatrix}
+
\begin{bmatrix}
\dfrac{i\mathbf{DQ}}{\sqrt{2}} \\[2mm]
\dfrac{-i\mathbf{DQ}}{\sqrt{2}} \\[2mm]
0 \\[2mm]
0
\end{bmatrix}. \tag{66}
$$

A real form of this equation is useful for purposes of numerical computation and is obtained from a unitary transformation of Eq. (65).

$$
\begin{bmatrix}
\dot{\mathbf{Z}}'(t) \\[2mm]
\dot{\mathbf{Z}}''(t) \\[2mm]
\tfrac{1}{2}\dot{\hat{\mathbf{X}}}(t) \\[2mm]
\tfrac{1}{2}\dot{\tilde{\mathbf{X}}}(t)
\end{bmatrix}
=
\begin{bmatrix}
\mathbf{R} & \mathbf{K}+\mathbf{K}' & 0 & 0 \\[2mm]
-\mathbf{K}+\mathbf{K}' & \mathbf{R} & 2\hat{\mathbf{d}} & 2\tilde{\mathbf{d}} \\[2mm]
0 & -2\hat{\mathbf{d}}^T & -\hat{\mathbf{W}} & 0 \\[2mm]
0 & -2\tilde{\mathbf{d}}^T & 0 & -\tilde{\mathbf{W}}
\end{bmatrix}
\begin{bmatrix}
\mathbf{Z}'(t) \\[2mm]
\mathbf{Z}''(t) \\[2mm]
\tfrac{1}{2}\hat{\mathbf{X}}(t) \\[2mm]
\tfrac{1}{2}\tilde{\mathbf{X}}(t)
\end{bmatrix}
+
\begin{bmatrix}
0 \\[2mm]
\mathbf{DQ} \\[2mm]
0 \\[2mm]
0
\end{bmatrix}. \tag{67}
$$

This expression is of the form given by Eq. (16). From the phenomenological theory one knows that a stimulated echo occurs in region VI of Fig. 7 at time $t_{\mathrm{SE}} = \tau_{p1} + \tau_{p2} + \tau_{p3} + 2\tau + T$.[4] Using the initial conditions

$$\mathbf{Z}(0) = \mathbf{Z}^*(0) = \hat{\mathbf{X}}(0) = \tilde{\mathbf{X}}(0) = \mathbf{0}, \tag{68}$$

we integrate Eq. (67) over each region and thus calculate $\mathbf{b}(t_{\mathrm{SE}})$ as a function of the rf. A plot of the sum of the $Z''^{(1,0)}(t_{\mathrm{SE}})$ for all the EPR transitions versus the radio frequency yields the ENDOR spin-echo spectrum.

3.3.2 $S = \frac{1}{2}$, $J_r = 1$ EXAMPLE

For the purposes of the present work we choose to study an electron spin $S = \frac{1}{2}$ radical coupled to two completely equivalent spin $I = \frac{1}{2}$ nuclei. The various partitioned matrices for this particular case are presented in reference 15. This system is not so simple as to lack in generality and yet maintains a manageable number of 36 equations to be solved. Typical ENDOR spin-echo spectral simulations are shown in Fig. 8. As in the

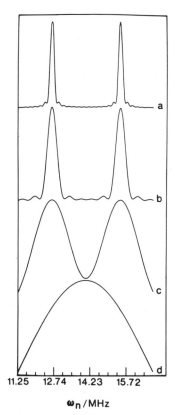

ω_n / MHz

Figure 8 Typical ENDOR spin-echo spectral simulations with rf pulse times T of: (a) 5.0×10^{-6} s, (b) 2.0×10^{-6} s, (c) 5.0×10^{-7} s, and (d) 2.0×10^{-7} s. The ripples found in the spectra are due to small Fourier components of the ENDOR transitions. All spectra were simulated with $\mathscr{A}_H = 2.80$ MHz, $g_s = 2.0030$, $T_{2e}^{-1} = T_{2x}^{-1} = 7.04 \times 10^5$ rad/s, $T_{2n}^{-1} = 8.80 \times 10^4$ rad/s, $W_e = W_x = W_n = 0.0$ rad/s, $B_0 = 3300$ G, $B_e = 10$ G, $B_n = 5$ G, and $\tau = 1.58 \times 10^{-7}$ s. A 90–90–90° pulse sequence was used. The zero values chosen for W_e, W_x, and W_n suppress any relaxation effects on the spectral line shapes and illustrate the effect of variation of rf pulse times T (From Stillman and Schwartz, reference 15).

conventional ENDOR spectrum for this hypothetical radical, one obtains a pair of resonances separated by the value of the isotropic hyperfine coupling constant.

To simplify the analysis we assume that the relaxation rates for the allowed spin transitions are the same for each transition. Moreover, it is assumed that the relaxation rates are the same for each of the mixed harmonic or cross transitions. We thus consider only the effects of single T_{2e}, T_{2x}, T_{2n}, W_e, W_x, and W_n where T_{2e}, T_{2x}, and T_{2n} are the respective transverse relaxation times of the EPR, mixed harmonic (cross transitions), and NMR transitions, and W_e, W_x, and W_n are the electron spin, mixed spin, and nuclear spin-flip rates, respectively. Although these must in fact be different for each transition,[19,27] this approximation is not considered to be too restrictive as the values of the relaxation rates for the various transitions of a given type are all similar in magnitude. Typical values of these parameters are given by Leniart et al.[40]

The most striking feature in Figs. 8 and 9 is that the linewidths of the

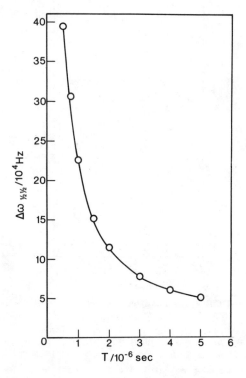

Figure 9 A plot of the half-width at half-height of an ENDOR spin echo versus the rf pulse time T. (From Stillman and Schwartz, reference 15).

ENDOR spin-echo spectra increase with decreasing rf pulse time T. Even if B_n is varied continuously so as to maintain a constant turning angle, this effect was manifest. The turning angle is defined by the relation $\theta_{\lambda_i} = \gamma_r B_n T J_{r-\lambda_i}$. A similar effect was found in the experiments of Liao and Hartmann[34] and attributed to the Heisenberg uncertainty principle.[41] We note that our calculated line widths are also consistent with the relation $\hbar \Delta\omega_{1/2,1/2} T \geq \hbar/2$ for computationally accessible rf pulse times T.[15b]

Although the dominant contribution to the linewidth results from the finite length of T, both T_{2n} and W_e also make nonnegligible contributions. As may be expected, the linewidth increases with decreasing T_{2n}. It was found, however, that the linewidths decrease with increasing W_e. Since there is the association that $T_{1e}^{-1} = 2W_e$ for a two-level system, this is a somewhat surprising result. A similar, albeit almost negligibly small "negative" linewidth effect was found for W_x and W_n. Neither T_{2e} nor T_{2x} appear to have any appreciable effect on the ENDOR spin-echo linewidth. However, both increased B_n fields and the inclusion of coherence effects result in additional linewidth contributions. Although both of these effects are interrelated, they do act independently of one another.

Figure 10 shows typical ENDOR spin echoes for various turning angles with and without inclusion of coherence effects. The spectra without coherence effects somewhat resemble those found by Liao and Hartmann.[34] The 2π pulse does not, however, completely split as they predict for a homogeneous rf field. The reason is that a 2π pulse is not simply a double inversion of population levels because there are degenerate ENDOR transitions for $J_r \geq 1$. The radiation field couples all the diagonal elements of the density matrix that are affected by the rf so that only a numerical solution of Eq. (66) can yield the effect of a 2π pulse when degenerate ENDOR transitions are present.

As seen in Fig. 10, additional splittings result when coherence effects are included. Such a dramatic effect is contrary to Liao and Hartmann's assumption that T_2's are short enough that the effects from the off-diagonal elements of the spin-density matrix may be neglected.[34] Although it is likely that most of the coherence terms do not manifest themselves because of the short length for the T_2's, the terms from double NMR transitions do prove to be important as they are directly coupled to the single-quantum NMR transitions. For broader ENDOR spin-echo lines the coherence effects are less dramatic.

Unfortunately, it is likely that the longitudinal relaxation times are too short for many free radicals in liquids to observe a well-resolved ENDOR spin-echo spectrum in spite of the negative linewidth effect observed for W_e. We therefore differ with Mims' assertion that an ENDOR spin echo should be observable providing only that a spin echo is detectable and

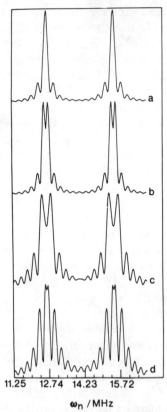

11.25   12.74   14.23   15.72

ω_n / MHz

Figure 10 ENDOR spin echoes with B_n varied for rf turning angles of (a) $\theta = \pi$ with no rf coherence, (b) $\theta = \pi$ with coherence, (c) $\theta = 2\pi$ without coherence, and (d) $\theta = 2\pi$ with inclusion of rf coherence. The rf pulse time T was 4×10^{-6} s for each of these plots. All other parameters are as in Fig. 8. (From Stillman and Schwartz, reference 15.)

that sufficient rf is applied.[32] One must also have longitudinal relaxation times long enough that a rf pulse of sufficient duration can be applied so as to resolve the ENDOR peaks.

4 APPLICATION OF GENERALIZED CUMULANT EXPANSIONS TO THE THEORY OF ELECTRON SPIN-ECHO PHASE MEMORY DECAY IN LIQUIDS

4.1 The Cumulant Expansion Method and Electron Spin Phase Memory Decay

Although the theory presented in Section 3 is "exact," some of the physical insight that is salient in Brown's approach[11] is made obscure by the nature of the formalism. It is therefore of interest to reexamine the basis of Brown's theory, namely, the work of Klauder and Anderson,[42] in light of our present understanding of electron spin relaxation in liquids.[2] Freed's generalized cumulant expansion theory for spin relaxation[43] proves to be useful for this purpose and its application to electron spin echoes is discussed below.

It is well known that the EPR line shape function is given by[1]

$$I(\omega) = \frac{4}{\pi} \int_0^\infty G(t) \cos{(\omega t)}\, dt, \tag{69}$$

where $G(t)$ is the autocorrelation function of the x component of the spin-angular momentum operator $\mathbf{S}_x(t)$:

$$G(t) = \mathrm{Tr}\,[\mathbf{S}_x(t)\mathbf{S}_x]. \tag{70}$$

$\mathbf{S}_x$ is an ensemble average over the various $\tilde{\mathbf{S}}_x$ for the individual spins. Thus

$$\mathbf{S}_x(t) = \langle \tilde{\mathbf{S}}_x(t) \rangle. \tag{71}$$

$\tilde{\mathbf{S}}_x(t)$ obeys the equation of motion

$$\frac{d}{dt}\tilde{\mathbf{S}}_x(t) = i[\mathscr{H}(t), \tilde{\mathbf{S}}_x]$$

$$\equiv i\mathscr{H}(t)^{\times}\tilde{\mathbf{S}}_x, \tag{72}$$

where the time-dependent spin Hamiltonian is given by

$$\hbar\mathscr{H}(t) = \hbar[\mathscr{H}_0 + \mathscr{H}_1(t)]. \tag{73}$$

$\mathscr{H}_0$ is the static zero-order Hamiltonian and $\mathscr{H}_1(t)$ is the time-dependent perturbation. In an interaction picture one has that

$$\frac{d}{dt}\tilde{\mathbf{S}}_x^{\ddagger}(t) = i\mathscr{H}_1^{\ddagger}(t)^{\times}\tilde{\mathbf{S}}_x^{\ddagger}, \tag{74}$$

where

$$\mathbf{U}^{\ddagger}(t) \equiv \exp{(-i\mathscr{H}_0 t)}\mathbf{U}(t)\exp{(i\mathscr{H}_0 t)}. \tag{75}$$

The formal solution to Eq. (74) is given by

$$\tilde{\mathbf{S}}_x^{\ddagger}(t) = \exp_o\left[i\int_0^t dt'\,\mathscr{H}_1^{\ddagger}(t')^{\times}\right]\tilde{\mathbf{S}}_x^{\ddagger}(0), \tag{76a}$$

so that Eq. (71) becomes

$$\mathbf{S}_x^{\ddagger}(t) = \left\langle \exp_o\left[i\int_0^t dt'\,\mathscr{H}_1^{\ddagger}(t')^{\times}\right]\right\rangle\mathbf{S}_x^{\ddagger}(0). \tag{76b}$$

The subscript o in Eq. (76) represents time ordering. Thus

$$\left\langle \exp_o \left[i \int_0^t dt' \mathcal{H}_1^\ddagger(t')^\infty \right] \right\rangle = \sum_{n=0}^\infty \mathbf{M}_n(t), \tag{77}$$

where

$$\begin{aligned}
\mathbf{M}_n(t) &= \frac{i^n}{n!} O \left\langle \left[\int_0^t dt' \mathcal{H}_1^\ddagger(t')^\infty \right]^n \right\rangle \\
&= i^n \int_0^t dt_1 \int_0^{t_1} dt_2 \cdots \int_0^{t_{n-1}} dt_n \\
&\quad \times \langle \mathcal{H}_1^\ddagger(t_1)^\infty \mathcal{H}_1^\ddagger(t_2)^\infty \cdots \mathcal{H}_1^\ddagger(t_n)^\infty \rangle \\
&= i^n \int_0^t dt_1 \int_0^{t_1} dt_2 \cdots \int_0^{t_{n-1}} \\
&\quad \times dt_n \mathbf{m}_n(t_1, t_2, \ldots, t_n), \tag{78a}
\end{aligned}$$

for $n > 1$,

$$\mathbf{M}_0(t) = \mathbf{1}, \tag{78b}$$

and where $t \geq t_i \geq t_j$ for $i > j$.[1,43-45]

Equation (76b) is analogous to the expression given by Klauder and Anderson[42] for the free-induction decay signal:

$$F(t) = \left\langle \exp \left[i \int_0^t dt' \omega(t') \right] \right\rangle. \tag{79}$$

Rather than considering the evolution of the spin system as being caused by random modulation of the operator $\mathcal{H}_1^\ddagger(t)^\infty$, Klauder and Anderson consider a random frequency $\omega(t)$. Being a scalar, $\omega(t)$ commutes with itself for arbitrary times while for the general case, $\mathcal{H}_1^\ddagger(t)^\infty$ does not. This is the reason for the necessity of the ordering prescription in Eq. (76). The commutivity of $\omega(t)$ in time results in considerable simplification in the theory; Mims[4,46] and Brown[11] consider $\omega(t)$ to be a scalar quantity that arises from random modulation of magnetic dipoles. The difficulty with this approach is that more complicated spin Hamiltonians do not in general commute in time, and hence one must utilize Eq. (76b) rather than Eq. (79) in any general theory of electron spin phase memory decay. One may, however, utilize the arguments of Klauder and Anderson to gain insight into the manner in which the microwave pulses affect $\mathbf{S}_x(t)$.

Consider the two-pulse experiment in which a 90° pulse is applied about the y-axis at $t = 0$ and a 180° pulse is applied about the x-axis at $t = \tau$. The effect of the second pulse is to flip each resonant spin by 180° about the x-axis. Thus the phase acquired during the time τ is reversed.

For times $t > \tau$ the acquired phase must therefore be given by[47]

$$\phi(t) = \int_\tau^t \omega(t') \, dt' - \int_0^\tau \omega(t') \, dt'$$

$$= \int_0^t S(t')\omega(t') \, dt', \tag{80}$$

where

$$S(t) = -1, \qquad 0 < t < \tau$$
$$S(t) = 1, \qquad \tau < t. \tag{81}$$

Other spin-echo experiments can be described by different functional forms for $S(t)$. Thus for a three-pulse stimulated echo one has

$$S(t) = -1, \qquad 0 < t < \tau$$
$$S(t) = 0, \qquad \tau < t < T \tag{82}$$
$$S(t) = 1, \qquad T < t,$$

where T is the time at which the third pulse is applied.

Utilizing these arguments, we augment the Liouville Eq. (72) so as to take into account the effect that the microwave pulses have on the accumulation of phase:[14,48]

$$\frac{d}{dt} \tilde{\mathbf{S}}_x(t) = iS(t)\mathcal{H}(t)^\times \tilde{\mathbf{S}}_x. \tag{83}$$

It is convenient to define an interaction picture by

$$\mathbf{U}^\dagger(t) = \exp\left[-i\mathcal{H}_0 \int_0^t dt' S(t')\right] \mathbf{U}(t) \exp\left[i\mathcal{H}_0 \int_0^t dt' S(t')\right]. \tag{84}$$

Note that at $t = 2\tau$ the echo has its maximum amplitude and

$$\mathbf{U}^\dagger(2\tau) = \mathbf{U}(2\tau). \tag{85}$$

One has in this interaction representation that

$$\frac{d}{dt} \tilde{\mathbf{S}}_x^\dagger(t) = i\mathcal{H}_1^\dagger(t)^\times \tilde{\mathbf{S}}_x^\dagger. \tag{86}$$

The formal solution to Eq. (86) is

$$\tilde{\mathbf{S}}_x^\dagger(t) = \exp_0\left[i\int_0^t dt' S(t')\mathcal{H}_1^\dagger(t')^\times\right] \tilde{\mathbf{S}}_x^\dagger(0), \tag{87}$$

so that

$$\mathbf{S}_x^\dagger(t) = \left\langle \exp_0\left[i\int_0^t dt' S(t')\mathcal{H}_1^\dagger(t')^\times\right]\right\rangle \mathbf{S}_x^\dagger(0). \tag{88}$$

This expression is analogous to that found by Klauder and Anderson:[42]

$$M(t) = \left\langle \exp\left[-i\int_0^t dt' S(t')\omega(t')\right]\right\rangle. \tag{89}$$

The problem that remains is the evaluation of Eq. (88). This may be accomplished by using a cumulant expansion method based on partial time ordering of the cumulant operators.[43–45] We utilize partial time ordering rather than total time ordering since it is generally more useful in the time domain.[49] Thus we seek a solution of the form

$$\left\langle \exp_o\left[i\int_0^t dt' S(t')\mathcal{H}_1^\dagger(t')\right]\right\rangle = \exp_o \mathbf{K}(t), \tag{90a}$$

where

$$\mathbf{K}(t) = \sum_{n=1}^\infty \mathbf{K}_n(t). \tag{90b}$$

It may be shown that

$$\mathbf{K}_n(t) = i^n \int_0^t dt_1 \int_0^{t_1} dt_2 \cdots \int_0^{t_{n-1}} dt_n$$
$$\times \langle O\mathcal{H}_1^\dagger(t_1)^\infty \mathcal{H}_1^\dagger(t_2)^\infty \cdots \mathcal{H}_1^\dagger(t_n)^\infty \rangle_c$$
$$\times S(t_1)S(t_2)\cdots S(t_n), \tag{91}$$

where O designates time ordering and the subscript c implies a type of cumulant rather than simple ensemble averaging.[44] On differentiating Eq. (88, 91) one has from Eq. (90a, b) that

$$\dot{\mathbf{S}}_x^\dagger(t) = \dot{\mathbf{K}}(t)\mathbf{S}_x^\dagger(t), \tag{92}$$

where

$$\dot{\mathbf{K}}_n(t) = i^n \int_0^t dt_1 \int_0^{t_1} dt_2 \cdots \int_0^{t_{n-2}} dt_{n-1}$$
$$\times \langle \mathcal{H}_1^\dagger(t)^\infty \mathcal{H}_1^\dagger(t_1)^\infty \cdots \mathcal{H}_1^\dagger(t_{n-1})^\infty \rangle_c$$
$$\times S(t)S(t_1)\cdots S(t_{n-1}) \tag{93a}$$
$$= i^n \int_0^t d\tau_1 \int_0^{t-\tau_1} d\tau_2 \cdots \int_0^{t-\sum_{j=1}^{n-2}\tau_j} d\tau_{n-1}$$
$$\times \left\langle \mathcal{H}_1^\dagger(t)^\infty \mathcal{H}_1^\dagger(t-\tau_1)^\infty \cdots \mathcal{H}_1^\dagger\left(t-\sum_{j=1}^{n-1}\tau_j\right)^\infty \right\rangle_c$$
$$\times S(t)S(t-\tau_1)\cdots S\left(t-\sum_{j=1}^{n-1}\tau_j\right). \tag{93b}$$

Equation (93b) is obtained from transforming the integrals by letting $\tau_i = t_{i-1} - t_i$.

Following Freed, we may write $\dot{\mathbf{K}}_n(t)$ as[43]

$$\dot{\mathbf{K}}_n(t) = i^n \mathbf{I}_n(t) \mathbf{A}_n \mathbf{P}_n \mathcal{H}_1^\dagger(t)^x \mathcal{H}_1^\dagger(t-\tau_1)^x \cdots \mathcal{H}_1^\dagger\left(t - \sum_{j=1}^{n-1}\tau_j\right)^x$$
$$\times S(t)S(t-\tau_1)\cdots S\left(t - \sum_{j=1}^{n-1}\tau_j\right), \tag{94}$$

where

$$\mathbf{I}_n(t) = \int_0^t d\tau_1 \int_0^{t-\tau_1} d\tau_2 \cdots \int_0^{t-\sum_{j=1}^{n-2}\tau_j} d\tau_{n-1}, \tag{95}$$

and $\mathbf{A}_n$ is an operator that performs the required averaging after $\mathbf{P}_n$ generates the sums of products of $\mathcal{H}_1^\dagger(t_i)^x$ necessary for the cumulant averages. Rewriting Eq. (94) in terms of matrix elements, one has that

$$\dot{\mathbf{K}}_n(t)_{\alpha\alpha',\beta\beta'} = i^n \mathbf{I}_n(t) \mathbf{A}_n \mathbf{P}_n \sum_{\gamma_1,\gamma_i} \mathcal{H}_1(t)^x_{\alpha\alpha',\gamma_1\gamma_i'}$$

$$\times \exp\left[-i(\omega_{\alpha\alpha'} - \omega_{\gamma_1\gamma_i'})\int_0^t dt'S(t')\right]\mathcal{H}_1(t-\tau_1)^x_{\gamma_1\gamma_1,\gamma_2\gamma_2'}$$

$$\times \exp\left[-i(\omega_{\gamma_1\gamma_i'} - \omega_{\gamma_2\gamma_2'})\int_0^{t-\tau_1} dt'S(t')\right] \cdots \mathcal{H}_1\left(t - \sum_{j=1}^{n-1}\tau_j\right)^x_{\gamma_{n-1}\gamma_{n-1}',\beta\beta'}$$

$$\times \exp\left[-i(\omega_{\gamma_{n-1}\gamma_{n-1}'} - \omega_{\beta\beta'})\int_0^{t-\sum_{j=1}^{n-1}\tau_j} dt'S(t')\right]$$

$$\times S(t)S(t-\tau_1)\cdots S\left(t - \sum_{j=1}^{n-1}\tau_j\right). \tag{96}$$

Note that for times $t < \tau$ one has that $\int_0^{t-\tau_i} dt'S(t') = -(t-\tau_i)$ and $S(t-\tau_i) = -1$. The cumulant average has the property that it vanishes if any of the $\mathcal{H}_1^\dagger(t_i)^x$ are uncorrelated. Thus if τ_c is a correlation time for $\mathcal{H}_1^\dagger(t)$, the only significant contributions to Eq. (96) come from times $\tau_i \lesssim \tau_c$. It follows that if $t > \tau + \tau_c$, then $t - \tau_i > \tau$ for those times τ_i for which the correlation functions are nonnegligible. We can therefore let $\int_0^{t-\tau_i} dt'S(t') = t - \tau_i - 2\tau$ and $S(t-\tau_i) = S(t) = 1$ for times $t > \tau + \tau_c$. The case for which $\tau < t < \tau + \tau_c$ is rather complex. We may, however, make use of the fact that $\dot{\mathbf{K}}_n(\tau) \cong \dot{\mathbf{K}}_n(\tau+\tau_c)$ for $\tau \gg \tau_c$ so that negligible error is introduced by assuming that $\int_0^{t-\tau_i} dt' S(t') = t - \tau_i - 2\tau$ and $S(t - \tau_i) = S(t) = 1$ for all $t > \tau$. Utilizing the preceding assumptions, one then has

$$\dot{K}_n(t)_{\alpha\alpha',\beta\beta'} = \exp\left[-i(\omega_{\alpha\alpha'} - \omega_{\beta\beta'}) \int_0^t dt' S(t')\right] i^n I_n(t) A_n P_n$$

$$\times \sum_{\gamma_i,\gamma_i} \mathcal{H}_1(t)_{\alpha\alpha',\gamma_1\gamma_1'}^{\infty} \mathcal{H}_1(t-\tau_1)_{\gamma_1\gamma_1',\gamma_2\gamma_2'}^{\infty}$$

$$\times \exp\left[-i(\omega_{\gamma_1\gamma_1'} - \omega_{\gamma_2\gamma_2'})S(t)\tau_1 \cdots \mathcal{H}_1\left(t - \sum_{j=1}^{n-1}\tau_j\right)_{\gamma_{n-1}\gamma_{n-1}',\beta\beta'}^{\infty}\right]$$

$$\times \exp\left[-i(\omega_{\gamma_{n-1}\gamma_{n-1}'} - \omega_{\beta\beta'})S(t) \sum_{j=1}^{n-1}\tau_j\right] S(t)^n. \tag{97}$$

For $t \gg \tau_c$ one may replace $I_n(t)$ with $I_n(t \to \infty)$ in Eq. (97) without introducing appreciable error. Thus $\dot{K}_n(t)$ is of the form

$$\dot{K}_n(t) \cong \exp\left[-i\mathcal{H}_0^{\infty} \int_0^t dt' S(t')\right] [R^{(n)'} - iS(t) R^{(n)''}]$$

$$\times \exp\left[i\mathcal{H}_0^{\infty} \int_0^t dt' S(t')\right] S(t)^n, \tag{98}$$

where $R^{(n)'}$ and $R^{(n)''}$ are time-independent operators obtained from one-sided cosine and sine Fourier transforms of the correlation functions, respectively.

Transforming back into the Schrödinger picture, one has that

$$\dot{S}_x(t) = \left[i\mathcal{H}_0^{\infty} S(t) + \sum_{n=1}^{\infty} (R^{(n)'} - iS(t)R^{(n)''})S(t)^n\right] S_x(t), \tag{99}$$

so that

$$S_x(2\tau) = \exp\left[2\tau \sum_{n=1}^{\infty} (R^{(2n)'} - iR^{(2n-1)''})\right] S_x(0). \tag{100}$$

For most cases of physical interest, $R^{(2n)'}$ is real whereas $R^{(2n-1)''}$ is imaginary. Thus the electron spin-echo decays as a superposition of exponentially decaying functions. If one utilizes the relation

$$\langle jk | A^{\infty} | j'k' \rangle = \delta_{kk'} \langle j | A | j' \rangle - \delta_{jj'} \langle k' | A | k \rangle \tag{101}$$

and assumes that the random modulation of $\mathcal{H}_1(t)$ is a stationary process, one then has to second order in the cumulant expansion that

$$R_{\alpha\alpha',\beta\beta'}^{(2)'} = J_{\alpha\beta,\beta'\alpha'}(\omega_{\alpha\beta}) + J_{\alpha\beta,\beta'\alpha'}(\omega_{\beta'\alpha'})$$

$$- \delta_{\alpha'\beta'} \sum_{\gamma} J_{\alpha\gamma,\gamma\beta}(\omega_{\gamma\beta}) - \delta_{\alpha\beta} \sum_{\gamma} J_{\beta'\gamma,\gamma\alpha'}(\omega_{\beta'\gamma}), \tag{102}$$

where

$$J_{\alpha\beta,\beta'\alpha'}(\omega) = \int_0^\infty d\tau_1 \cos{(\omega\tau_1)}[\langle\mathcal{H}_1(t)_{\alpha\beta}\mathcal{H}_1(t-\tau_1)_{\beta'\alpha'}\rangle$$
$$-\langle\mathcal{H}_1(t)_{\alpha\beta}\rangle\langle\mathcal{H}_1(t)_{\beta\alpha}\rangle]. \tag{103}$$

Thus $\mathbf{R}^{(2)'}$ is the real part of the familiar relaxation matrix and the phase memory time for a two-level spin system is given by $T_2 = -1/R^{(2)'}_{\alpha\alpha',\alpha\alpha'}$. Note that $\mathbf{R}^{(2)''}$ gives rise to dynamic frequency shifts in cw EPR spectra.[50] It is seen from Eq. (100) that it does not contribute to the electron spin-echo decay signal, however.

Although the preceding results are strictly valid only in the limit of negligibly short microwave pulse widths, our previous work has shown that finite pulse times do not appreciably affect phase memory times in the motionally narrowed regime.[13] The cumulant expansion method does offer the advantage of providing a prescription for calculating the relaxation matrix to arbitrary order.[43,45] This is important since the rate of convergence of the cumulant expansion depends on the relative magnitude of $|\mathcal{H}_1(t)\tau_c|$. Thus higher-order terms are expected to become important for $|\mathcal{H}_1(t)\tau_c| \gtrsim 1$.

4.2   Phase Memory Decay in the Slow-Motional Region

It is instructive to consider the slow-motional behavior of a simple line resulting from the Markovian modulation of a secular g-tensor Hamiltonian due to isotropic Brownian rotational diffusion of a spin $S = \frac{1}{2}$ radical. Freed has shown in this case that [43,51]

$$R^{(2)'}_{\alpha\alpha',\alpha\alpha'} = -(\tfrac{2}{15})B_0^2\beta_e^2\hbar^{-2}\tau_2\sum_i(g_i^2 - g_s^2), \tag{104a}$$

and

$$R^{(4)'}_{\alpha\alpha',\alpha\alpha'} = -(\tfrac{13}{35})\tau_2[R^{(2)'}_{\alpha\alpha',\alpha\alpha'}]^2. \tag{104b}$$

The predicted phase memory decay to fourth order in the cumulants may be compared with that calculated from the time-dependent stochastic Liouville theory discussed in Section 3.2. One finds that for $|\mathcal{H}_1(t)\tau_2| < 1$ the phase memory decay is well represented by Eq. (104a, b). For longer correlation times, however, the calculated phase memory decay rate is smaller than that predicted to fourth order in the cumulants. It appears therefore that higher-order cumulants are required in this case and that there must be positive contributions from these cumulants so as to be consistent with the calculated results.

As discussed in Section 3.2, cw EPR line shapes in the slow-motional region are known to be sensitive to the particular model chosen for the dynamics of rotational diffusion. Clearly since electron spin echoes always decay as a superposition of exponentials for $2\tau \gg \tau_c$, the shape of the decay signal cannot be expected to be very sensitive to the nature of the molecular dynamics. As demonstrated in Section 3.2, however, phase memory decay times do display a model dependence in the slow-motional region. From the present viewpoint this results from the different magnitudes as well as signs that the higher-order cumulants ($n > 2$) may have depending on the choice for the stochastic diffusional model. For example, the secular g-tensor Hamiltonian for an electron spin $S = \frac{1}{2}$ radical undergoing strong jump rotational diffusion has $R^{(4)'}_{\alpha\alpha',\alpha\alpha'}$ given by [cf. Eq. (104b)]

$$R^{(4)'}_{\alpha\alpha',\alpha\alpha'} = +(\tfrac{1}{7})\tau_2[R^{(2)'}_{\alpha\alpha',\alpha\alpha'}]^2. \tag{105}$$

Brown maintains that the phase memory decay rate is a more sensitive function of the correlation time than is the cw EPR spectrum.[10,11] This premise is supported by considering the properties of a convolution of Lorentzian lines and is demonstrated in Fig. 6. For a two-level spin system in the long time limit, the electron spin-echo decays in general as a single exponential and thus its Fourier transform is Lorentzian. The width of this line is a function of the correlation time. In contrast, the slow-motional cw EPR spectrum consists of a superposition of Lorentzian lines whose widths and positions are functions of the correlation time.[29] For correlation times longer than that associated with the phase memory time minimum, the component lines narrow with little change in their positions. The envelope width becomes much larger than the component line widths so that little change is observed for the overall line shape. The cw EPR spectrum thus achieves a rigid-limit appearance. This problem is similar to that encountered for the cw EPR spectra of inhomogeneously broadened radicals undergoing Heisenberg exchange.[6] Another difficulty arises from unresolved intra- and intermolecular electron-nuclear dipolar couplings. These produce an orientation dependent contribution to the cw line shape that becomes more important as the motion slows and may tend to mask subtle changes in the overall envelope shape. They also affect the echo envelope by producing a beat pattern (i.e., the nuclear modulation effect).[4] The rate of decay must still be exponential for intermediate rotational correlation times, however.[52] The main point is that the frequency shifts and positions of the component lines have no effect on the echo envelope decay rate (cf. Section 4.3). Thus phase memory times may be expected to be sensitive to the motion of the radicals even though the cw EPR spectrum appears to be rigid. For even longer rotational correlation times, solid-state contributions to the phase memory decay rate become important and nonexponential behavior may be observed.[4,46]

4.3 Finite Time Effects on the Electron Spin-Echo Phase Memory Decay

It is also possible to examine $\mathbf{K}_n(t)$ for finite t using the cumulant theory discussed in Section 4.1. This is especially important if the condition $2\tau \gg \tau_c$ is not fulfilled. In this case one retains $\mathbf{I}_n(t)$ for finite times in Eq. (97). Thus for $\tau \gtrsim \tau_c$ one obtains

$$\dot{\mathbf{K}}_n(t) \cong \exp\left[-i\mathcal{H}_0^x \int_0^t dt' S(t')\right][\mathbf{R}^{(n)'}(t) - iS(t)\mathbf{R}^{(n)''}(t)]$$

$$\times \exp\left[i\mathcal{H}_0^x \int_0^t dt' S(t')\right]S(t)^n, \tag{106}$$

where $\mathbf{R}^{(n)'}$ and $\mathbf{R}^{(n)''}$ are now *time dependent*.

It is instructive to consider the simple case for which $\mathcal{H}_1(t)$ is secular with $\langle \mathcal{H}_1(t) \rangle = 0$. We assume that

$$m_2(\tau_1) = \Delta^2 \exp\left(-\frac{|\tau_1|}{\tau_c}\right), \tag{107}$$

and

$$m_3(\tau_1, \tau_2) = \Delta^3 \exp\left[-\frac{(|\tau_1| + |\tau_2|)}{\tau_c}\right], \tag{108}$$

where

$$\Delta^2 \equiv \langle |\mathcal{H}_1(t)_{\alpha\alpha} - \mathcal{H}_1(t)_{\alpha'\alpha'}|^2 \rangle. \tag{109}$$

One then has that

$$\dot{\mathbf{K}}_2(t)_{\alpha\alpha',\alpha\alpha'} = -\int_0^t dt' m_2(t')_{\alpha\alpha',\alpha\alpha'}$$

$$= -\Delta^2 \tau_c \left[1 - \exp\left(-\frac{|t|}{\tau_c}\right)\right], \tag{110}$$

while

$$\dot{\mathbf{K}}_3(t)_{\alpha\alpha',\alpha\alpha'} = -i\int_0^t d\tau_1 \int_0^{t-\tau_1} d\tau_2 m_3(\tau_1, \tau_2)_{\alpha\alpha',\alpha\alpha'} S(t)$$

$$= -i\,\Delta^3 \tau_c^2 \left[1 - \exp\left(-\frac{|t|}{\tau_c}\right) - \left(\frac{|t|}{\tau_c}\right)\exp\left(-\frac{|t|}{\tau_c}\right)\right]S(t), \tag{111}$$

so that

$$K_2(2\tau)_{\alpha\alpha',\alpha\alpha'} = -\Delta^2\tau_c\left[2\tau + \tau_c \exp\left(-\frac{2\tau}{\tau_c}\right) - \tau_c\right], \qquad (112)$$

and

$$K_3(2\tau)_{\alpha\alpha',\alpha\alpha'} = -i\,\Delta^3\tau_c^2\left[2\tau_c - (4\tau_c + 2\tau)\exp\left(-\frac{\tau}{\tau_c}\right)\right.$$

$$\left. + 2(\tau + \tau_c)\exp\left(-\frac{2\tau}{\tau_c}\right)\right]. \qquad (113)$$

From the form of Eq. (112) it is clear that $\exp[K_2(2\tau)_{\alpha\alpha',\alpha\alpha'}]$ is not a simple exponential decay. Moreover, the presence of an imaginary $K_3(2\tau)_{\alpha\alpha',\alpha\alpha'}$ term implies that the echo envelope decays in an oscillatory manner. This oscillatory contribution decays quickly in time, however, with a concomitant asymptotic approach of $\exp[K_2(2\tau)_{\alpha\alpha',\alpha'\alpha'}]$ to simple exponential behavior. Thus for $\tau \gg \tau_c$ the results of Section 4.1 are recovered. One finds in general that for $\tau \gtrsim \tau_c$ the echo envelope decay function is rather complex but asymptotically approaches exponential behavior in the long time limit.

Freed has shown that the inclusion of finite time effects in the cumulant expansion yields weak subsidiary lines that produce non-Lorentzian behavior in the wings of motionally narrowed cw spectra.[43] Its effect for electron spin echoes is often negligible, however, since the condition $\tau \gg \tau_c$ is nearly always satisfied except for very long correlation times. The reason is that there is a characteristic "dead" time for electron spin-echo spectrometers that is required so that the signal resulting from cavity ringing, after the pulses, has decayed to zero.[37] This dead time is often much larger than τ_c and places a practical lower limit on τ. Thus the short time behavior has decayed by the time that the echo is observed.

This has an important consequence for Fourier transform EPR spectroscopy. The cw EPR line shape is given by the Fourier transform of the free-induction decay signal.[1] One may use the theory presented in Section 4.1 to describe the free-induction decay (FID). The difference for this experiment is that $S(t) = 1$ for all times t. Using this definition, one recovers the results obtained by Freed.[43] In the absence of any inhomogeneous line broadening, the FID signal of a simple line decays in general as a complex exponential in the long time limit. Hence its Fourier transform is Lorentzian. Slow-motional EPR spectra are not Lorentzian, however.[29] This implies that the subsidiary lines resulting from finite time effects are very important in determining slow-motional cw EPR line shapes. Since the finite time terms of the cumulant expansion may have decayed during the dead time of the pulse spectrometer, the observed Fourier transform of the free-induction decay signal may not resemble the cw EPR slow-motional spectrum.

5 Summary

The theory of electron spin echoes in liquids has been examined by use of both density matrix-relaxation operator and generalized cumulant expansion formalisms. Although intimately related, each of these methods offers their own particular advantages; the former proves to be more useful for computational purposes, whereas the cumulant theory provides considerable insight into the nature of time domain relaxation behavior.

If $\tau \lesssim \tau_c$, the echo envelope is shown to decay in a complex, oscillatory manner. It is found that for pulse times $\tau \gg \tau_c$, the two-pulse electron spin-echo envelope decays exponentially and for motionally narrowed systems, the phase memory decay rate is given by the elements of the Redfield relaxation matrix. The effects of Heisenberg spin exchange are readily incorporated into the theory and are found to be additive to the phase memory decay rate in the absence of exchange. Since the obtained results are independent of any source of inhomogeneous line broadening, it is clear that the desired "intrinsic" linewidth is directly determined in this experiment, whereas it can only be determined from cw EPR line shapes by computer simulation methods. These are *strongly* dependent on accurate knowledge of the hyperfine model for the nuclei that give rise to the inhomogeneous line broadening. As this can be difficult to determine and since the computer calculations can become quite expensive, the value of the electron spin-echo technique for motionally narrowed systems is apparent.

For radicals undergoing slow-rotational reorientation, the phase memory time is shown to be sensitive to both the rotational correlation time and the assumed dynamical model for rotational diffusion even for correlation times so long that the cw EPR spectra appear to be rigid. Thus the time scale of EPR detectable dynamics may be extended by use of electron spin-echo methods. This may also be accomplished by saturation transfer spectroscopy.[53] However, the electron spin-echo technique may still be expected to enjoy its own advantages arising from the lack of complications due to inhomogeneous line broadening and its different mechanism.

The theoretical methods employed in the present work may also be applied to other time domain experiments such as stimulated electron spin echoes, inversion recovery, and ENDOR spin-echo spectroscopy. In an investigation of the latter for liquid solutions, it is found that the duration of the rf pulse is very important in determining the ENDOR linewidth and that the strength of the rf field, the various relaxation times, and the inclusion of rf coherence effects also contribute to the line shape in a complicated manner. Although the analysis is rigorously valid only for the case of motional narrowing, it is expected that the calculated line shapes demonstrate the gross features of relaxation and coherence effects found in rigid-limit ENDOR spin-echo spectra.

There are presently both computational and experimental difficulties with electron spin-echo methods. Although cumulant expansions are useful for obtaining some understanding of phase memory decay behavior, the great difficulty in calculating the higher-order terms limits its applicability for slow-motional systems. The computation time for using the stochastic Liouville theory discussed in Section 3.2 can become quite expensive unless it is programmable on a minicomputer. Also, experimentally it is found that the signal-to-noise ratio can become troublesome especially near the phase memory time minimum. It is hoped that the present work will serve as an impetus for the solution of both of these problems.

REFERENCES

1. A. Abragam, *The Principles of Nuclear Magnetism*, Oxford University Press, London, 1961.

2. L. T. Muus and P. W. Atkins, Eds., *Electron Spin Relaxation in Liquids*, Plenum, New York, 1972.

3. T. C. Farrar and E. D. Becker, *Pulse and Fourier Transform NMR*, Academic, New York, 1971.

4. W. B. Mims, in S. Geschwind, Ed., *Electron Paramagnetic Resonance*, Plenum, New York, 1972.

5. G. Poggi and C. S. Johnson, Jr., *J. Mag. Resonance*, **3,** 436 (1970).

6. A. E. Stillman and R. N. Schwartz, *J. Mag. Resonance*, **22,** 269 (1976).

7. K. M. Salikov, A. G. Semenov, and Yu. D. Tsvetkov, *Electron Spin Echoes and Their Applications*, Science, Novosibirsk, 1976.

8. R. Brändel, G. J. Krüger, and W. Müller-Warmuth, *Z. Naturforsch.*, **25A,** 1 (1970).

9. (a) A. D. Milov, K. M. Salikov, and Yu. D. Tsvetkov, *Chem. Phys. Lett.*, **8,** 523 (1971); (b) A. D. Milov, A. B. Mel'nik, and Yu. D. Tsvetkov, *Theor. Exp. Chem.*, **11,** 790 (1975).

10. I. M. Brown, *Chem. Phys. Lett.*, **17,** 404 (1972).

11. I. M. Brown, *J. Chem. Phys.*, **60,** 4930 (1974).

12. F. Köksal and G. J. Krüger, *Z. Naturforsch.*, **30A,** 883 (1975).

13. A. E. Stillman and R. N. Schwartz, *Mol. Phys.*, **32,** 1045 (1976). The footnote on p. 1054 of this reference is in error; the relation $T_2^{-1}(\text{ex}) \approx T_2^{-1}(0) + \omega_{\text{ex}}$ is in general valid.

5 REFERENCES

14. A. E. Stillman and R. N. Schwartz, *J. Chem. Phys.*, **69**, 3532 (1978).

15. (a) A. E. Stillman and R. N. Schwartz, *Mol. Phys.*, **35**, 301 (1978). (b) Reference 15a states that the energy uncertainty relation was not obeyed for large *T*. A more refined calculation, however, demonstrates that this statement is not in fact the case. Also, we note that the times in the figures of this reference should have units of seconds rather than sec/rad.

16. It is this distribution of resonance frequencies that is responsible for the inhomogeneous broadening in the cw spectrum.

17. R. H. Dicke, *Phys. Rev.*, **93**, 99 (1954).

18. J. D. Macomber, *The Dynamics of Spectroscopic Transitions*, Wiley-Interscience, New York, 1976, pp. 170–237.

19. J. H. Freed, *J. Chem. Phys.*, **43**, 2312 (1965); J. H. Freed, in L. T. Muus and P. W. Atkins, Eds., *Electron Spin Relaxation in Liquids*, Plenum, New York, 1972, Chap. 18.

20. J. H. Freed, *J. Phys. Chem.*, **78**, 1155 (1974); see also Chapter 2 of this book.

21. J. H. Freed, D. S. Leniart, and H. D. Conner, *J. Chem. Phys.*, **58**, 3089 (1973).

22. P. W. Atkins, K. A. McLauchlan, and P. W. Percival, *Mol. Phys.*, **25**, 281, 1973.

23. C. B. Moler and C. F. van Loan, "Nineteen Ways to Compute the Exponential of a Matrix," Cornell University Computer Science Report TR76-283, 1976.

24. C. W. Gear, *Numerical Initial Value Problems in Ordinary Differential Equations*, Prentice-Hall, New York, 1971.

25. We found that the Bulirsch–Stoer rational extrapolation method (see reference 24) was best suited for our purpose. It has the fastest computational time of all the methods tested and also required only matrix storage space for **R**, **K**, and $\hat{\mathbf{d}}$ rather than the entire supermatrix.

26. C. C. Whisnant, S. Ferguson, and D. B. Chesnut, *J. Phys. Chem.*, **78**, 1410 (1974).

27. J. H. Freed and G. K. Fraenkel, *J. Chem. Phys.*, **39**, 326 (1963).

28. H. C. Torrey, *Phys. Rev.*, **76**, 1059 (1949).

29. (a) J. H. Freed, G. V. Bruno, and C. F. Polnaszek, *J. Phys. Chem.*, **75**, 3385 (1971); (b) J. H. Freed, in L. J. Berliner, Ed., *Spin Labeling Theory and Applications*, Academic, New York, 1976.

30. D. Garfinkel, C. B. Marbach, and N. Z. Shapiro, *Ann. Rev. Biophys. Bioeng.*, **6**, 525 (1977).

31. J. S. Hwang, R. P. Mason, L. P. Hwang, and J. H. Freed, *J. Phys. Chem.*, **79**, 489 (1975).

32. W. B. Mims, *Proc. Roy. Soc.* **A283**, 452 (1965).

33. I. M. Brown, D. J. Sloop, and D. P. Ames, *Phys. Rev. Lett.*, **22**, 324 (1969).

34. P. F. Liao and S. R. Hartmann, *Phys. Rev.*, **B8**, 69 (1973).

35. E. R. Davies, *Phys. Lett.*, **A47**, 1 (1974).

36. L. Kevan and L. D. Kispert, *Electron Spin Double Resonance Spectroscopy*, Wiley-Interscience, New York, 1976.

37. W. E. Blumberg, W. B. Mims, and D. Zuckerman, *Rev. Sci. Instrum.*, **44**, 546 (1973).

38. J. H. Freed, D. S. Leniart, and J. S. Hyde, *J. Chem. Phys.*, **47**, 2762 (1967).

39. P. Narayana, M. K. Bowman, D. Becker, L. Kevan, and R. N. Schwartz, *J. Chem. Phys.*, **67**, 1990 (1977).

40. D. S. Leniart, H. D. Conner, and J. H. Freed, *J. Chem. Phys.*, **63**, 165 (1975).

41. A. Messiah, *Quantum Mechanics*, North-Holland, Amsterdam, 1958, pp. 135–138.

42. J. R. Klauder and P. W. Anderson, *Phys. Rev.*, **125**, 912 (1962).

43. J. H. Freed, *J. Chem. Phys.*, **49**, 376 (1968); J. H. Freed, in L. T. Muus and P. W. Atkins, Ed., *Electron Spin Relaxation in Liquids*, Plenum, New York, 1972, Chap. 8.

44. R. Kubo, *J. Phys. Soc. Jap.*, **17**, 1100 (1962).

45. (a) N. G. van Kampen, *Physica*, **74**, 215 (1974); (b) N. G. van Kampen, *Physica*, **74**, 239 (1974).

46. W. B. Mims, *Phys. Rev.*, **168**, 370 (1968).

47. As it appears to be more transparent, we have chosen our definition for $S(t)$ to be opposite in sign than that used by Klauder and Anderson, reference 42.

48. The time domain signal intensity in this formalism is given by $I(t) =$ Re $\{\int_0^\infty d\omega g(\omega) \exp[+i\omega \int_0^t dt' S(t')] \text{Tr}[\mathbf{S}_x(t)\mathbf{S}_x(0)]\}$, where $g(\omega)$ is a normalized function for the inhomogeneous distribution of frequencies. Thus for a two-pulse spin echo, $I(2\tau) = \text{Re}\{\text{Tr}[\mathbf{S}_x(2\tau)\mathbf{S}_x(0)]\}$ so that the time dependence of the echo intensity is determined by knowledge of $\mathbf{S}_x(t)$.

49. B. Yoon, J. M. Deutsch, and J. H. Freed, *J. Chem. Phys.*, **62**, 4687 (1975).

50. G. K. Fraenkel, *J. Chem. Phys.*, **42**, 4275 (1965).

51. If nonsecular terms are retained in the spin Hamiltonian, there will also be a contribution to the phase memory decay resulting from $R^{(3)''}_{\alpha\alpha',\alpha\alpha'}$.

52. If $2\tau \lesssim \tau_c$ nonexponential behavior may be observed (cf. Section 4.3). In this case one can define the phase memory time as the time for which the echo has decayed to $1/e$ of its initial value. The phase memory time can still then be measured by comparison to that calculated from stochastic Liouville theory and may be expected to be sensitive to the dynamics of the radical.

53. L. R. Dalton, B. H. Robinson, L. A. Dalton, and P. Coffey, *Adv. Mag. Resonance*, **8**, 149 (1976).

6 ELECTRON SPIN-ECHO STUDIES OF RELAXATION PROCESSES IN MOLECULAR SOLIDS

Ian M. Brown

McDonnell Douglas Research Laboratories
McDonnell Douglas Corporation,
St. Louis, Missouri

1   RELEVANT BACKGROUND INFORMATION

1.1   Introduction

In this chapter we are concerned with the generation and decay of electron spin echoes in molecular solids containing free radicals. In particular, we describe the results of some pulsed electron paramagnetic resonance (EPR) experiments that provide information about the paramagnetic relaxation processes in rigid glass matrices containing low concentrations of randomly oriented free radicals and also in nonrigid matrices where the radicals are undergoing molecular motions. Some of the advantages of using electron spin-echo techniques to study these relaxation processes are illustrated.

The EPR line resulting from a dilute collection of randomly oriented free radicals in a rigid glass matrix is inhomogeneously broadened usually because of g-anisotropy and unresolved hyperfine splittings. Consider the application of a resonant microwave pulse to such a free radical system at $t = 0$ under the condition that ω_1 is less than the EPR linewidth, where $\omega_1 = \gamma B_1$ and B_1 is the amplitude of the microwave magnetic field. If spectral diffusion is slow enough, the pulse will selectively excite the spin system and "burn" a hole of half-width ω_1 in the line. Following the application of a second resonant microwave refocusing pulse at a time $t = \tau$, an echo will form at approximately $t = 2\tau$ providing that spectral diffusion and electron spin-lattice relaxation processes are not prohibitively fast. Fortunately, in most molecular solids containing low concentrations of free radicals, the spectral diffusion rates are sufficiently fast to allow the excitation to spread to adjacent parts of the EPR line in a time comparable to the electron spin-lattice relaxation time but are slow enough to allow electron spin echoes to be detected easily with modern pulsed EPR spectrometers. Consequently, electron spin echoes have been observed in a large variety of cation, anion, and neutral radicals randomly oriented in rigid glass matrices.

1.2   Phase Memory Time

As τ, the interpulse time, is increased, the echo height decreases, and the overall two-pulse echo envelope decay is characterized by a time called the phase memory time T_M.

In discussing the mechanisms that cause the two-pulse echo envelope decay, it is convenient to divide the electron spin system into two types of spins, A and B.[1,2] The A spins are directly excited by the resonant microwave pulses and hence contribute to the echo signal, whereas the B

spins are the remainder of the electron spin system. The A spins can be responsible for echo decay if spin-lattice relaxation removes A spins from precessional motion and hence their contribution to the echo at $t = 2\tau$. In this case, one observes $T_M = T_1$. However, it is more common for the echo envelope decay to be determined by the local fields at the A spin sites arising from dipolar interactions with the B spins. The static contributions to the local fields results in reversible dephasings that are nullified at $t = 2\tau$ by the refocusing effects of the second microwave pulse. On the other hand, local field fluctuations produce variations in the Larmor frequencies and thus a randomization of the precessional phases that result in echo attenuation.

The Fourier transform of the echo envelope decay can be used to define the homogeneous line shape of an individual spin packet with a half-width at half-intensity that corresponds to the reciprocal of the phase memory time.[1] Since this homogeneous spin packet width can be several orders of magnitude smaller than the width of the inhomogeneously broadened EPR line, it is more sensitive to time-dependent local fields. For example, in Section 3 we describe a free radical system where, as a result of molecular motions, there is a distinctive temperature dependence of the homogeneous spin packet width, whereas there is no significant change in the EPR line shape or linewidth.

In a rigid lattice the local field fluctuations can be determined by spin flips of the B electron spins or the surrounding nuclear spins. These electron spin flips are the result of electron spin-lattice relaxation processes (T_1-induced) or electron spin-spin relaxation processes (T_2-induced), whereas the nuclear spin flips are usually a consequence of nuclear spin-spin relaxation. We summarize each of these mechanisms in turn.

1.3 T_1-Induced Dipolar Field Fluctuations

The situation where the local field fluctuations are caused by spin-lattice relaxation of unlike electron spins acting as B spins is the case that has received the most attention both theoretically[3-5] and experimentally,[4] principally because it is tractable to mathematical analysis.

Although fluctuations in the dipolar fields cause a random time dependence of the spin packet frequencies, an average behavior can be predicted using a conditional probability distribution function $K(\omega - \omega_0, t')$, which is sometimes referred to as the diffusion kernel.[4] It is defined as the probability that a spin whose Larmor frequency is ω_0 at $t = 0$ will have a frequency ω at $t = t'$. The exact mathematical form of this function depends on the particular model chosen to represent the behavior of the spin system.

Two models have been proposed: the sudden-jump model of Klauder and Anderson[3] and the Gauss-Markovian model of Mims.[4] In the sudden-jump model, the z-component of the B-spin magnetic moment can undergo random sudden jumps between the values $\mu_z = \pm\frac{1}{2}g_B\beta_e$ at an average rate $R(=1/T_1)$. In a time $t_1 \ll 1/R$, a number of B spins $\mathcal{N}_B R t_1$ will reverse their z-component values, where $\mathcal{N}_B$ is the number of B spins per unit volume. The local field change at the A spin site is the same as if $\mathcal{N}_B R t_1$ moments, each with $\mu_z = g_B\beta_e$, were randomly distributed around the A spin. Just as the statistical theory of line shapes[6] predicts a Lorentzian form for the EPR line shape for a dilute random distribution of magnetic moments, so it also predicts that a spin with frequency ω_0 will at $t = t_1$ broaden out into a Lorentzian distribution.

$$K(\omega - \omega_0, t_1) = \frac{2R\,\Delta\omega_{\text{dip}}t_1/\pi}{(\omega - \omega_0)^2 + (2R\,\Delta\omega_{\text{dip}}\,t_1)^2}, \tag{1}$$

where $\Delta\omega_{\text{dip}}$ is the half-width attributable to dipolar broadening from a random distribution of static B spins. For B spins with $S = \frac{1}{2}$, one finds from the statistical theory that[4]

$$\Delta\omega_{\text{dip}} = 2.53\,\gamma_A g_B\beta_e\mathcal{N}_B, \tag{2}$$

where γ_A is the gyromagnetic ratio for the A spins, g_B is the g-value for the B spins, and β_e is the Bohr magneton.

Following the same argument, in the next time interval t_2, $\mathcal{N}_B R t_2$ randomly distributed spin flips will cause each frequency in Eq. (1) to broaden out to a corresponding Lorentzian distribution with a half-width $2R\,\Delta\omega_{\text{dip}}t_2$. Thus, during the time $t_2 \ll 1/R$, the local field values experienced by a collection of A spins can be considered as a diffusion process characterized by a Lorentzian diffusion kernel whose width increases linearly with t_2.

For long times the appropriate diffusion kernel is stationary and of the form

$$K(\omega - \omega_0, t) = \frac{\Delta\omega_{\text{dip}}(1 - e^{-2Rt})/\pi}{(\omega - \omega_0)^2 + [\Delta\omega_{\text{dip}}(1 - e^{-2Rt})]^2}. \tag{3}$$

Klauder and Anderson[3] have shown that in the limit $R\tau \ll 1$, the two-pulse echo envelope decay $E(2\tau)$ is given by[7]

$$E(2\tau) = \exp\left[-\left(\frac{2\tau}{T_M}\right)^2\right], \tag{4a}$$

with

$$T_M = 1.4(R\,\Delta\omega_{\text{dip}})^{-1/2}. \tag{4b}$$

In the Gauss–Markovian model, Mims[4] assumed that the magnetic moments of the B spins behaved as Gaussian random variables with the Markovian correlation function

$$\langle \mu(t_1)\,\mu(t_2)\rangle = \tfrac{1}{4}g_B^2\beta_e^2\,\exp\left(-R|t_2-t_1|\right). \tag{5}$$

By averaging first over an ensemble of A spins having identical B spins and then over all A spin environments, Mims showed that in the limit $R\tau\ll1$, the Gauss–Markovian model predicts

$$E(2\tau)=\exp\left[-\left(\frac{2\tau}{T_M}\right)^{3/2}\right], \tag{6a}$$

with

$$T_M=1.9[R(\Delta\omega_{\text{dip}})^2]^{-1/3}, \tag{6b}$$

whereas when $R\tau\gg1$, it predicts

$$E(2\tau)=\exp\left[-\left(\frac{2\tau}{T_M}\right)^{1/2}\right], \tag{7a}$$

with

$$T_M=\frac{0.56R}{(\Delta\omega_{\text{dip}})^2}. \tag{7b}$$

The diffusion kernel for the Gauss–Markovian model has also been derived[4] and is of the form

$$K(\omega-\omega_0,t)=\frac{\Delta\omega_{\text{dip}}[1-\exp\left(-2Rt\right)]^{1/2}/\pi}{(\omega-\omega_0)^2+(\Delta\omega_{\text{dip}})^2[1-\exp\left(-2Rt\right)]}. \tag{8}$$

This kernel also has a Lorentzian form but with a width that is proportional to $(\text{time})^{1/2}$.

It is appropriate to introduce the stimulated or three-pulse echo at this point since its decay can be directly related to the diffusion kernel.[4] If three $\pi/2$ pulses are applied at $t=0$, $t=\tau$, and $t=\tau+T$, the stimulated echo appears at $t=2\tau+T$. The first two pulses form a serrated pattern of z-magnetization that can be recalled as a stimulated echo by the third pulse. Spectral diffusion occurring during the time interval T smooths out this serrated pattern and hence produces an attenuation of the stimulated echo. If the overall decay factor for the stimulated echo, $D(\tau,T)$, can be written as

$$D(\tau,T)=D_\tau(\tau)D_T(\tau,T), \tag{9}$$

where $D_\tau(\tau)$ is the echo decay attributable to the loss of phase coherence occurring in the two intervals τ, and $D_T(\tau,T)$ is the echo decay caused by spectral diffusion occurring in the interval T, then in the limit $R\tau\ll1$,

$D_T(\tau, T)$ is the Fourier transform of the diffusion kernel. Thus in the sudden-jump model where the Lorentzian form of the diffusion kernel shown in Eq. (3) is appropriate, the form of $D_T(\tau, T)$ is given by[3]

$$D_T(\tau, T) = \exp\left[-2R\,\Delta\omega_{dip}\,\tau T\right], \tag{10}$$

when $R\tau \ll 1$, whereas the corresponding expression for the Gauss–Markovian model[4] is

$$D_T(\tau, T) = \exp\left[-(2R)^{1/2}\,\Delta\omega_{dip}\tau T^{1/2}\right]. \tag{11}$$

Mims has discussed the physical validity of both models[1,4] and has expressed some reservations about using Eq. (10) and Eq. (11) for the stimulated echo decay because of a spin sorting effect. A spins located in regions containing large numbers of nearby B spins experience larger than average dipolar field fluctuations and are selectively removed during the time interval τ before they can contribute to the serrated pattern of z-magnetization formed at $t = \tau$. The remaining A spins that contribute to the stimulated echo experience local fields dominated by a large number of B spins at intermediate distances. From the central limits theorem, the diffusion kernel might then be expected to have a Gaussian instead of a Lorentzian form. Furthermore, when T_1 is determined by an Orbach process, it may not be realistic to consider $\mu(t)$ as a Gaussian random variable; hence the results derived from the Gauss-Markov model in the limit $R\tau \ll 1$ may not be applicable.

Although the measurements in single crystals of $CaWO_4$ doped with either manganese and erbium or cerium and erbium, agreed reasonably well with the predictions of the Gauss–Markov model,[4] recent considerations by Hu and Hartmann[8] have shown that the agreement with the predictions of a sudden-jump model are also remarkably good. In the more rigorous calculations of Hu and Hartmann,[5] expressions were derived for the free-induction decay and the two-pulse and the three-pulse echo envelope decays using an uncorrelated sudden-jump model to account for the T_1-induced dipolar field fluctuations. The solutions were obtained by solving the problem exactly and are valid for all values of $R\tau$. The limiting behavior of the form of the two-pulse echo envelope decay reduces to Eq. (4a) for $R\tau \ll 1$ and Eq. (7a) for $R\tau \gg 1$, and the three-pulse echo decay is of the form shown in Eq. (10) for $R\tau \ll 1$. Moreover, a minimum value of T_M is predicted to occur at $R = 1/T_1 = 2\Delta\omega_{dip}$.

1.4 T_2-Induced Dipolar Field Fluctuations

There is little experimental data on and no rigorous theoretical treatment of the case where the echo envelope decays are caused by electron spin flips resulting from electron spin-spin relaxation processes. In solids there are four times that can characterize the spin-spin relaxation processes: (1) the time associated with the reciprocal of the inhomogeneously broadened linewidth, sometimes designated T_2^*, (2) the time associated with the reciprocal of the dipolar broadened spin-packet width $(\Delta\omega_{dip})^{-1}$, (3) the mean spin-spin flip time, T_f, and (4) the experimentally determined phase memory time T_M.

In molecular solids containing low concentrations of free radicals, usually the values of these times will be such that

$$T_2^* < (\Delta\omega_{dip})^{-1} < T_M < T_f. \tag{12}$$

The phase memory time can be less than the mean spin-spin flip time since the surrounding electron spins can act as B spins to produce the fluctuations in the dipolar fields. This is illustrated by substituting $R = 1/T_f$ into Eq. (4b), where one obtains $T_M < T_f$ if $T_f > (2\Delta\omega_{dip})^{-1}$. For example, at radical concentrations of $10^{-3}M$, where $\mathcal{N} = 6 \times 10^{17}$ spins cm^{-3} and $(\Delta\omega_{dip})^{-1} = 2\ \mu s$, we calculate a value of $T_M = 63\ \mu s$ if we assume $T_f = 1$ ms.

It appears likely that in most radical systems T_f is longer than $(\Delta\omega_{dip})^{-1}$ since T_f depends on only the $S_{i+}S_{j-}$ terms in the dipolar interaction, whereas $\Delta\omega_{dip}$ depends on the $S_{iz}S_{jz}$ terms. Hence any static radical hyperfine interaction greater than $\Delta\omega_{dip}$ can cause a large enough shift in the Larmor frequencies of the neighboring electron spins to inhibit the mutual spin-flip processes. The electron spins are said to be detuned from one another by the hyperfine interaction.[1]

At low enough radical concentrations the values of the phase memory time are independent of the radical concentration and, in many cases, depend on the matrix. The matrix nuclei can then play the role of the B spins and the local field fluctuations at the A spin sites are usually caused by mutual spin flips that result from nuclear spin-spin relaxation processes. The nuclei on the radical and on the matrix molecules in shells close to the radical are excluded from this mechanism. Shifts in the Larmor frequencies of these nuclei that arise because of static electron-nuclear dipolar interactions result in a suppression of the mutual spin flips. These nuclear spins are said to be detuned from one another by the dipolar interaction with the unpaired electron.

We follow essentially the same argument as Mims[1] to obtain the dependence of the phase memory time on the nuclear environment. Only nuclei in shells outside a radius r_s undergo mutual spin flips, where r_s is defined as the distance from the electron at which the electron-nuclear interaction equals the interaction causing the mutual spin-flip processes, that is,

$$\frac{\frac{1}{2}\hbar\gamma_n\gamma_e|1-3\cos^2\theta|}{r_s^3} = \Delta\omega_{nn}, \tag{13}$$

where $\Delta\omega_{nn}$ is the mutual spin-flip rate and is of the order of $1/T_2$, the homogeneous nuclear magnetic resonance (NMR) linewidth. If the line shape is Gaussian, then[9]

$$\Delta\omega_{nn} \approx \frac{1}{3}(M_2)^{1/2}, \tag{14}$$

where M_2 is the second moment of the NMR line. Hence for a collection of randomly oriented cubic lattice sites,[10] $\Delta\omega_{nn}$ reduces to

$$\Delta\omega_{nn} \approx 0.75\gamma_n^2\hbar\mathcal{N}[I(I+1)]^{1/2}, \tag{15}$$

where $\mathcal{N}$ is the number of nuclei per cubic centimeter each having a nuclear spin I.

We assume that the effect of a mutual spin flip at a distance r from the electron is to produce a frequency shift of the electron Larmor frequency $\delta\omega_{ne}$, given by

$$\delta\omega_{ne} = \frac{\frac{1}{2}\hbar\gamma_e\gamma_n(1-3\cos^2\theta)}{r^3}. \tag{16}$$

Fluctuations in I_z arising from the mutual spin flips render these frequency shifts time dependent with a rate of about $\Delta\omega_{nn}$. If the fluctuations in I_z are uncorrelated, the square of the homogeneous broadening of a spin packet, $(\Delta\omega_e)^2$, can be obtained by a summation of contributions of the form $(\delta\omega_{ne})^4/(\Delta\omega_{nn})^2$ from all the shells of surrounding nuclei with radii greater than r_s. If this summation is replaced with an integration over a continuous distribution of shells, the expression for the phase memory time determined by the mutual spin flips can be reduced to the form[1]

$$(T_{MN})^{-1} = \Delta\omega_e = 0.49\left(\frac{\gamma_e}{\gamma_n}\right)^{1/2}\frac{\Delta\omega_{nn}}{[I(I+1)]^{1/4}}. \tag{17}$$

For simplicity we consider a cubic lattice so that Eq. (17) reduces to

$$(T_{MN})^{-1} = 0.37\gamma_e^{1/2}\gamma_n^{3/2}\mathcal{N}\hbar[I(I+1)]^{1/4}. \tag{18}$$

If we assume a cubic lattice for a sulfuric acid matrix and substitute $\mathcal{N} = 2.2\times10^{22}$ protons/cm^3, the value for sulfuric acid, into Eq. (18), we obtain a value of $T_{MP} = 6.8\ \mu s$, whereas the measured value for the anthracene cation in sulfuric acid at 77 K is 10.4 μs. This is good agreement considering the crudeness of the calculation.

We are also interested in evaluating the ratio T_{MD}/T_{MP} where T_{MD} and T_{MP} are the phase memory times of a radical in a deuterated and protiated matrix, respectively. From Eq. (17), this ratio should be

$$\frac{T_{MD}}{T_{MP}} = 1.28\left(\frac{\gamma_D}{\gamma_P}\right)^{1/2}\frac{\Delta\omega_{PP}}{\Delta\omega_{DD}}, \tag{19}$$

where $\Delta\omega_{PP}$ and $\Delta\omega_{DD}$ are the mutual spin-flip rates in the protiated and deuterated matrix, respectively. For a matrix of deuterium nuclei with cubic or higher symmetry, where there is no nuclear quadrupole splitting, $\Delta\omega_{DD}$ is given by Eq. (15), and the ratio T_{MD}/T_{MP} is 13.0. As is described in Section 2.1, the measured ratio for the radical cations of anthracene-d_{10} in sulfuric acid-d_2 and anthracene in sulfuric acid matrices at 77 K is ≈ 19.2. The difference between the observed value and that calculated from Eq. (19) could be due to the quadrupole interaction, which is effectively a static interaction producing an inhomogeneous broadening of the matrix deuterium NMR line. The resultant detuning renders $\Delta\omega_{DD} < (M_{2D})^{1/2}/3$, where M_{2D} is the second moment of the deuterium NMR line. From Eq. (19), it therefore appears that for most radicals the effect of using a deuterated matrix instead of a protiated matrix is to increase the phase memory time by at least an order of magnitude.

1.5 Instantaneous Diffusion

In addition to the spectral diffusion arising from the B spins, the microwave pulses themselves can cause an instantaneous diffusion that is independent of R, the time characterizing the B spin-flip rate.[3] Specifically, this diffusion is brought about by the local field changes resulting from the spin flips induced by the second pulse.

Consider $\mathcal{N}_R$ spins per cubic centimeter to be reoriented by the second pulse and randomly distributed throughout the sample. It can be seen from the statistical theory of EPR line shapes[1,6] that as a result of the $\mathcal{N}_R$ spin flips, a spin packet with frequency ω_0 will spread into a distribution of frequencies given by[1,3]

$$K(\omega - \omega_0) = \frac{\Omega/\pi}{(\omega - \omega_0)^2 + \Omega^2}, \tag{20a}$$

where

$$\Omega = 5.1\gamma g\beta_e\mathcal{N}_R. \tag{20b}$$

The echo envelope then contains a decay factor that is the Fourier transform of $K(\omega - \omega_0)$ and hence is of the form

$$E(2\tau) = E_0 \exp\left(\frac{-2\tau}{T_M}\right), \tag{21a}$$

where

$$T_M = \frac{2}{\Omega},$$ (21b)

and E_0 is the echo envelope decay in the absence of instantaneous diffusion.

If the second pulse is a $2\pi/3$ pulse for the spins on resonance and $\omega_1 \ll$ linewidth, then $\mathcal{N}_R$ is given by $\mathcal{N}_R = \mathcal{N}_A$, where $\mathcal{N}_A (=\mathcal{N}_T\delta)$ is the number of A spins per cubic centimeter excited by the first pulse and $\mathcal{N}_T$ is the total number of electron spins (A and B) per cubic centimeter in the sample.

Substituting this value of $\mathcal{N}_R$ in Eq. (20b) and using the expression for $\Delta\omega_{dip}$ given in Eq. (2), one obtains

$$\Omega = \frac{2\mathcal{N}_A}{\mathcal{N}_T}\Delta\omega_{dip}.$$ (22)

As ω_1 is increased, T_M decreases until $\mathcal{N}_A = \mathcal{N}_T$ where it reaches a value $T_{MA} = 1/(\Delta\omega_{dip})$ when $\omega_1 \gg$ linewidth and a 90–180° pulse sequence is used. For a radical concentration of $10^{-3}M$, T_{MA} is given by $T_{MA} = 2\ \mu s$.

It can be seen from Eqs. (21) and (22) that instantaneous diffusion effects are easily distinguishable from the B spin effects by the τ dependence of the two-pulse echo envelope decay and by the dependence of the phase memory time on the A spin concentration. It is possible to change the A spin concentration in three ways: by changing the radical concentration, by changing the magnetic field position in the EPR line at which the A spins are excited,[11] and by changing ω_1 and hence the width of the A spin excitation. Figure 1 illustrates the latter two ways of varying the A spin concentration.

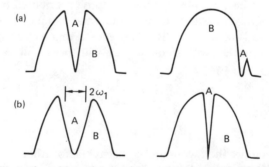

Figure 1   An illustration of how the A spin concentration can be changed (a) by varying the position of the excited region in the EPR line and (b) by varying the width of the excited region. The A and B spin concentrations are proportional to the indicated areas under the EPR line. Here ω_1/γ is the amplitude of the microwave magnetic field.

2 EXPERIMENTS WITH RADICALS IN RIGID MATRICES

In this section we describe the behavior of the two-pulse and three-pulse echo envelope decays, the electron spin-lattice relaxation time, and the echo-height line shape spectra that have been observed in selected systems of randomly oriented radicals in rigid glass matrices. The experimental results that have been chosen show features that can be expected from many radicals and matrices.

2.1 Observed Values of T_M

The values of the phase memory time for several types of radicals in different matrices are listed in Table 1. The measurements were made at 77 K in samples containing low concentrations of radicals (i.e., $\leq 3 \times 10^{-3} M$).[12,13] At the pulse powers and widths used (400 mW and 200 ns), most of the two-pulse echo envelopes exhibited no modulation, and their decays fitted reasonably well to an $\exp(-2\tau/T_M)^2$ relationship. The values of T_M shown in Table 1 were evaluated from the best-fits to this function. An inspection of Table 1 leads to the overall conclusion that for the radicals and matrices listed, the value of the phase memory time is

Table 1   Values of the Phase Memory Time T_M at 77 K.

Radical	Matrix	T_M (μs)
Naphthalene anion	MTHF	3.4
Naphthalene-d_8 anion	MTHF	3.2
1,3,5-triphenylbenzene anion	MTHF	3.2
Triphenylene anion	MTHF	3.2
DPPH	MTHF	3.2
DPPH	Fluorolube (FS-5)	10.0
Perylene cation	Sulfuric acid	10.4
Anthracene cation	Sulfuric acid	10.4
Naphthacene cation	Sulfuric acid	10.4
Thianthrene cation	Sulfuric acid	10.0
Anthracene-d_{10} cation	Sulfuric acid-d_2	200.0
Anthracene cation	Boric acid	14.6
Biphenyl cation	Boric acid	14.6
p-terphenyl cation	Boric acid	14.2
Naphthalene cation	Boric acid	13.2
1,3,5-triphenylbenzene cation	Boric acid	13.8
Coronene cation	Boric acid	10.0
Triphenylene cation	Boric acid	9.8
Thianthrene cation	Boric acid	10.0

independent of the number and type of radical nuclei, but it is dependent on the matrix nuclei. The local field fluctuations that determine the values of the phase memory time arise from the mutual spin flips of the nuclei located in shells with radii greater than r_s. [See Eq. (13).] Thus the phase memory times in both the sulfuric acid and the 2-methyltetrahydrofuran (MTHF) matrices are independent of the nature of the radical. In the boric acid matrix the typical value of the phase memory time is $T_M \approx 14.6 \mu s$. Notable exceptions are the cations of coronene, triphenylene, and thianthrene. What sets these radicals apart are the values of their electron spin–lattice relaxation times. In solution, radicals such as the anions of coronene and triphenylene with orbitally degenerate ground states have greater electron spin-lattice relaxation rates than those with nonorbitally degenerate ground states.[14] Measurements of T_1 for the cations of coronene, triphenylene, and thianthrene in boric acid matrices at 77 K show that they are several orders of magnitude smaller than the value for the anthracene cation.[12]

As was mentioned in Section 1.3, the electron spin-lattice relaxation rates can shorten the phase memory time by causing fluctuations of the dipolar fields produced by the surrounding electron B spins. If we substitute $R = (T_1)^{-1} = 2.85 \times 10^3 \, s^{-1}$ obtained from the value of T_1 for the thianthrene cation in Table 2, and $\Delta\omega_{dip} = 0.49 \times 10^6$ rad/s corresponding to a radical concentration of $10^{-3} M$ into Eq. (4) derived from the sudden-jump model, we obtain $T_M = 37.5 \mu s$. If we assume $(T_{Mobs})^{-1} = (T_{Mtyp})^{-1} + (T_{MR})^{-1}$, where T_{Mobs} is the observed value, T_{Mtyp} is the typical value of T_M in boric acid, and T_{MR} is the contribution to T_M from the electron B spins, and substitute the values $T_{Mtyp} = 14.6 \mu s$ and $T_{MR} = 37.5 \mu s$, we obtain a value $T_{Mobs} = 10.6 \mu s$ that is in good agreement with the measured value of $10.0 \mu s$.

More detailed and informative results concerning the dependence of the phase memory time on the radical concentration are provided by a set of experiments performed in the rigid glass matrices of sulfuric acid and sulfuric acid-d_2 containing the radical cations of anthracene and anthracene-d_{10}, respectively.[13] All measurements were made at 77 K.

The shapes of the two-pulse echo envelope decays differed markedly in the two systems studied. The envelope decay from the radical cations of anthracene in rigid sulfuric acid matrices showed good agreement with the relation $E(2\tau) = \exp(-2\tau/T_M)^n$, where $n \approx 2.5$. All the two-pulse echo envelope decay shapes from the radical cations of anthracene-d_{10} in sulfuric acid-d_2 matrices showed excellent agreement with $n = 1$. No modulation pattern was observed in either system at the pulse powers and widths used (400 mW and 200 ns). Typical examples of the echo envelope shapes are shown in Fig. 2.

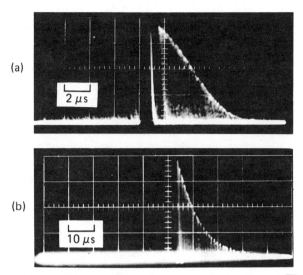

Figure 2 Typical shapes of the two-pulse echo envelope decays observed at 77 K from (a) the radical cations of anthracene in rigid matrices of sulfuric acid and (b) the radical cations of anthracene-d_{10} in rigid matrices of sulfuric acid-d_2.

As illustrated in Fig. 3, the values of the phase memory time for the radical cations of anthracene in sulfuric acid matrices show only a slight dependence on the radical concentration. This implies that the local field fluctuations determining the value of the phase memory time are caused by mutual spin flips of the proton spins on the matrix molecules located in shells with $r > r_s$, where r_s is given by Eq. (13).

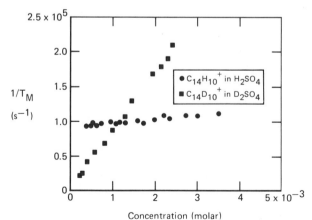

Concentration (molar)

Figure 3 The measured values of the reciprocal of the phase memory time (T_M^{-1}) at 77 K as a function of the radical concentration for the radical cations of anthracene in rigid matrices of sulfuric acid and the radical cations of anthracene-d_{10} in sulfuric acid-d_2.

On the other hand, as is shown in Fig. 3, the phase memory times for the deuterated system depend strongly on the radical concentration. More specifically, the phase memory times in the deuterated system depend on the A spin concentration. In Fig. 4 the values of $1/T_M$ are plotted as a function of the position of the excitation in the EPR line normalized to the line center. These values decrease on going out from the line center, on both the high and the low side, down to a limiting value, $1/T_{ML}$, where T_{ML} is the value of the phase memory time at zero A spin concentration. Also shown in Fig. 4 are the values of $1/T_M$ at different pulse powers with an additional 5 and 10 dB attenuation. At each pulse power value the pulse widths were adjusted to maximize the echo height. The reciprocal values of T_M again decrease on going out from the line center down to a limiting value that is the same for all pulse powers. Measurements were also made for different pulse powers at the line center for the same sample, and the observed values of $1/T_M$ are plotted in Fig. 5 against (microwave pulse power attenuation)$^{1/2}$ normalized to zero attenuation. The latter parameter is proportional to the amplitude of the microwave magnetic field. As the pulse powers decrease, the phase memory times increase up to a limiting value T_{ML} that was the same (16 μs) as the T_{ML} value evaluated from Fig. 4. We therefore conclude that in the deuterated system the phase memory time depends on the position in the

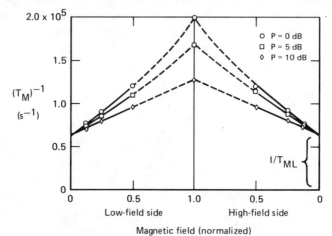

Figure 4 The values of the reciprocal of the phase memory time (T_M^{-1}) measured at different positions in the EPR line of the sample containing the radical cations of anthracene-d_{10} in a sulfuric acid-d_2 matrix at a radical concentration $\sim 2.5 \times 10^{-3}$ M. The abscissa values were measured from the two-pulse echo height with $\tau = 2$ μs and were normalized to the line center. The limiting value of the phase memory time, T_{ML}, was evaluated by extrapolating to magnetic field values which gave zero echo height. The measured values of T_M^{-1} at pulse powers with 5 and 10 dB attenuation are also plotted.

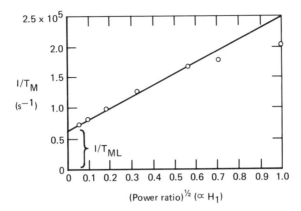

Figure 5   The reciprocal of the phase memory time (T_M^{-1}) as a function of (the microwave pulse power attenuation)$^{1/2}$ normalized to zero attenuation. The abscissa values are proportional to H_1, the amplitude of the microwave magnetic fields. The measurements were made in a sample containing the radical cations of anthracene-d_{10} in sulfuric acid-d_2 at 77 K. This sample was the same as used to obtain the values shown in Fig. 4. A limiting value of the phase memory time, T_{ML}, was evaluated by extrapolating to zero-pulse powers.

EPR line and is proportional to the width of the A spin excitation. This implies that the phase memory time depends on δ, the fraction of spins excited. Furthermore, the results of the measurements of T_M at a fixed power level (0 dB attenuation) shown in Fig. 3 indicate that in the deuterated system, $1/T_M$ is proportional to the total radical concentration $\mathcal{N}_T$. These observations together with the single exponential shape of the two-pulse echo envelope decay are exactly the predictions [Eq. (21)] for instantaneous diffusion brought about by the microwave pulses. The limiting value T_{ML} is the intrinsic phase memory time for the anthracene-d_{10} radical cations in the sulfuric acid-d_2 matrix, and at any given radical concentration it is determined by the B spins. The question arises whether the relevant B spins are the surrounding B electron spins, the nuclear spins on the radical, or the nuclear spins on the surrounding matrix molecules. This question was answered by measuring T_{ML} at different radical concentrations. These T_{ML} values were evaluated by extrapolating to zero pulse powers; the results are plotted in Fig. 6. For the concentration range 2×10^{-4}–$2.5 \times 10^{-3} M$, an approximate linear dependence of $1/T_{ML}$ on the radical concentration was observed.

The concentration-dependent contribution to T_{ML} (15.4 μs at $2 \times 10^{-3} M$) is caused by local field fluctuations arising from mutual spin flips of the electron spins on surrounding radicals, which act as B spins. An estimate of the spin-spin-flip time T_f can be obtained by assuming that Eq. (4), derived from the sudden-jump model, is valid when $R^{-1} = T_f$.

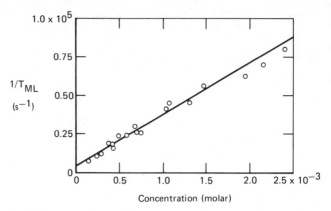

Figure 6   The reciprocals of the limiting values of the phase memory time $(T_{ML})^{-1}$ plotted against the radical concentration for the radical cations of anthracene-d_{10} in rigid matrices of sulfuric acid-d_2 at 77 K. The values of $(T_{ML})^{-1}$ were evaluated by extrapolating to zero pulse powers, as is shown in Fig. 5.

Substituting $T_M = 15.4$ μs and $\Delta\omega_{dip} = 0.99 \times 10^6$ rad/s, the value corresponding to a concentration of $2 \times 10^{-3}M$, we obtain $T_f = 119$ μs. Since the value of T_f and also T_M is larger than the reciprocal of the dipolar width (1.0 μs), there is some detuning of the Larmor frequencies of the B spins by an inhomogeneous broadening, probably the deuterium hyperfine interaction and g-anisotropy. Furthermore, the observed concentration dependence of $(T_{ML})^{-1}$ over the range shown in Fig. 6 implies that T_f is also linearly proportional to the radical concentration over this range.

The contributions to the local field fluctuations from spin flips of the B spins as a result of electron spin-lattice relaxation processes appear to be too small to account for the observed values of T_{ML} plotted in Fig. 6. This is apparent when the values of T_1 shown in Fig. 11 are substituted into Eq. (4). For example, at a concentration of $2 \times 10^{-3}M$, where $R = 1/T_1 = 80$ s^{-1} and $\Delta\omega_{dip} = 0.99 \times 10^6$ rad/s, T_{ML} is estimated to be 158 μs, whereas the observed value is $T_{ML} = 14$ μs. This implies that at $2 \times 10^{-3}M$, the contribution to T_M from the T_1-induced dipolar field fluctuations is only $\approx 10\%$ of the measured value. Moreover, since the electron and nuclear spin-lattice relaxation times in rigid solids are usually temperature dependent, whereas the spin-spin relaxation times are usually temperature independent, measurements of the phase memory times at different temperatures can indicate the importance of these relaxation processes in determining the local field fluctuations. In the case of the radical cations of anthracene in sulfuric acid, measurements at 27 and 77 K showed exactly the same values of the phase memory time. On the other hand, in samples of the deuterated system ($\sim 2 \times 10^{-3} M$), T_{ML} at 27 K increased by approximately 20% of its value at 77 K.

By extrapolating to zero radical concentration in Fig. 6, we obtain a value of $T_{ML} \approx 200 \ \mu s$. Unfortunately, the concentration dependence of T_{ML} was not obtained at 27 K, so it is uncertain whether the spin-lattice relaxation of the deuterium spins on the radical or in the matrix make any contribution to the concentration-independent value of T_{ML} at 77 K. It seems likely, however, that this value is caused primarily by mutual spin flips of the deuterium spins on the surrounding matrix molecules. Thus the measured value of T_{MD}/T_{MP} is ≈ 19.2 where T_{MD} and T_{MP} are the values of T_{ML} in the deuterated and protiated systems, respectively. As discussed in Section 1.4, the difference between this value and that predicted by Eq. (19) (i.e., $T_{MD}/T_{MP} = 13.0$) could be due to a detuning of the matrix deuterium spins by the deuterium quadrupole interaction.

2.2 Stimulated Echo Decays

The three-pulse echo decays in the protiated and deuterated anthracene cation–sulfuric acid systems were obtained by varying T, the time between the second and third pulses, at several fixed values of τ, the time between the first and second pulses. The results shown in Fig. 7 for the protiated system and Fig. 8 for the deuterated both indicate that $D_T(\tau, T)$ can be written as

$$D_T(\tau, T) = \exp \left[-f(\tau) T^{1/2} \right] \tag{23a}$$

$$= \exp \left[-\alpha \tau T^{1/2} \right], \tag{23b}$$

where α is independent of radical concentration in the protiated system and dependent on radical concentration in the deuterated system. Thus

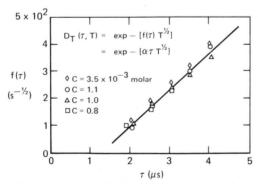

Figure 7 A plot of $f(\tau)$, the function in the three-pulse echo decay factor $D_T(\tau, T)$ versus τ at different concentrations (C) of anthracene radical cations in rigid sulfuric acid matrices at 77 K.

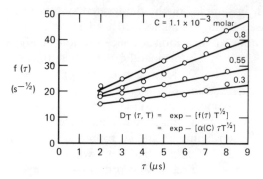

Figure 8 A plot of $f(\tau)$, the functions in the three-pulse echo decay factor $D_T(\tau, T)$ versus τ at different concentrations (C) of anthracene-d_{10} radical cations in rigid sulfuric acid-d_2 matrices at 77 K.

the three-pulse echo decays indicate that over the range of radical concentrations studied $(\sim3.5\times10^{-3}\text{–}3\times10^{-4}M)$ the spectral diffusion rates are independent of radical concentration in the former system but are dependent on radical concentration in the latter. This conclusion is consistent with the two-pulse echo data discussed in Section 2.1.

The dependence of $D_T(\tau, T)$ on τ and $T^{1/2}$ in both systems is the same as that predicted by the Gauss–Markov model [Eq. (11)] using the Lorentzian diffusion kernel given by Eq. (8). It is difficult, however, to justify applying this model to the protiated system since this would imply that the proton magnetic moments located in shells with $r > r_s$ can be considered Gaussian random variables as a result of the mutual spin flips. Similarly, it is difficult to rationalize the application of the Gauss–Markov model and the Lorentzian diffusion kernel of Eq. (8) to the deuterated system where we know from the two-pulse echo data that at radical concentrations $\geqslant10^{-3}M$ the local field fluctuations are determined primarily by mutual spin flips of the surrounding B electron spins.

2.3   Observed Values of T_1

It can be seen from Eq. (4) that the electron spin-lattice relaxation time can be an important factor in determining the value of the phase memory time even when it is over one order of magnitude larger. Consequently, in identifying the mechanism for T_M, it is important to know the value of T_1 in the sample.

The electron spin-lattice relaxation time can be measured by perturbing the electron spin system and monitoring the recovery of the z-magnetization at a later time with the two-pulse echo.[15] Three perturbing methods have been used in free radical studies[13]: (1) a single pulse saturation with pulse widths of approximately $1 \mu s$ to 1 ms, (2) an adiabatic fast passage,[15] and (3) a π-pulse with pulse widths of 40–400 ns depending on the value of the pulse powers available.

Figure 9 shows the typical recoveries of the two-pulse echo height observed at 77 K from radical cations of anthracene in sulfuric acid using these three methods of perturbing the spin system. All three curves show deviations from a single exponential recovery. This is not surprising following the π-pulse and the wide pulse because of spectral diffusion. In the adiabatic fast passage, however, all the spin packets are brought to the same negative spin temperature, and hence the recovery of the z-magnetization should be a single exponential with a time constant $1/T_1$. We explain deviations from this form by a distribution of T_1 values.

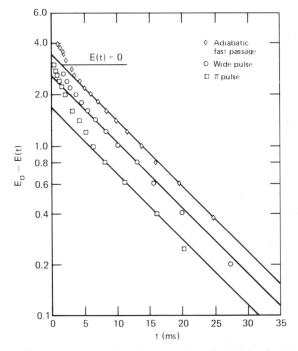

Figure 9 The difference between E_0, the two-pulse echo height ($\tau = 2 \mu s$) when the electron spin system is in thermal equilibrium, and $E(t)$, the two-pulse echo height at a time t following a π pulse, a wide pulse (pulse width 1 ms) and an adiabatic fast passage. The measurements were made in a sample containing the radical cations of anthracene in sulfuric acid at 77 K. The value of $[E_0 - E(t)]$ in which $E(t) = 0$ is indicated. The echoes associated with the points above this line differed from those for the points below the line by a 180° phase change.

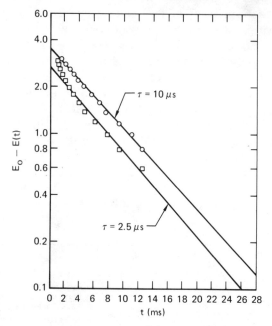

Figure 10   The difference between E_0, the two-pulse echo height when the electron spin system is in thermal equilibrium, and $E(t)$, the two-pulse echo height at time t following an adiabatic fast passage for the values of $\tau = 2.5\ \mu s$ and $\tau = 10 \mu s$. The measurements were made in a sample containing the radical cations of anthracene-d_{10} in sulfuric acid-d_2 at 77 K.

This conclusion was substantiated by the results in the following experiment. The recoveries of the two-pulse echo height following an adiabatic fast passage were measured for different values of τ, the interpulse time. Typical forms of the recoveries observed from the radical cations of anthracene-d_{10} in sulfuric acid-d_2 for $\tau = 2.5\ \mu s$ and $\tau = 10\ \mu s$ are shown in Fig. 10, where it can be seen that the latter approximates a single exponential. Nonuniform distributions of radicals may be present in the samples as a result of the volume changes produced when the sulfuric acid solidifies. Radicals in a high local concentration region have a smaller than average T_1 value because of a concentration-dependent mechanism. The values of T_{ML} for radicals in this high concentration region are also less than the average value because of B electron spin effects. It would appear, therefore, from the results plotted in Fig. 10, that by increasing τ it is possible to select the radicals with the slower T_{ML} values and measure their value of T_1. This implies that the radicals in the high concentration region relax independently of those in the low-concentration region and no cross relaxation occurs. The shape of the recoveries observed from the anthracene radical cations in sulfuric acid did not change markedly when the τ value was increased, partly because the longest τ value that could be used was $\tau \approx 5 \ \mu s$ and partly because the T_M values were determined primarily by the matrix nuclear spins.

The concentration dependence of T_1 was investigated for both the protiated and deuterated anthracene–sulfuric acid systems. The longest time in the recovery was taken to be T_1. The measured values for the anthracene radical cations in sulfuric acid are plotted in Fig. 11 and fit the expression $1/T_1 = 1.2 \ (\pm 0.15) \times 10^7 \ C^2 + 28 \ (\pm 5)$, where T_1 is in seconds and C is the radical concentration in moles per liter. The T_1 values for the deuterated system showed reasonable fits to the concentration-dependent part and excellent agreement with the concentration-independent term.

In identifying the mechanisms determining the electron spin-lattice relaxation at 77 K, we deal first with the concentration-dependent term. The observed dependence of T_1 on the (radical concentration)^2 agrees with that predicted by Waller for a mechanism involving a modulation of the electron magnetic dipole-dipole interaction by the lattice vibrations.[16] For a Raman process in a temperature region where $T \geq \theta_D$, the Debye temperature for the matrix, the theoretical expression for the electron spin-lattice relaxation rate, can be written as[16,17]

$$(T_{1\text{dip}})^{-1} = 1.9 \frac{\beta^4 h^3 \mathcal{N}^{10/3}}{r_0^6 k^3 \rho^2 \theta_D^3} \frac{\exp(\theta_D/T)}{[\exp(\theta_D/T) - 1]^2}, \tag{24}$$

where r_0 is the average distance between radicals, ρ is the density of the matrix, and $\mathcal{N}$ is the number of matrix molecules per cubic centimeter.

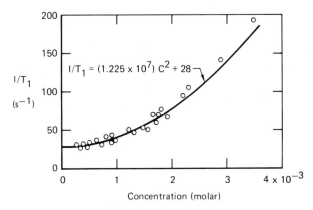

Figure 11 The electron spin-lattice relaxation rates, T_1^{-1}, plotted against the radical concentration for the radical cations of anthracene in rigid matrices of sulfuric acid at 77 K.

Specific heat[18] and relaxation data[19] indicate that the Debye temperature for organic solids can be less than 77 K. In fact, if $\mathcal{N} = 1.1 \times 10^{22}$ cm^{-3}, the value for the sulfuric acid matrix, is substituted into the equation[17]

$$\frac{\theta_D}{v_0} = \frac{h}{k}\left(\frac{3\mathcal{N}}{4\pi}\right)^{1/3},\qquad (25)$$

along with $v_0 = 10^5$ cm·s^{-1}, where v_0 is the acoustic velocity in the matrix, one obtains $\theta_D = 66$ K. If the values $T = 77$ K, $\rho = 1.8$g cm^{-3}, $\theta_D = 66$ K, and $r = 1.2 \times 10^{-6}$ cm, corresponding to a radical concentration of $10^{-3}M$, are substituted into Eq. (24), one obtains a predicted value of $(T_{1dip})^{-1} = 2.1 \times 10^{-8}$ s^{-1}. Even if θ_D is as small as 10 K, the predicted value of $(T_{1dip})^{-1}$ is only 2.8×10^{-4} s^{-1}, whereas the measured rate is $(T_{1dip})^{-1} \approx 10$ s^{-1}.

Furthermore, measurements of free radicals in rigid matrices below 4.2 K show a temperature and concentration dependence of T_1 expected from a single phonon Waller mechanism[20] (i.e., $T_1^{-1} \propto T/r_0^6$). Estimates of $(T_{1dip})^{-1}$ were also too small to account for the observed values.

This discrepancy between the observed and calculated values does not rule out the Waller mechanism completely since the simple Debye continuum that was assumed for the phonon spectrum in deriving Eq. (24) may be an inadequate assumption.[17] However, the following alternative explanation would also account for the observed concentration dependence of T_1. Pairs or triads of radicals may be present in which the distance between the radicals is less than the average intermolecular radical distance in the sample.[21,22] These pairs or triads relax faster than the average rate associated with the monoradicals because of the Waller mechanism. The overall spin-lattice relaxation of the radical system then involves a cross relaxation between the monoradicals and the pairs or triads, but the observed values of T_1^{-1} are still proportional to (radical concentration)^2 [20–22].

A spin-orbit, orbit-lattice mechanism must dominate in determining the concentration-independent contribution to the electron spin-lattice relaxation rates shown in Fig. 11 since the values for the protiated and deuterated systems are the same (28 ± 5 s^{-1}). Furthermore, we can infer that the contribution to the spin-lattice relaxation rates in the anthracene and anthracene-d_{10} cations at 77 K from a modulation of the radical hyperfine interactions by the lattice vibrations is less than 5 s^{-1}.

Alexsandrov[23] has derived expressions for the electron spin-lattice relaxation rate for free radicals in solids when it is determined by the spin-orbit, orbit-lattice mechanism.

Table 2 Values of the Electron Spin-lattice Relaxation Time T_1 at 77 K

Radical	Matrix	T_1 (ms)	$(g_{is}-g_e) \times 10^5$
Anthracene cation	Sulfuric acid	36±6.0	24.8
Anthracene-d$_{10}$ cation	Sulfuric acid	36±6.0	(24.8)[a]
Thianthrene cation	Sulfuric acid	0.3±0.05	578
Anthracene cation	Boric acid	32±6.0	24.8
Naphthalene cation	Boric acid	20±4.0	(43.3)[b]
Biphenyl cation	Boric acid	19±4.0	(45)[b]
Thianthrene cation	Boric acid	0.35±0.04	578
Anthracene anion	MTHF	24±4.0	39.2
Anthracene-d$_{10}$ anion	MTHF	23±4.0	(39.2)[a]

[a] The value for the corresponding protiated radical.
[b] The value for the corresponding radical anion.

If $T \gtrsim \theta_D$, the expression for the Raman process can be reduced to the form

$$(T_{1so})^{-1} = 0.83 \left(\frac{\Delta g}{g}\right)^2 \frac{h^5 \nu^2 \mathcal{N}^{10/3}}{k^3 \rho^2 \theta_D^3} \frac{\exp(\theta_D/T)}{[\exp(\theta_D/T)-1]^2}, \qquad (26)$$

where Δg is a measure of the g-anisotropy and ν is the Larmor frequency. On substituting $(\Delta g/g) = 2.5 \times 10^{-4}$, $\theta_D = 10$ K, $T = 77$ K into Eq. (26), we obtain $(T_{1so})^{-1} = 3.4 \text{ s}^{-1}$. Thus, with reasonable values of the parameters, Eq. (26) gives a value that is large enough to account for the observed value of $(T_{1so})^{-1} = 28 \text{ s}^{-1}$ to within an order of magnitude.

Further evidence of the importance of the spin-orbit coupling in the spin-lattice relaxation rates of organic radicals in solids is shown in Table 2, which lists the T_1 values for dilute concentrations of radicals ($\leqslant 3 \times 10^{-3} M$) in different matrices and the value $(g_{is} - g_e)$, which is a measure of the spin-orbit coupling, where g_{is} is the isotropic g-value and g_e is the free electron g-value.[24,25] The T_1 values were measured at 77 K by monitoring the recovery of the two-pulse echo following saturation by either a saturating pulse (1–10 μs) or an adiabatic fast passage.

2.4 Electron Spin-Echo Spectroscopy

Calculations by Mims[26] of the echo waveforms following two $2\pi/3$ pulses have shown that if $\omega_1 \ll$ linewidth, the echo height is determined by a peak precessing magnetization of $1.7 \, \omega_1 M_0 S(\omega_A)$, where ω_1/γ is the microwave magnetic field amplitude, M_0 is the total static magnetic moment in the line, and $S(\omega_A)$ is the normalized line shape function due to an inhomogeneous broadening evaluated at the applied microwave frequency. The echo height should therefore be a good measure of the

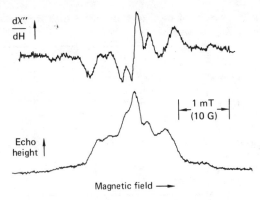

Figure 12   The conventional EPR derivative spectrum and echo height spectrum for the radical cation of naphthalene in a boric acid matrix at 293 K.

z-magnetization at various points in a broad EPR line. Hence it is possible to perform spectroscopy with the two-pulse echo in the following manner. The echo height is maximized by adjusting the pulse widths to provide $2\pi/3$ pulses. The boxcar integrator is set with the gate width less than the echo width and triggered to open at $t \approx 2\tau$. The echo height is then recorded as the two-pulse sequences are repeated at a convenient repetition rate $(\ll 1/T_1)$, and the magnetic field is slowly swept.

The echo height spectrum obtained in this manner[12] from the naphthalene cation in boric acid is shown in Fig. 12 along with the corresponding derivative of the EPR spectrum. It is evident from Fig. 12 that the echo spectrum shown approximates the integral of the EPR derivative and hence the absorptive part of the magnetic susceptibility.

In this method two types of line-broadening effects can lead to severe distortions of the line shapes. First, since the echo height is determined by all the spin packets within a distribution with a half-width at half-height of ω_1, the spectral resolution is limited. No structure with a splitting $\lesssim \omega_1$ will be resolved. Second, large ω_1 values together with large radical concentrations in the sample can lead to sizable contributions to the phase memory time from instantaneous diffusion effects. As described in Section 1.5, this will result in the phase memory time measured in the wings being longer than that measured at the center of the line. Hence the ratio of the echo height in the wings to the echo height at the center of the line will depend on the value of τ used in recording the spectra. The half-width at half-height for the spectra then increases as the value of τ is increased.

If a system contains two free radicals with widely different T_M values T_{M1} and T_{M2}, where $T_{M1} \gg T_{M2}$, it is possible to record selectively an echo height spectrum of only the species with the T_{M1} value by choosing τ such that $T_{M1} \gtrsim \tau \gg T_{M2}$.

We have used this form of time domain spectroscopy to analyze spectra of the spin probe 2,2,6,6-tetramethyl-4-piperidinol-1-oxyl (Tanol) in block copolymers of dimethylsiloxane and bisphenol-A carbonate.[27] In this polymer the EPR spectra above 220 K appear to be the superposition of a broad-line spectrum and three narrow lines; that is, the spin probe can exist in essentially two environments with two different motional correlation times. Alternatively, there is the possibility that the spectra can be assigned to Tanol molecules having only one motional correlation time, and the complicated shape is due to an incomplete averaging of the hyperfine and g-anisotropy brought about by an anisotropic motion. We have obtained the following results to establish that the former explanation is correct.

Figure 13a is the EPR absorption at 293 K showing the broad-line and narrow-line spectra. Figure 13b is the echo height spectrum obtained

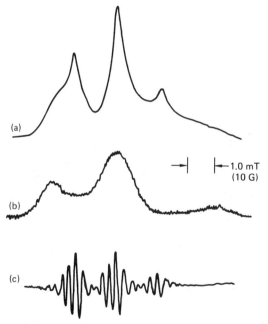

Figure 13 (a) The EPR absorption spectrum resulting from an integration of the EPR derivative spectrum of Tanol in a block copolymer of dimethylsiloxane and bisphenol–A carbonate at 293 K. (b) The echo height spectrum for the same sample with pulse widths = 20 ns, $\tau = 800$ ns. (c) The free induction decay signal from the same sample with the boxcar gate set at 150 ns after the $\pi/2$ pulse.

using a two-pulse sequence, with 20 ns pulse widths and a value of $\tau = 800$ ns. There is no indication of the narrow lines in this spectrum, which indicates that the value of T_M for these lines is less than 800 ns. Furthermore, using a single $\pi/2$ pulse, it was possible to detect the free-induction decay signals from each of the three narrow lines and record these signals with a boxcar integrator set at a time 150 ns after the $\pi/2$ pulse. These are shown in Fig. 13c. The free-induction decay signals from the broad-line spectrum were not recorded in Fig. 13c because the decay time is given by approximately $1/\gamma B_1 = 8$ ns ($B_1 = 0.7$ mT in these experiments). Thus using a two-pulse sequence the echo height will respond only to the broad-line spectrum. On the other hand, using a single pulse the free-induction decay signal responds only to the narrow-line spectrum. These experiments indicate that the broad-line spectrum is inhomogneously broadened, whereas the narrow lines are homogeneously broadened (or nearly so). Thus the broad-line and narrow-line spectra have different characters.

This technique of recording echo height spectra at different values of τ should prove useful in studying (1) overlapping spectra each of which are associated with a different T_M value and are a consequence of a distribution of motional correlation times and (2) small variations in the value of T_M across an inhomogeneous line.

3 RADICALS UNDERGOING MOTION IN NONRIGID MATRICES

Anisotropic interactions that are rendered time dependent by molecular motions can lead to a randomization of the Larmor frequencies and a dephasing of the individual vectors in the precessing magnetization. The two-pulse echo envelope decays should therefore contain important information about molecular motional processes. In this section we describe the behavior of the two-pulse echo envelope decay and electron spin-lattice relaxation times that have been measured in free radical systems where the radical and the matrix molecules are undergoing molecular motions.

3.1 Experimental Results

3.1.1 ANTHRACENE-D_{10} ANION IN METHYLTETRAHYDROFURAN

At low temperatures where the solvent 2-methyltetrahydrofuran forms a rigid or semirigid glass matrix, an amplitude modulation pattern consisting of several maxima and minima was observed in the echo envelope.[28]

This modulation can be attributed to time-dependent local fields at the electron which arise from anisotropic hyperfine couplings with deuterium nuclei on the radical and protons on nearby matrix molecules.[29] As the temperature was increased, the shape of the echo envelope and the phase memory time showed significant changes. Some of the observed echo envelope shapes are shown in Fig. 14, and the temperature dependence

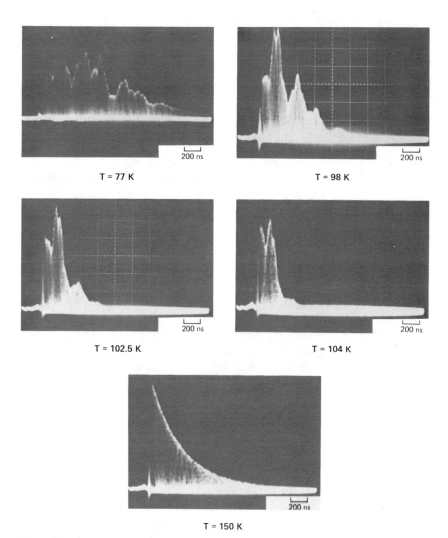

Figure 14   The two-pulse echo envelope decays for the anthracene-d_{10} radical anion in MTHF observed at 77, 98, 102.5, 104, and 150 K.

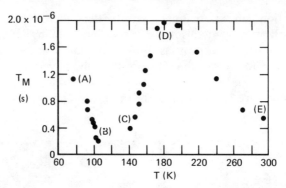

Figure 15   The measured values of the phase memory time T_M for the anthracene-d_{10} radical anion in MTHF as a function of temperature.

of the phase memory time is plotted in Fig. 15. The echo envelope shapes shown in Fig. 14 were obtained with 20 ns, 500 W pulses, whereas the data in Fig. 15 were obtained with 80 ns, 25 W pulses.

In the region AB in Fig. 15 where the amplitude modulation was observed in the echo envelope, the phase memory time decreased with increasing temperature. No echo was observed in the region between B and C. When the echo reappeared at C, the echo envelope had a smooth shape (the example obtained at 150 K shown in Fig. 14 is typical), and the phase memory time increased with increasing temperatures from C to D. At D the phase memory time reached a maximum and decreased with further increases in temperature up to 293 K. This behavior of the phase memory time implies that the width of the homogeneous spin packet defined by the Fourier transform of the echo envelope decay first increases between A and B, decreases between C and D, and then increases again between D and E. On the other hand, the continuous wave (cw) EPR absorption showed only a single line at all temperatures between 77 and 293 K with a width that changed from 0.37 ± 0.03 mT below 95 K to 0.28 ± 0.03 mT above 175 K. It was difficult, therefore, to obtain any information about radical rotational rates from these line width changes.

The temperature dependence of the electron spin-lattice relaxation time for the anthracene-d_{10} anion in MTHF was determined by monitoring the recovery of the two-pulse electron spin echo following saturation by a 1 μs pulse; 20 ns, 500 W pulses were used to generate the echoes. By this method, measurements could be made from 95 to 293 K except over a small region in the vicinity of the T_M minimum where no echo was observed. The measured values of T_1 are plotted in Fig. 16, where it can be seen that the form of the temperature dependence is different from

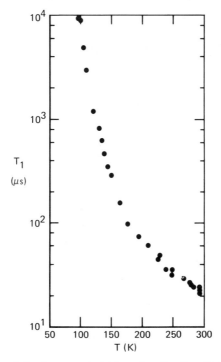

Figure 16 The values of the electron spin-lattice relaxation time T_1 for the anthracene-d_{10} radical anion in MTHF as a function of temperature.

that of T_M shown in Fig. 15. This difference would imply that the values of T_M for temperatures below D in Fig. 15 are not determined by the values of T_1.

At low temperatures ($\sim$95 K), the values of T_1 are determined by the lattice phonons, and a Raman process involving the spin-orbit coupling probably dominates. At high temperatures ($\sim$293 K), since the radicals are undergoing fast rotational and translational motions, the values of T_1 are probably primarily determined by a rotational modulation of either the g-anisotropy or anisotropic hyperfine coupling but may also include a contribution from Heisenberg exchange.[30]

3.1.2 TANOL IN METHYLTETRAHYDROFURAN

The measured values of T_M for 2,2,6,6-tetramethyl-4-piperidinol-1-oxyl (Tanol) in 2-methyltetrahydrofuran (MTHF) are plotted in Fig. 17. The temperature dependence is essentially the same as that shown in Fig. 15 in that there is a minimum and a maximum value of T_M. Pulse powers of

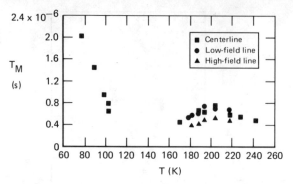

Figure 17 The measured values of the phase memory time T_M for Tanol in MTHF as a function of temperature.

only 25 W and pulse widths of 80 ns were used, and no proton modulation of the echo envelope was observed at any temperature. Measurements on the low-temperature side of the minimum showed that the phase memory times were longer at the extrema positions of the spectrum than at the center. However, the T_M values approached the same value as the pulse powers reduced, indicating that the differences were due mainly to instantaneous diffusion. We therefore conclude that measurements made at different parts of the EPR line showed no significant differences in the intrinsic T_M value.

Between 170 and 240 K, three partially resolved lines were observed in the EPR spectrum arising from the isotropic nitrogen hyperfine interaction. As is shown in Fig. 17, the T_M value measured at the high-field line was shorter than those measured at either of the other two lines, but all showed a maximum at $T = 205$ K.

3.1.3 PADS IN A GLYCEROL-WATER MATRIX

The radical peroxylamine disulfonate (PADS) was investigated in a mixed matrix of glycerol and water, 85 : 15 wt %. This system was chosen since extensive studies already had been made, and the correlation time had been deduced from the EPR line shape changes over a wide temperature range.[31] The temperature dependence of the T_M values measured at the center of the EPR spectrum is shown in Fig. 18. The maximum pulse power used was 25 W, and only a poorly resolved modulation pattern was observed in the initial stages of the echo envelope at temperatures below 210 K. When the echo reappeared at temperatures above 248 K, its amplitude was comparable to that of the free-induction decay following the second microwave pulse, and the determination of T_M in the presence

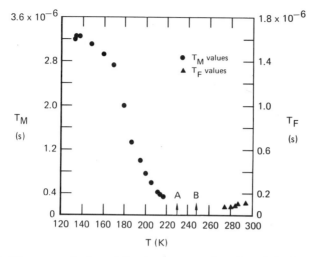

Figure 18 The measured values of the phase memory time T_M and the time characterizing the free-induction decay T_F as a function of temperature. The measurements were made in a sample of PADS in a glycerol-water matrix (85:15 wt%). The arrow at A indicates the temperature below which the two-pulse echo could be detected; the arrow at B indicates the temperature above which the free-induction decay could be detected.

of this large free-induction decay signal was difficult. We therefore include in Fig. 18 the values of the time T_F which were evaluated from a fit of the free-induction decay shape for the center line to the form $\exp(-t/T_F)$. The arrows at A and B in Fig. 18 correspond to the temperatures below which the echo could be observed and above which the free-induction decay could be detected, respectively. Both sets of data in Fig. 18 suggest that a minimum value of T_M lies between A and B. According to the data of Goldman et al.[31] obtained from line shape studies, the rotational correlation time for the PADS radical has the value $\tau_r = 3.5 \times 10^{-8}$ s at A (229 K) and $\tau_r = 6 \times 10^{-9}$ s at B (248 K).

3.2 Discussion

At low temperatures where the matrix is rigid, the phase memory times are often determined primarily by the local field fluctuations resulting from nuclear spin flips in the surrounding matrix. As described in Section 1.4, these spin flips are usually due to mutual spin flips involving nuclei located in shells that are at distances greater than r_s from the unpaired electron, where r_s is given by Eq. (13). As the temperature is increased and the matrix molecules begin to tumble, the mean spin-spin flip time for the matrix nuclei increases. One could therefore expect the phase memory time to increase with increasing temperature if it were determined solely by this mechanism. Other effects, however, can determine the phase memory time as the temperature increases. Anisotropic interactions that are static at low temperatures become time dependent as the radical and matrix molecules tumble. On the low-temperature side of the T_M minimum (e.g., the region AB in Fig. 15), the inequality $\tau \ll \tau_c$ holds where τ_c is the correlation time for the motion. The time-dependent anisotropic interactions then produce random variations in the Larmor frequencies and hence additional dephasings of the individual vectors in the precessing magnetization. This would account for the decrease of T_M with increasing temperature. This mechanism can involve electron-nuclear dipolar interactions with matrix protons in shells with $r \lesssim r_s$, anisotropic hyperfine interactions with radical nuclei, or an anisotropic g-tensor. All of these time-dependent interactions probably play a role in determining the value of T_M in the slow-motion region and each may dominate, depending on the temperature and specific system studied. For example, because of the smaller values of the g-anisotropy[24] and hyperfine coupling constants,[32] it might be expected that these interactions would be less important in the anthracene-d_{10} radical anions than in the nitroxides.[33] However, because of the lack of decisive experimental data, we are unable positively to identify the most important term in the three systems discussed here.

Echo envelope modulation has been observed from several paramagnetic metal ions[34] and trapped electrons[35] in glassy matrices. Computer simulations of the envelope modulation showed reasonable agreement with the experimental observations when the modulation was attributed to shells of surrounding matrix nuclei. However, no successful analysis has yet been made of the echo envelope modulation patterns involving radical nuclei where the radicals are randomly oriented in a rigid matrix. Furthermore, no attempts have been made to calculate the changes produced in the echo modulation pattern when the radicals are' undergoing slow tumbling. It should be noted, however, that the depth of the modulation depends on the magnitude of the anisotropic hyperfine interaction, and the period depends on the nuclear Larmor frequencies.[29] It is therefore clear from the periods in the modulation of the echo envelopes in Fig. 14 that deuterium nuclei on the radical and protons on the matrix nuclei both contribute to the modulation patterns. Consequently, rotational diffusion of the radical is the motion that will exclusively influence the deuterium modulation pattern.

At the low end of the temperature range between A and B in Fig. 15, no changes in the period or depth of modulation could be detected in either the deuterium or proton modulation pattern after the echo envelopes were corrected for phase memory decay, whereas reductions in the phase memory time were observed. From this observation we conclude the following for the system of anthracene-d_{10} anions in MTHF in the slow-tumbling region. If the phase memory time is determined by a rotational modulation of the g-tensor or the anisotropic deuterium hyperfine interaction, it is more sensitive to the radical rotations than is the echo envelope modulation pattern. On the other hand, the phase memory time can be determined by a rotational modulation of the electron-nuclear dipolar coupling with the matrix nuclei. The absence of changes in the modulation pattern at temperatures where the phase memory time has markedly decreased is then not surprising if the respective mechanisms involve different matrix nuclei. This would imply that the proton modulation pattern is due mainly to protons located in the first few shells of matrix molecules, whereas the phase memory time is primarily determined by matrix protons located between these shells and $r \sim r_s$. [See Eq. (13).] This does not rule out the possibility that a modulation of the g-tensor and/or anisotropic hyperfine interactions with radical nuclei can still determine the value of T_M at higher temperatures.

The shapes of the echo envelopes observed in the region CD in Fig. 15 indicate that the radical rotational rates are fast enough to average out the anisotropic hyperfine interactions responsible for the echo envelope modulation. This is consistent with our estimates for the value of the radical correlation time ($\tau_r < 6 \times 10^{-9}$ s) for the PADS radical at temperatures above the T_M minimum in Fig. 18. The radical correlation times are also fast enough in the region CD to average the local field fluctuations that determine T_M and result in an increase of T_M with increasing temperature.

For temperatures above the T_M maximum in Figs. 15 and 17, another mechanism besides that involving rotational motions is responsible for the time-dependent local fields determining T_M. This is the Heisenberg exchange, which also produces a concentration dependence of T_M in the region DE in Fig. 15. The spin exchange is the result of radical-radical collisions arising from an increase of the radical translational diffusion rates with increasing temperature.

In the calculations of Stillman and Schwartz,[36] it was shown that for a nitroxide radical in the fast-tumbling region (i.e., $\tau \gg \tau_r$) the phase memory time is given by $1/T_M = 1/T_2 + \omega_{ex}$, where $1/T_2$ is the homogeneous linewidth in the absence of Heisenberg exchange and ω_{ex} is the exchange frequency. This dependence is indicated in Fig. 17 where the T_M value measured at the high-field nitrogen hyperfine line differs from that measured at the center-field line because the homogeneous linewidths depend on the nuclear quantum number.[37]

With improvements in the sensitivity and response times ($<10^{-7}$ s) of pulsed EPR spectrometers, electron spin-echo techniques should become increasingly important for relaxation studies of free radicals both in molecular solids and in solution when the cw absorption lines are inhomogeneously broadened.

ACKNOWLEDGMENT

This research was conducted under the McDonnell Douglas Independent Research and Development Program.

REFERENCES

1. W. B. Mims, *Electron Paramagnetic Resonance*, Plenum, New York, 1972, p. 263.
2. B. Herzog and E. L. Hahn, *Phys. Rev.*, **103**, 148 (1956).
3. J. R. Klauder and P. W. Anderson, *Phys. Rev.*, **125**, 912 (1962).
4. W. B. Mims, *Phys. Rev.*, **168**, 370 (1968).
5. P. Hu and S. R. Hartmann, *Phys. Rev.*, **9**, 1 (1974).
6. A. Abragam, *Principles of Nuclear Magnetism*, Oxford University Press, London, 1961, p. 126.
7. Unless otherwise stated in the text, the definition of T_M used in this chapter is the time measured from the first microwave pulse for a $1/e$ attenuation of the echo. In some of the work cited, the definition of T_M was one-half of this value.
8. P. Hu and S. R. Hartmann, *J. Mag. Resonance*, **15**, 226 (1974).
9. The factor $\frac{1}{3}$ is included to account for the fact that we require only the mutual spin-flip contribution to the line broadening, that is, the $I_{i+}I_{j-}$ term rather than the $I_{iz}I_{jz}$ term.
10. A. Abragam, *Principles of Nuclear Magnetism*, Oxford University Press, London, 1961, p. 112.
11. One can investigate the A spin contribution to the measured value of T_M in this way providing the intrinsic value of T_M does not depend on the position of the excitation in the line.
12. I. M. Brown, *J. Chem. Phys.*, **55**, 2377 (1971).
13. I. M. Brown, *J. Chem. Phys.*, **58**, 4242 (1973).
14. M. Townsend and S. I. Weissman, *J. Chem. Phys.*, **32**, 309 (1960).
15. A. Kiel and W. B. Mims, *Phys. Rev.*, **161**, 386 (1967).
16. I. Waller, *Z. Phys.*, **79**, 370 (1932).
17. G. Pake, *Paramagnetic Resonance*, Benjamin, New York, 1962, p. 120.
18. J. P. Goldsborough, M. Mandel, and G. E. Pake, *Conference on Low Temperature Physics*, Toronto, University of Toronto Press, Toronto, 1961, p. 702.
19. I. M. Brown, D. J. Sloop, and D. P. Ames, *Chem. Phys. Lett.*, **1**, 167 (1967).
20. I. M. Brown and D. J. Sloop, *Chem. Phys. Lett.*, **1**, 579 (1968).

21. J. C. Gill and R. J. Elliot, *Advances in Quantum Electronics*, Columbia University Press, New York, 1961, p. 399.

22. E. A. Harris and K. S. Yngvesson, *Phys. Lett.*, **21**, 252 (1966).

23. I. V. Aleksandrov, *Teor. Eksp.* Khim. **1**, 221 (1966).

24. B. G. Segal, M. Kaplan, and G. K. Fraenkel, *J. Chem. Phys.*, **43**, 4191 (1965).

25. G. Vincow, in E. T. Kaiser and L. Kevan, Ed., *Radical Ions*, Interscience, New York, 1968, p. 151.

26. W. B. Mims, *Rev. Sci. Instrum.*, **36**, 1472 (1965).

27. I. M. Brown, to be published.

28. I. M. Brown, *J. Chem. Phys.*, **60**, 4930 (1974).

29. L. G. Rowan, E. L. Hahn, and W. B. Mims, *Phys. Rev.*, **A137**, 61 (1965).

30. M. P. Eastman, R. G. Kooser, M. R. Das, and J. H. Freed, *J. Chem. Phys.*, **51**, 2690 (1969).

31. S. A. Goldman, G. V. Bruno, C. F. Polnaszek, and J. H. Freed, *J. Chem. Phys.*, **56**, 716 (1972).

32. J. Bolton and G. Fraenkel, *J. Chem. Phys.*, **40**, 3307 (1964).

33. L. J. Berliner, *Spin Labeling: Theory and Applications*, Academic, New York, 1976, p. 565.

34. W. B. Mims, J. Peisach, and J. L. Davis, *J. Chem. Phys.*, **66**, 5536 (1977).

35. L. Kevan, M. K. Bowman, P. A. Narayana, R. K. Boekman, Y. F. Yudanov, and Y. D. Tsvetkov, *J. Chem. Phys.*, **63**, 409 (1975).

36. A. E. Stillman and R. N. Schwartz, *Mol. Phys.*, **32**, 1045 (1976).

37. D. Kivelson, in L. T. Muus and P. W. Atkins, Ed., *Electron Spin Relaxation in Liquids*, Plenum, New York, 1972, p. 213.

7 ELECTRON SPIN-ECHO STUDIES OF SPIN-SPIN INTERACTIONS IN SOLIDS

K. M. Salikhov and Yu. D. Tsvetkov

Institute of Chemical Kinetics and Combustion
Siberian Branch—USSR Academy of Sciences
Novosibirsk, USSR

1   INTRODUCTION

Studies of spin-spin interactions in solids are of great importance from
various points of view. They are necessary for the general development of
the physics of paramagnetism and, in particular, of electron spin resonance (ESR) spectroscopy. Indeed, spin-spin interactions determine such
phenomena as spin diffusion, cross relaxation, and intermediate quasi-
stationary states of spin systems that are important for the dynamic
behavior of paramagnetic systems. In addition to providing detailed
information concerning the interaction between the individual spins,
spin-spin interactions are widely used to solve various physical and
chemical problems associated with structural chemistry, photochemistry,
radiation chemistry, molecular physics, and the physics of solids.

In this chapter we consider problems of electron paramagnetic relaxation associated with two types of spin-spin interactions in magnetically
dilute solids: dipolar interactions, designated as S-S interactions, between
the unpaired electrons of the paramagnetic centers and superhyperfine
interactions, referred to as S-I interactions, between the unpaired electrons of the paramagnetic centers and the magnetic nuclei of the matrix
molecules. In electron spin-echo (ESE) experiments the S-I interactions
with nearest molecules are manifested in modulation of the ESE signal
decay. Analysis of these modulation effects yields unique information
about the structure of the environment surrounding the paramagnetic
centers. A detailed discussion of this problem is given by L. Kevan in
Chapter 8 of this book. In this chapter we consider in detail the contribution made by magnetic nuclei of remote matrix molecules to the electron
spin phase relaxation of paramagnetic centers.

In stationary ESR spectroscopy it is difficult to obtain information on
the contribution of S-S and S-I interactions to spin relaxation because

the spectral lines are inhomogeneously broadened—for example, by g-tensor anisotropy and intramolecular hyperfine interactions with magnetic nuclei of the paramagnetic particle. The pulse ESE method allows one to study these considerably weaker contributions to the spin dynamics and to remove the masking effects of the inhomogeneous broadening. The principles of this method are described in the literature and in other chapters in the present volume.

Modern microwave and electronic technology make it possible to observe ESE signals for spin systems that are characterized by paramagnetic relaxation times of the order of 10^{-7} s. Therefore paramagnetic ions cooled to liquid helium temperatures or free radicals stabilized in solid matrices (i.e., spin systems that interact comparatively weakly with the lattice) are the most typical objects for ESE studies. In such systems S-S and S-I interactions are of primary importance for the spin dynamics. For example, these make the main contribution to the process of irreversible dephasing of the Larmor precession of electron spins (phase relaxation process).[1-14] These interactions determine the ESE signal decay and the analysis of the signal decay kinetics gives information about the spin-spin processes and the magnitude of S-S and S-I interactions.

The present chapter reviews the main results of the theoretical and experimental ESE studies of the influence of S-S and S-I interactions on the dynamics of electron spins of magnetically dilute solids and the applications of the ESE method for investigating the spatial distributions of paramagnetic centers in various systems.

2 PHASE RELAXATION MECHANISMS DUE TO SPIN-SPIN INTERACTIONS

The amplitude of the microwave magnetic field in the cavity of an ESE spectrometer of moderate power corresponds typically to a spectral width of $\sim$5 G. Hence in systems that exhibit strong inhomogeneously broadened lines, as is most often encountered in ESE studies, not all spins can be excited. According to the basic papers,[3-5] those spins excited by microwave pulses and taking part in the formation of the ESE signal will be called A spins and the other spins of the system will be designated B spins.

The dipolar interactions of the A spins with each other and with the B spins affect differently the ESE signal decay of the A spins. To make it clear, we consider a two-pulse ESE (i.e., the spin system response to two microwave pulses that turn the A spins through 90 and 180°).[6] The microwave pulses leave the local magnetic field produced by the B spins unchanged at the location of some A spin. The local dipolar field from the B spins is, in reality, a kind of external field for the A spins. If during the time of the echo signal observation the B spins do not change their orientation, the resonant frequency shift of the A spins, resulting from the B spins, becomes one of the additional sources of inhomogeneous broadening and thus like any inhomogeneous broadening, does not contribute to the echo signal decay. If the B spins have enough time to change their spin orientations, the frequency shift of the A spins does not remain constant with respect to time and fluctuations or migrations of the A spin's frequency occur. These migrations of spin frequency are called *spectral diffusion*. Spectral diffusion results in an irreversible dephasing of the spin precession. Changes in spin orientation occur because of either spin-lattice interactions or mutual spin flip-flops. If spectral diffusion is induced by random modulation of the S-S spin interactions by spin-lattice relaxation, then these systems are usually called T_1-type samples. If spin flip-flops are more effective, these systems are called T_2-type samples.[5] At sufficiently high temperatures, where T_1 is short, or at comparatively low concentrations of paramagnetic centers, where flip-flops are infrequent, the T_1-type mechanism is realized. On the other hand, spectral diffusion of the T_2-type is observed at lower temperatures and higher concentrations of paramagnetic particles.

The S-S interaction between A spins is reflected in the echo signal decay in another way. A $180°$ pulse changes the sign of the local magnetic field created by the A spins. The microwave pulse thus causes an "instantaneous" resonant frequency shift of the spins equal to the duration of the pulse. As a result, the S-S interaction of the A spins leads to spin dephasing by a mechanism called *instantaneous diffusion*.[3-5] However, A spin flips, either due to T_1 or flip-flops, modulate in a random manner the resonant frequency shift resulting from the interaction of the A spins with each other. Therefore the mechanism of instantaneous diffusion works only when the A spin flips are sufficiently infrequent and cannot be observed within the time of the ESE signal decay observation $(\tau \sim 10^{-7}$–10^{-5} s). If during this time the A spins change their spin orientation several times, then one additional flip caused by the $180°$ pulse is of no importance against the background of these random flips. As a result, if the A spin-flip rates are comparatively fast, the interaction between the A spins will contribute to the echo signal decay not by the instantaneous but by the spectral diffusion mechanism, acting like B spins.

The preceding mechanisms of spin dephasing are associated with the adiabatic part of the S-S interaction. It is this part that results in spin-resonant frequency shifts. There is also a nonadiabatic contribution from the S-S interaction that results in mutual spin flips. Only the adiabatic part of the S-S interaction influences the ESE signal decay by the instantaneous diffusion mechanism. The spectral diffusion in T_1-type samples is also associated with this part of the S-S interaction. But in T_2-type samples both frequency shifts and mutual spin flips are important. The latter induce random migration of the frequency within the limits of the dipolar width of ESR lines. It is, however, not the only effect that mutual spin flips have on the kinetics of the ESE signal decay. A mutual flip of two or more spins results in spin-excitation transfer from one particle to another. The spin-excitation transfer from the A-type spins, which are the ones excited by the microwave pulses and which participate in the formation of the echo signal, to the unexcited B-type spins, decreases the number of spins participating in the echo signal formation, and thus results in some additional decay of this signal. In this case the excitation transfer is a kind of additional spin-lattice relaxation mechanism in which the B spins act as the *lattice*.

Note that in the literature various terms are used when discussing the effects of spin-excitation transfer. In experiments on double resonance or on the recovery of the stationary ESR signal after pulse saturation of some part of the spectrum, it is important that the excitation transfer represents the energy transfer from spins with a particular resonant frequency to those with another frequency, that is, the excitation is transferred from one part of the spectrum to another. This energy transfer between spins, which is manifested as migration of spin excitation over the ESR spectrum, is also often called spectral diffusion. However, such energy migration differs from the spectral diffusion mechanism considered earlier that results from random adiabatic changes of the resonant frequency of *each* spin. We shall use the term *spectral diffusion* only to denote the frequency migration of each spin. When energy is transferred between spins with different Zeeman frequencies, it is important to find out the source of compensation of the difference between the Zeeman spin energies. When this aspect of the problem of spin-excitation transfer is considered, the term *cross relaxation* is usually used if the difference between the Zeeman spin energies is compensated for by the reservoir of dipolar interactions between paramagnetic centers.[15] Another aspect of the energy transfer process due to mutual spin flips is that the energy transfer leads to a spatial diffusion of spin excitation or spin diffusion.[16] It is this spin diffusion that results in the random changes of the local dipolar field at the location of a given spin.

It has been shown in the earliest ESE investigations[17] that an essential contribution to the ESE signal decay is made by the S-I interaction of unpaired electrons with magnetic matrix nuclei. The magnetic nuclei behave as B-type spins. Therefore they cause dephasing of the electron spins by the spectral diffusion induced by the random modulation of the local magnetic field resulting from the nuclear spin flips at the location of the paramagnetic centers. These flips are caused by nuclear spin diffusion.

3 THE KINETICS OF ELECTRON SPIN-ECHO DECAY CAUSED BY SPIN-SPIN INTERACTIONS

3.1 Spectral Diffusion in T_1-Type Samples

Let us first discuss the contribution made by the B spins to the ESE signal decay due to spectral diffusion that results from the random modulation of the S-S interaction by spin-lattice relaxation. Much attention has been paid to this problem in recent theoretical studies of the ESE signal decay.

Experimental and theoretical investigations of the contributions made by the mechansims of instantaneous and spectral diffusions to the ESE signal decay kinetics were made by Mims et al.[4] and Klauder and Anderson.[5] Further development of the theory and its comparison with experiments on model systems have been recently provided by several groups.[7-14] The resonant frequency shift due to the S-S interaction of a given A spin with B spins is

$$\Delta\omega_p(t) = \gamma_A \gamma_B \hbar \sum_k^{(B)} r_{pk}^{-3}(1 - 3\cos^2\theta_{pk})m_k(t), \tag{1}$$

where m_k is the z-projection of the kth B spin, r_{pk} is the distance between the A and the kth B spin, θ_{pk} is the angle between $\mathbf{r}_{pk}$ and the direction of the external magnetic field (Z-axis), γ_A and γ_B are the gyromagnetic ratios of the A and B spins, and the summation in Eq. (1) is carried out over all B spins. In all theories of spectral diffusion in T_1-type samples it is assumed that the projections of the different spins vary independently of each other.

The contribution made by the B spins to the decay of the two-pulse and stimulated ESE signals is described by[5]

$$V(t) = V_0 \left\langle \left\langle \exp\left(i \int_0^t s(t)\,\Delta\omega_p(t)\,dt\right)\right\rangle_t \right\rangle_p, \tag{2}$$

where $s(t) = 1$ over the time interval $(0, \tau)$ and $s(t) = -1$ over the interval $(\tau, 2\tau)$ for the two-pulse ESE signal, and $s(t) = 1, 0, -1$ over $(0, \tau)$, $(\tau, \tau + T)$, $(\tau + T, 2\tau + T)$, respectively, for the stimulated ESE signal. Here $\langle \cdots \rangle_t$ means averaging over all realizations of the random process $m_k(t)$, and $\langle \cdots \rangle_p$ represents the average over all possible spatial distributions of B spins surrounding the A spins.

To carry out the averaging in Eq. (2) it is necessary to know the random process of changing the spin projections and the statistics of the spin distribution over the lattice sites of a solid. To simplify this complex problem it is assumed that the paramagnetic particles are distributed over the lattice sites in a random manner. In this case there are no preferable locations of the spins and they are assumed to be uniformly distributed over the lattice. It is then possible to average over the B spin distribution in the lattice for an arbitrarily chosen random process $m_k(t)$. For this purpose we introduce the random variable

$$X(t) = \int_0^t s(t)m(t)\,dt, \tag{3}$$

where the subscript k has been dropped from $m_k(t)$. Using Eq. (1) through (3) and averaging over the various possible spatial distributions of the B spins by the well-known Markov method,[18] one obtains[9]

$$V(t) = V_0 \exp\left[-\frac{8\pi^2}{9\sqrt{3}}\gamma_A\gamma_B \hbar C_B \langle |X| \rangle_t\right]. \tag{4}$$

Here $\langle |X| \rangle_t$ is the mean value of $|X|$, and C_B is the concentration of B spins. Equation (4) has an interesting physical interpretation: the exponent represents the mean value of the modulus of the phase that the A spins will have at the moment of the echo signal observation. This phase is accumulated in the local magnetic field of an effective B spin that is located approximately $C_B^{1/3}$ from the spin under observation.

Using Eq. (4) the effects of spectral diffusion on the kinetics of the ESE signal decay can be elucidated. First, consider the case when the time of the ESE signal observation t exceeds the characteristic time of B spin reorientation τ_c. According to the central limit theorem in the probability density theory of the sum distribution, at $t > \tau_c$ Eq. (3) tends asymptotically to its normal form,

$$\phi_t(X)\,dX = \frac{dX}{\sqrt{2\pi\langle X^2\rangle}}\exp\left(-\frac{X^2}{2\langle X^2\rangle}\right), \tag{5}$$

where

$$\langle X^2\rangle = \int_0^t\int_0^t dt_1\,dt_2 s(t_1)s(t_2)\langle m(0)m(|t_1-t_2|)\rangle_t. \tag{6}$$

Substituting Eq. (5) into Eq. (4), one obtains

$$V(t) = V_0 \exp\left(-\frac{8\pi\sqrt{\pi}}{9\sqrt{3}}\gamma_A\gamma_B \hbar C_B \sqrt{2\langle X^2\rangle}\right). \tag{7}$$

If the correlation function for $m(t)$ has the form $\langle m(0)m(t)\rangle_t = \frac{1}{3}S_B(S_B+1)\exp(-t/\tau_c)$, then the amplitudes of the two-pulse and the stimulated ESE signals are

$$V(2\tau) = V_0 \exp[-K'\sqrt{B(\tau, 0)}], \tag{8a}$$

$$V(2\tau+T) = V_0 \exp[-K'\sqrt{B(\tau, T)}], \tag{8b}$$

where

$$K' = \frac{16\pi\sqrt{\pi}}{27}\sqrt{S_B(S_B+1)}\,\gamma_A\gamma_B\hbar C_B, \tag{8c}$$

and

$$B(\tau, t) = 2\tau\tau_c + 2\tau_c^2\left[\exp\left(-\frac{\tau}{\tau_c}\right)-1\right]$$
$$-\tau_c^2\exp\left(-\frac{t}{\tau_c}\right)\left[1-\exp\left(-\frac{\tau}{\tau_c}\right)\right]^2. \tag{8d}$$

When τ, $t > \tau_c$, $B(\tau, t)\approx 2\tau\tau_c$. Thus in the limit when $\tau, t > \tau_c$ the ESE signal must behave as follows:

$$V(2\tau) \approx V(2\tau+T) = V_0 \exp(-K'\sqrt{2\tau_c\tau}), \tag{9}$$

for an arbitrary process $m(t)$.

In the other limiting case, when $\tau, T < \tau_c$, the kinetics of the echo signal decay is strongly dependent on the nature of the random $m(t)$ process. It is thus possible to obtain only an estimate of the upper decay rate via the correlation function $\langle m(0)m(t)\rangle_t$. For this purpose one can use the inequality

$$\langle |X|\rangle_t \le (\langle X^2\rangle)^{1/2}. \tag{10}$$

For example, using Eq. (4), (6), and (10) it is possible to show[9] that for any process $m(t)$ with a correlation function $\langle m(0)m(t)\rangle_t \approx \exp(-t/\tau_c)$, the two-pulse echo signal decay at short times will be described by

$$V(2\tau) = V_0 \exp\left[-2b\tau\left(\frac{\tau}{\tau_c}\right)^{\mathcal{H}}\right], \tag{11}$$

where $\mathcal{H} \ge \frac{1}{2}$, $b \approx \Delta\omega_{1/2}$; $\Delta\omega_{1/2}$ is the ESR linewidth resulting from $S\text{-}S$

interactions with B spins:

$$\Delta\omega_{1/2} = \frac{8\pi^2}{9\sqrt{3}} \gamma_A \gamma_B \hbar C_B \langle |m| \rangle_t ; \tag{12a}$$

$$\langle |m| \rangle_t = \frac{2S_B + 1}{4}, \qquad \text{if } S_B \text{ is a half-integer;} \tag{12b}$$

and

$$\langle |m| \rangle_t = \frac{S_B(S_B + 1)}{2S_B + 1}, \qquad \text{if } S_B \text{ is an integer.} \tag{12c}$$

The results given in Eq. (8), (9), (11), and (12) were obtained previously for the case $S_B = \frac{1}{2}$.[9] Here we have generalized these expressions for arbitrary spin S_B.

An important result is, in fact, that among random $m(t)$ processes with the correlation function given approximately as $\exp(-t/\tau_c)$, it is impossible to find a model that results in a two-pulse ESE signal decay described by the exponential law

$$V(2\tau) = V_0 \exp(-2b\tau), \tag{13}$$

for comparatively short times, $\tau < \tau_c$. If experimentally the spectral diffusion follows Eq. (13), then one should employ a more sophisticated model to interpret this observation. As it was shown by Milov et al.,[9] the kinetics of the echo signal decay for small τ obeys the exponential form Eq. (13) if the correlation function for $m(t)$ is described by the expression $\exp(-(t/\tau_c)^n)$, where $n < 1$. ESE signal decay kinetics governed by Eq. (13) can also be expected for $m(t)$ processes with an exponential correlation function $\langle m_k(0)m_k(t) \rangle \approx \exp(-t/\tau_{ck})$, if there is a proper distribution of correlation times for the B spins, τ_{ck}. A more detailed discussion of this problem is presented in Section 3.2.

The preceding results were obtained by using general assumptions for the random process $m(t)$. It is quite natural that for this situation an explicit law for the echo signal decay can be obtained only in the limit of either large or small observation times t.

For a complete analysis of the echo signal decay kinetics it is necessary to describe in detail the process $m(t)$. In the literature two models have been considered. Mims[7] discusses the situation where $m(t)$ is described by a normal (Gaussian) process. In this case any linear combination of $m(t)$ is also a normal process. Therefore the distribution of $X(t)$ is given by Eq. (5) and the ESE signal decay kinetics at any τ and T are described by Eq. (7) and (8), assuming that $\langle m(0)m(t) \rangle \approx \exp(-t/\tau_c)$. As expected, it follows that in the limit when $\tau, T > \tau_c$ the signals are again described by Eq.

(9). For $m(t)$ described by a normal process and in the time region $\tau < \tau_c$, one obtains

$$\langle X^2(2\tau) \rangle \approx \frac{\frac{4}{9} S_B (S_B + 1) \tau^3}{\tau_c}, \tag{14a}$$

and

$$\langle X^2(2\tau + T) \rangle \approx \frac{2}{3} S_B (S_B + 1) \tau^2 \left[1 - \exp\left(-\frac{T}{\tau_c} \right) \right]. \tag{14b}$$

On substituting Eq. (14a, b) into (7) one finds that the echo signal decay at $\tau < \tau_c$ is given by

$$V(2\tau) \approx V_0 \exp\left(-K' \sqrt{\frac{2}{3} \frac{\tau^3}{\tau_c}} \right), \tag{15a}$$

and

$$V(2\tau + T) \approx V_0 \exp\left[-K'\tau \left[1 - \exp\left(-\frac{T}{\tau_c} \right) \right]^{1/2} \right]. \tag{15b}$$

For the normal process $m(t)$ various values of the spin projection must be randomly distributed according to the Gaussian law. The higher the spin S_B, the more valid is such a model. However, for low S_B the model of random jumps between various discrete values of m is more valid. For example, if $S = \frac{1}{2}$, the m projection can be $\pm\frac{1}{2}$. Up to now calculations have been carried out only for spin systems with $S = \frac{1}{2}$. Random spin jumps between states with different m values can be described by a Poisson random process and such a model has been considered in detail by several workers.[5,8,9,12] Assuming that the spin projections vary with the mean frequency τ_c^{-1}, then the probability that n spin jumps occur during a time t is

$$P_n(t) = \frac{1}{n!} \left(\frac{t}{\tau_c} \right)^n \exp\left(-\frac{t}{\tau_c} \right). \tag{16}$$

For $m(t)$ described by a Poisson process[9]

$$\langle |X| \rangle = \sum_{n=0}^{\infty} K_n \frac{\tau^{n+1}}{\tau_c^n} \exp\left(-\frac{2\tau}{\tau_c} \right), \tag{17}$$

where $K = 0$, $K_1 = 1$, $K_2 = \frac{1}{3}, \ldots$. Substituting Eq. (17) into (4) and taking into account that for small times, $\tau < \tau_c$, the sum in (17) may be reduced to a term that is proportional to τ^2, hence one obtains

$$V(2\tau) \approx V_0 \exp(-m\tau^2), \tag{18}$$

where

$$m = \frac{8\pi^2}{9\sqrt{3}} \gamma_A \gamma_B \hbar C_B \tau_c^{-1} = 2\, \Delta\omega_{1/2}\tau_c^{-1}.$$

This result was obtained for the first time by Klauder and Anderson.[5] Numerical calculations given by Hartmann and Hu[12] show that for $\tau/\tau_c \approx 1$ the echo signal decay is well approximated by the formula

$$V(2\tau) \approx V_0 \exp(-2\,\Delta\omega_{1/2}\tau). \tag{19}$$

Thus for the case of a uniform distribution of paramagnetic centers over the sites of the solid lattice, the contribution of the B spins to the ESE signal decay has been investigated quite extensively for T_1-type samples.

3.2 Spectral Diffusion in T_2-Type Samples

The ESE signal decay due to spectral diffusion in T_2-type samples was first discussed by Klauder and Anderson.[5] They assumed that the ESE signal decay in T_2-type samples can be described by the same relations as those used for T_1-type samples and that in T_2-type samples some mean effective time of flips for spins participating in spin diffusion can be used as the characteristic correlation time τ_c. In reality, the problem of spectral diffusion in T_2-type samples turns out to be much more complicated. One unfavorable circumstance is that in magnetically dilute solids the spins have nearest environments that differ from point to point in the lattice and thus, their spin orientations change with different frequencies during spin diffusion. As a result, there is a set of spin flip-flop rates. If the rate of mutual flip-flops is calculated using perturbation theory, this rate will be $W_{pk} \approx r_{pk}^{-6}$. The distribution of rates of spin flip-flops will then be given by[16,19]

$$\phi(W)\, dW = \sqrt{\frac{3W_{max}}{2\pi W^3}} \exp\left(-\frac{3W_{max}}{2W}\right) dW, \tag{20}$$

where W_{max} is the most probable rate of spin flip-flops.

Numerical calculations of the kinetics of the two-pulse ESE signal decay due to spectral diffusion in T_2-type samples for spins with $S_B = \frac{1}{2}$ were carried out by Milov et al.[9] It was assumed that the flip-flops of each spin are described by the Poisson process and that all spins are distributed with respect to W according to Eq. (20). The ESE signal decay is then expressed by the equation

$$V(2\tau) = V_0 \exp\left[-4\pi C_B Q(\tau)10^{-19}\right]. \tag{21}$$

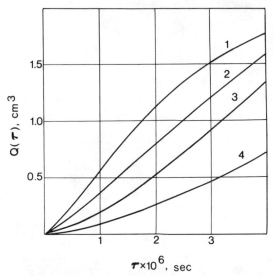

Figure 1 The function $Q(\tau)$ for various W_{max} (s^{-1}): (1) 1.07×10^5; (2) 1.5×10^4; (3) 3.25×10^3; (4) 6×10^2.

Here C_B is the concentration of B spins per cm^{-3}. The results of the calculations of $Q(\tau)$ for various values of W_{max} are given in Fig. 1. The analysis of these curves shows that when $W_{max} < 10^3 s^{-1}$, the condition $W\tau < 1$ holds for the majority of the B spins, and, according to the results for spectral diffusion in T_1-type samples, the two-pulse ESE signal decay is described by

$$V(2\tau) \approx V_0 \exp(-m\tau^2), \tag{22}$$

with

$$m \approx \Delta\omega_{1/2} W_{max}.$$

In this case the kinetics of the ESE signal decay due to B spins is the same in T_1- and T_2-type samples [see Eq. (18)], and the approach proposed by Klauder and Anderson[5] for analysis of the spectral diffusion mechanism in T_2-type samples is valid. Note that though the time dependences in Eq. (18) and (22) are similar, the parameter m depends on the B spin concentration in different ways: in T_1-type samples $m \approx C_B$; in T_2-type samples $m \approx C_B^2$ or even higher orders of C_B.

The fraction of spins for which $W\tau \gtrsim 1$ increases with W_{max}. At $10^4 \lesssim W_{max} \lesssim 10^5 s^{-1}$, the quantity $Q(\tau)$ can be approximated by the linear function $Q(\tau) = 4 \times 10^5 \tau$. For larger values of W_{max}, $Q(\tau)$ becomes proportional to $\sqrt{\tau}$.

Thus the analysis shows that for the intermediate case, when

$$10^4 \lesssim W_{max} \lesssim 10^5 \text{ s}^{-1}, \tag{23}$$

the two-pulse ESE signal decay can be described by the equation

$$V(2\tau) = V_0 \exp(-5 \times 10^{-13} C_B \tau), \tag{24}$$

where C_B and τ are in cm^{-3} and sec, respectively. This result is rather unexpected: There is a region where the spin dephasing due to spectral diffusion is weakly dependent on the value of the most probable rate of spin flip-flops. The exponent in Eq. (24) is linearly dependent on the spin concentration and the time interval between the pulses forming the echo signal. In Eq. (24) the phase relaxation rate is

$$b = 2.5 \times 10^{-13} C_B \text{ s}^{-1}. \tag{25}$$

3.3 Instantaneous Diffusion

Klauder and Anderson[5] have shown that instantaneous diffusion results in a two-pulse signal decay that obeys the equation

$$V(2\tau) = V_0 \exp(-2 \Delta\omega_{1/2} p\tau), \tag{26}$$

where $\Delta\omega_{1/2}$ is the dipolar broadening and p is the fraction of A spins, that is, the fraction of the spins participating in the formation of the ESE signal. A more detailed analysis shows that the contribution of the instantaneous diffusion to the ESE signal decay is described by the formula[11]

$$V(2\tau) = V_0 \exp(-2b_{M.g.}\tau), \tag{27}$$

where

$$b_{M.g.} = \Delta\omega_{1/2}\left\langle \sin^2 \frac{\theta}{2} \right\rangle_g.$$

Here $\theta(\omega)$ is the angle, with respect to the Z-axis, which the second microwave pulse rotates spins with frequency ω and $\langle \cdots \rangle_{g(\omega)}$ denotes an average over the Larmor spin-frequency distribution associated with the linewidth of the ESR spectrum.

The angle θ depends on the amplitude of the microwave magnetic field B_1. This allows the contribution of the instantaneous diffusion to the ESE signal decay to be determined experimentally by varying the amplitude of the microwave field.

Spin dephasing by instantaneous diffusion results from the static spread of the Larmor spin frequencies due to S-S interactions. A spin flip-flops during either spin-lattice relaxation or spin diffusion reduces the contribution of instantaneous diffusion to the ESE signal decay. Consequently, the A spins begin to contribute to the ESE signal decay by the mechanism of spectral diffusion.

For the general case, where the ESR spectrum is excited by microwave pulses, the two-pulse ESE signal decay resulting from S-S interactions is described by[2,11]

$$V(2\tau) = V_0 \exp\left[-\frac{8\pi^2}{9\sqrt{3}} \gamma^2 \hbar C \langle\langle |X| \rangle_t \rangle_g\right], \qquad (28)$$

with

$$X(\omega) = \int_0^t S(\omega, t)m(t)\,dt.$$

Here $S(\omega, t) = \cos\theta(\omega)$ over the interval $(0, \tau)$ and $S(\omega, t) = -1$ over $(\tau, 2\tau)$. If $m(t)$ represents a normal process, then

$$V(2\tau) = V_0 \exp\left[-K'\sqrt{B(\tau)}\right], \qquad (29a)$$

where

$$B(\tau) = [1 + \langle\cos^2\theta(\omega)\rangle_g]\left\{\tau\tau_c + \tau_c^2\left[\exp\left(-\frac{\tau}{\tau_c}\right) - 1\right]\right\}$$

$$- \langle\cos\theta(\omega)\rangle_g \tau_c^2\left[\exp\left(-\frac{\tau}{\tau_c}\right) - 1\right]^2. \qquad (29b)$$

Hence, for the two limiting situations, $\tau < \tau_c$ and $\tau > \tau_c$, we have

$$B(\tau) \approx \tfrac{1}{2}\tau^2\left(1 - \frac{\tau}{3\tau_c}\right)(1 + \langle\cos^2\theta\rangle_g) - \tau^2\left(1 - \frac{\tau}{\tau_c}\right)\langle\cos\theta\rangle_g, \qquad (30a)$$

and

$$B(\tau) \approx \tau\tau_c(1 + \langle\cos^2\theta\rangle_g), \qquad (30b)$$

respectively. If we assume that all A spins rotate through $180°$ because of the microwave pulse, then $\cos\theta = -1$. According to Eq. (29) and (30), when $\tau > \tau_c$, the A spins make the same contribution to the echo signal decay as the B spins [see Eq. (9)]. If $\tau < \tau_c$, then

$$V(2\tau) \approx V_0 \exp\left[-K'\sqrt{2}\,\tau\left(1 - \frac{2\tau}{3\tau_c}\right)\right]. \qquad (31)$$

It is seen from Eq. (31) that spin flip-flops reduce the contribution of instantaneous diffusion to the echo signal decay.

Numerical calculations of the two-pulse ESE signal decay for a Poisson process of $S = \frac{1}{2}$ spin flip-flops for various ratios of the amplitude of the pulsed microwave field to the width of the ESR spectrum were made by Raitsimring et al.[11] The total contribution of all spins (A and B) to the two-pulse echo signal decay is given by Eq. (21). In the general case $Q(\tau)$ is a nonlinear function of τ. However, the time interval of the ESE signal decay measurements is rather limited and lies usually within 3×10^{-7}–3×10^{-6} s. Over this range $Q(\tau)$ can be approximated by a straight line whose slope determines the effective relaxation rate b, and $V(2\tau) \approx V_0 \exp(-2b\tau)$. According to Eq. (32), the effective rate can be represented as a sum of two addends,

$$b = \alpha_c(W)C + \alpha_M(W)C\left\langle \sin^2 \frac{\theta}{2} \right\rangle_g, \qquad (32)$$

which corresponds to the contributions of spectral and instantaneous diffusions. Plots of the coefficients α_c and α_M versus the spin flip-flop rate W are given in Fig. 2. For $W < 10^4 \, \text{s}^{-1}$, $\alpha_M(\omega) \cdot C \approx \Delta\omega_{1/2}$, and therefore the spectral diffusion makes a negligible contribution to the phase relaxation rate. On the other hand, when $W \approx 10^4$–$10^5 \, \text{s}^{-1}$, the contribution of the instantaneous diffusion decreases sharply, and that due to spectral diffusion grows. Finally, for $W > 10^5 \, \text{s}^{-1}$

$$b \approx \alpha_c C \approx 10^{-13} C \, \text{s}^{-1}. \qquad (33)$$

Summarizing the preceding results, we conclude that at present the basic quantitative and qualitative trends of the dependence of the phase relaxation rate on the S-S interactions between paramagnetic centers are well characterized. In general, the kinetics of the ESE signal decay depend on the concentrations of the paramagnetic particles, on the amplitude of the pulsed microwave field, and on the electron spin-lattice relaxation time.

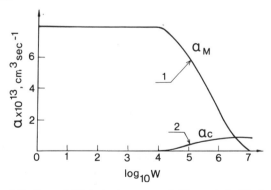

Figure 2 The contributions of (1) instantaneous and (2) spectral diffusions to the phase relaxation rate versus the spin flip-flop rate W.

3.4 The Contribution of Magnetic Matrix Nuclei to the ESE Signal Decay

Let us consider spectral diffusion due to the random modulation of the S-I interaction between the unpaired electrons and the matrix magnetic nuclei by the spin-diffusion process. This situation is, in fact, similar to the preceding case of spectral diffusion in T_2-type samples. There are, however, some distinct differences. Magnetic nuclei form, as a rule, not dilute but concentrated spin systems. Therefore the mode of averaging the spatial distribution of nuclear spins differs from that for magnetically dilute systems.[2] In addition, account must be taken of the diffusion barrier that surrounds the electron spins. Spin diffusion can take place only outside the diffusion barrier. In spin diffusion the spin pairs vary their spin orientations in a correlated manner. Therefore when considering spectral diffusion in T_2-type samples, one must not assume that the spin orientations vary independently. The correlation of spin flip-flops in T_2-types samples has not been investigated. For magnetic nuclei spin correlation can be taken into account in the following way. Electron spin-resonant frequency shifts due to a flip-flop of the kth nuclear pair can be approximated as

$$\Delta\omega_k(t) \approx 3r_n\gamma_n\gamma_e\hbar r_k^{-4}\varepsilon(t). \tag{34}$$

Here r_n is the distance between neighboring magnetic nuclei, r_k is the distance between the electron spin and the kth nuclear pair, γ_n and γ_e are gyromagnetic ratios for nuclei and electrons, and $\varepsilon(t)$ takes on the values $\pm\frac{1}{2}$ in a random way and changes with the mean nuclear flip-flop frequency, W_n.[20]

The ESE signal decay resulting from phase relaxation due to the random modulation of the S-I interaction by nuclear spin diffusion obeys the expression[10]

$$V(2\tau) \approx V_0 \exp(-m\tau^{\mathcal{H}}), \tag{35}$$

where $2 \leq \mathcal{H} \leq 3$. In the region of comparatively small τ, $\mathcal{H} = 3$. In the limit of large τ, the kinetics of the ESE signal decay takes the form

$$V(2\tau) \approx V_0 \exp(-m\tau^{7/4}), \tag{36a}$$

$$V(2\tau + T) = V_0 \exp[-m\tau^{3/4}(\tau + T)], \tag{36b}$$

where

$$m \approx 2\pi W_n C_n (3r_n\gamma_n\gamma_e\hbar)^{3/4}, \tag{36c}$$

and C_n is the concentration of magnetic nuclei and the nuclear flip-flop rate is given by[20]

$$W_n \approx \frac{\gamma_n^2 \hbar}{10} C_n. \tag{3.7}$$

It is seen from Eq. (36a, b, c) and (37) that for the nuclei $m \approx C_n^2$.

3.5 Spin Excitation Transfer

The spin-spin interaction between paramagnetic centers may result in ESE signal decay due to excitation transfer from A spins (excitation donors) to B spins (excitation acceptors).

A peculiarity of magnetically dilute solids is that different spins have different spin environments. The rate of excitation transfer from a given spin depends on its acceptor environment. Therefore there is a distribution of spins with energy transfer rate to acceptors, W. This distribution is given by Eq. (20) and the energy transfer rate is described by the formula

$$p(t) = \int \exp(-Wt)\phi(W)\,dW = \exp(-a\sqrt{t}), \tag{38}$$

where

$$a^2 = 6W_{max}.$$

The most probable energy transfer rate depends on many factors: the acceptor concentration, the difference between Zeeman frequencies of the acceptors and the donors, and the interaction reservoir, which guarantees conservation of energy in the process of energy transfer. The simplest results are obtained when spins with sufficiently short paramagnetic relaxation times, T_{1A} and T_{2A}, are acceptors.[2,21] If $(\omega_A + \omega_D)T_{1,2} \ll 1$ and $T_{1A} = T_{2A}$, then

$$W_{max} \approx 2 \times 10^{15} \mu_A^2 T_{1A} C_A^2, \tag{39}$$

where μ_A is the magnetic moment of the acceptors. The energy transfer rate is independent of the difference between the Zeeman frequencies of donors and acceptors, ω_D and ω_A. If the paramagnetic relaxation of the acceptors is less effective, $(\omega_A - \omega_D)T_{2A} > 1$, then

$$W_{max} \approx 0.1\gamma_D^2 \mu_A^2 C_A^2 T_{2A}^{-1}(\omega_A - \omega_D)^{-2}, \tag{40}$$

and the energy transfer rate is proportional to the inverse square of the difference between the Zeeman frequencies of the donors and acceptors.

A more complicated case occurs when the Zeeman energies in the excitation transfer process are compensated for by the S-S interaction reservoir. Free radicals stabilized in a solid matrix serve as an example of such a system. Estimations made using perturbation theory for particles with spin $\frac{1}{2}$ give[2,19]

$$W_{max} \approx \gamma^4 \hbar^2 C_k^2 \frac{\Delta\omega_{1/2}}{\Delta\omega_k^2} \left[1n \left(\frac{\Delta\omega_k}{2\,\Delta\omega_{1/2}} \right) \right]^2. \tag{41}$$

Here C_k is the concentration of spins belonging to the kth component of the ESR spectrum with linewidth $\Delta\omega_k$. Thus the kinetics of the ESE signal decay due to spin excitation transfer obey the following expression

$$V(2\tau) \approx \exp\left(-\sqrt{12\,W_{max}\tau}\right). \tag{42}$$

In summary, there currently is a theory of relaxation decay of the ESE signal due to the interaction of paramagnetic centers with each other and with the matrix magnetic nuclei. Here we have described only the basic results of the theory that will be useful for discussing the applications of the ESE method for investigating the spatial distributions of paramagnetic species. A more detailed discussion of all these problems is found in the papers of Mims[1] and Salikhov et al.[2]

4 THE CONTRIBUTION OF SPIN-SPIN INTERACTIONS TO THE ESE SIGNAL DECAY

The preceding theory of relaxation due to dipolar interactions between paramagnetic centers implies a random uniform distribution of spins in the matrix. Therefore a direct comparison of theory and experiment may be done only for such systems where a uniform spatial distribution of paramagnetic species can be realized. Glassy solutions of VO^{2+} ions, stable DPPH radicals, and 2,4,6-tri-*tert*-butylphenoxyl radicals (TBP) are model systems of this type.[2,22] The homogeneity of such systems has been verified by ESR data[23] and the concentrations of the paramagnetic centers in the sample as determined by ESR and by other methods (e.g., by weighing, in the case of stable radicals or by spectrophotometric methods) are in agreement within experimental error. Irradiated substances can also serve as model systems with a uniform distribution of paramagnetic centers. However, the irradiation dose must be sufficiently high to avoid effects that are possible in the region of small doses. As is shown below, for γ-irradiated substances, the spatial distribution of radicals is uniform for doses exceeding approximately 0.5 Mrad.

The S-S interaction between paramagnetic centers leads to the ESE signal decay as a result of phase relaxation by the mechanisms of instantaneous and spectral diffusion. The relative contributions of these mechanisms depends on the concentration of paramagnetic species, the degree of excitation of the ESR spectrum by the microwave pulses, and the rate of spin flip-flops resulting in the random modulation of the S-S interaction. Therefore a complete analysis of the phase relaxation associated with the S-S interactions between paramagnetic centers requires various types of measurements. To distinguish the contribution of the interaction between spins during their dephasing, one must study the kinetics of the ESE signal decay at varying concentrations of paramagnetic species. The contribution of S-S interactions to the phase relaxation by the mechanisms of instantaneous and spectral diffusions can be separated on the basis of the dependence of the kinetics of ESE signal decay on the amplitude of the pulsed microwave magnetic field. Studies of the temperature dependence of phase relaxation allows one to choose between T_1- and T_2-type samples and to identify the effective mechanism of spin flip-flops.

Detailed phase relaxation investigations have been carried out for hydrogen atoms trapped in various matrices, for free radicals, and for paramagnetic ions. ESE data for a number of systems follow and are compared with the phase relaxation theory.

4.1 Phase Relaxation of Hydrogen Atoms in Solid Matrices

Hydrogen atoms trapped in various matrices are convenient model systems for investigating the phase relaxation induced by S-S and S-I interactions. A series of detailed investigations of spin-lattice relaxation of hydrogen atoms trapped in amorphous quartz at 77 K[14] and in glassy water solutions of sulfuric acid at 4.2 and 77 K have been made.[19,21] At sufficiently low hydrogen atom concentrations the spin-lattice relaxation kinetics are described by an exponential dependence of the form $\exp(-a\sqrt{t})$. This means that there is a distribution of spin-lattice relaxation times for this system. The most probable value of T_1 for hydrogen atoms at 77 K is 2×10^{-2} s and about 100 s at 4.2 K. Because of the weak interaction between the hydrogen atoms and the lattice one expects that instantaneous spectral diffusion should be the main mechanism of the two-pulse ESE signal decay. Estimations using Eq. (41) show that when the hydrogen atom concentration exceeds 10^{19} cm^{-3}, mutual spin flip-flops must effectively modulate the local dipolar fields, and the spectral diffusion mechanism resulting from a random modulation of S-S interactions by the spin-diffusion process begins to contribute to the ESE signal decay.

The two-pulse signal decay for hydrogen atoms trapped in glassy quartz at 77 K has been studied at various concentrations of hydrogen atoms and for various amplitudes of the pulsed microwave magnetic field B_1. Hydrogen atoms were obtained in glassy quartz by γ-irradiation of the samples at 77 K. The width of the ESR spectral line for hydrogen atoms trapped at 77 K is 0.3 G. The conditions for complete excitation of this line are satisfied over a wide range of B_1 values. Experimentally the kinetics of the ESE signal decay for hydrogen atoms are described by the simple exponential function given in Eq. (13). Figure 3a represents plots of the phase relaxation rate b *versus* B_1, which is accurately described by the equation

$$b = b_0 + \tfrac{1}{2}b_M \sin^2 \left(\tfrac{1}{2}\gamma B_1 t_p\right),\tag{43}$$

where B_1 and t_p are the amplitude of the microwave field and the duration of the microwave pulse, respectively. As is seen from Fig. 3b,

$$b_M = \alpha_M C_H = 8 \times 10^{-13} C_H \text{ s}^{-1},\tag{44}$$

where C_H is the hydrogen atom concentration (cm^{-3}). Comparison of these results with theory shows that in the case of hydrogen atoms trapped in quartz and complete excitation of the ESR spectral linewidth, S-S interactions, as was expected, cause the ESE signal to decay only by the instantaneous diffusion mechanism.

The problem of the influence of other paramagnetic species and magnetic matrix nuclei on phase relaxation was thoroughly studied for hydrogen atoms stabilized in γ-irradiated glassy solutions of sulfuric[21] and phosphoric acids.[11] In addition to hydrogen atoms, SO_4^- and PO_4^{2-} anion radicals are also stabilized in these matrices.

The decay rate of the two-pulse ESE signal for hydrogen atoms at 77 K in irradiated solutions of sulfuric acid is described by the expression[10]

$$V(2\tau) = V_0 \exp\left(-2b\tau - m\tau^2\right).\tag{45}$$

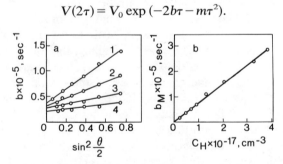

Figure 3 (a) B_1 dependence of the phase relaxation rate of hydrogen atoms for various concentrations of hydrogen atoms (cm^{-3}): (1) 3.8×10^{17} (2) 2.0×10^{17}; (3) 0.9×10^{17}; (4) 0.5×10^{17}. (b) Plots of the phase relaxation rate b_M versus the hydrogen atom concentration.

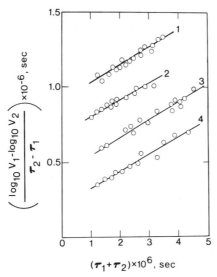

Figure 4 The phase relaxation kinetics (linear anamorphoses) of hydrogen atoms in irradiated sulfuric acid at 77 K versus the hydrogen atom concentration (cm^{-3}): (1) 2.4×10^{18}; (2) 2×10^{17}; (3) 2.2×10^{17}; (4) 2×10^{17}. V_1 and V_2 are the amplitudes of the ESE signal corresponding to the time intervals τ_1 and τ_2.

It is seen that the experimental data in Fig. 4, plotted in these special coordinates, yields a straignt line, which is good evidence for the validity of Eq. (45). The values of m were determined from the slopes of the straight lines and the values for b were obtained from the intercepts.

An ESE signal decay proportional to $\exp(-m\tau^2)$ is indicative that the interaction between the hydrogen atoms and the matrix protons contributes significantly to the phase relaxation. The parameter m is independent of the concentration of the paramagnetic species and equal to $2.6 \times 10^{11}\,\mathrm{s}^{-1}$. This is in good agreement with the value calculated using Eq. (36a, c). Detailed investigations have been made on the dependence of m on the proton concentration C_p in the matrix.[10] It was observed that for glassy sulfuric acid solutions with proton concentrations ranging from 2.8×10^{22} to $6 \times 10^{22}\,\mathrm{cm}^{-3}$ the parameter m depends strongly on C_p,

$$m = AC_p^2; \qquad A = 7 \times 10^{-35}\,\mathrm{cm}^6/\mathrm{s}^2. \qquad (46)$$

This dependence and the value of A are in good agreement with the theoretical analysis of the contribution of the matrix magnetic nuclei to the ESE signal decay.

The concentration of hydrogen atoms and SO_4^- radicals in glassy sulfuric acid solutions increases with the irradiation dose. The dependence of the phase relaxation rate b on the SO_4^- concentration is represented in Fig. 5.

The presence of the term depending on the concentration of paramagnetic species and linear with respect to τ in the exponent of Eq. (45) is associated with the interaction of the hydrogen atoms with each other and with SO_4^- ion radicals. An analysis of the dependence of the relaxation rate b on the amplitude of the pulsed microwave field B_1 shows that the interaction of the hydrogen atoms with each other leads to an exponential decay of the two-pulse ESE signal according to Eq. (45) by the mechanism of instantaneous diffusion.

To determine the contribution of the SO_4^- radicals to the phase relaxation rate of the hydrogen atoms, the following experiments were performed. Irradiated samples were annealed by storing at the boiling temperature of liquid oxygen. After annealing the concentration of the hydrogen atoms was reduced to less than 2×10^{17} cm^{-3}, whereas that of SO_4^- remained practically the same. It appeared that the phase relaxation kinetics in these samples are described by Eq. (45) with the same value of m but different values of b. The dependence of b on the SO_4^- concentration in the samples is given in Fig. 5 (curve 2) and has the form $b = \alpha C_i$, where $\alpha = 1.2 \times 10^{-13}$ cm^3/s and C_i is the concentration of SO_4^- radicals.

The difference in the slopes of curves 1 and 2 in Fig. 5 results from the interaction between hydrogen atoms. The magnitude of this difference is in agreement with that calculated using the theory of instantaneous diffusion.

SO_4^- ion radicals can influence the hydrogen atom phase relaxation either through the process of SO_4^- spin-lattice relaxation or through SO_4^-

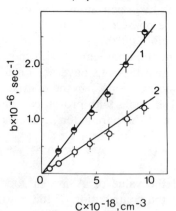

$C \times 10^{-18}$, cm^{-3}

Figure 5 Plots of the phase relaxation rate versus the SO_4^- ion-radical concentration for hydrogen atoms in (1) nonannealed and (2) annealed at 90 K samples.

spin flip-flop processes. Experimental results obtained at 4.2 K for the hydrogen two-pulse signal decay enables one to distinguish between the two mechanisms of spectral diffusion and to establish which mechanism of random spin reorientation is responsible for the modulation of the S-S interactions between hydrogen atoms and SO_4^- ion radicals.[14]

The kinetics of the ESE signal decay at 4.2 K are described, as for the case at 77 K, by expression (45). Unlike the data obtained at 77 K, it was found experimentally that at 4.2 K the phase relaxation rate b is independent of the SO_4^- concentration. Thus when the temperature decreases from 77 to 4.2 K, the spectral diffusion caused by the modulation of the S-S interactions of the hydrogen atoms with the SO_4^- ion radicals by random flip-flops of the SO_4^- spins "freezes." Taking into account that the rate of mutual SO_4^- flip-flops has a negligible temperature dependence, whereas the SO_4^- spin-lattice relaxation rate decreases by several orders with decreasing temperature from 77 to 4.2 K, one can conclude that at 77 K spectral diffusion is controlled by the SO_4^- spin-lattice relaxation and not by the mutual SO_4^- flip-flops. These results support the conclusion that at 77 K the system of SO_4^- ions with $C_i < 10^{19}$ cm^{-3} is an example of a T_1-type sample. Similar results were obtained by an ESE investigation of hydrogen atoms in glassy water solutions of phosphoric acid at 77 K.[11]

Recently Raitsimring et al. have investigated spin diffusion and phase relaxation of hydrogen atoms at liquid helium temperatures over a wide range of hydrogen atom concentrations.[13] Quite high concentrations of hydrogen atoms were achieved in irradiated water solutions of perchloric acid ($C_H \approx 2 \times 10^{19}$ cm^{-3}) and it was observed that at these high concentrations the dipolar shift is modulated sufficiently fast by the spin-diffusion process. Consequently, the phase relaxation mechanism for the hydrogen atoms in this system changes with increasing hydrogen atom concentration. In the interval $10^{18} \lesssim C_H < 10^{19}$ only the instantaneous diffusion mechanism is operative, whereas with increasing concentration, $C_H \gtrsim 10^{19}$ cm^{-3}, spectral diffusion, resulting from the random modulation of the S-S interaction of hydrogen atoms by the spin-diffusion process, makes a contribution to the phase relaxation of the hydrogen atoms.

4.2 Phase Relaxation of Molecular Radicals in Solid Matrices

Among the spin systems that have been studied by the ESE method, hydrogen atoms stand apart because of the vast amount of available experimental data. Unfortunately, for the majority of the other systems of interest, there is a paucity of experimental data on spin-lattice relaxation, spin diffusion, and B_1 dependence of the contribution of S-S interactions to the ESE signal decay. Nevertheless, it is still possible to give a

Table 1 Parameters Determining the Dependendence of the Two-Pulse ESE Signal Decay on the Spin Concentration

$$V(2\tau) = V_0 \exp\left(-\alpha_c C 2\tau - \alpha_M C\left\langle \sin^2\frac{\theta}{2}\right\rangle 2\tau - m\tau^2\right)$$

Sample	$T\,K$	$\alpha_M \times 10^{13}$ (cm^3/s)	$\alpha_c \times 10^{13}$ (cm^3/s)	m (s^{-2})	Reference
TBP in	4.2	3.9	0.9	0	14
toluene	77	2.8	0.7	0	11
CH$_2$OH in	4.2	3.7	0.9	0	14
methyl alcohol	77	2.7	0.6	0	11
TMOPO in					
methyl alcohol	77	5.5	0	$4 \times 10^{-27} C^2$	11

satisfactory interpretation of the existing experimental data.

First of all, we shall discuss the experimental data on the phase relaxation for the radicals listed in Table 1. It is typical for all of them that the *S-S* interaction contributes to the two-pulse ESE signal decay through the instantaneous and spectral diffusion mechanisms. Moreover, at 77 K these systems appear to be T_2-type samples over the concentration range studied.

For solutions of the iminoxyl radical 2,2,6,6-tetramethyl-4-oxypiperidine-1-oxyl (TMOPO) in methyl alcohol the two-pulse ESE signal decay is governed by Eq. (45), where $b = b_M \langle \sin^2 \theta/2 \rangle$, $b_M = 5.5 \times 10^{-13} C\,s^{-1}$ (see Fig. 6), and m increases with the spin concentration as (see Fig. 7)

$$m = 4 \times 10^{-27} C^2\,s^{-2}. \tag{47}$$

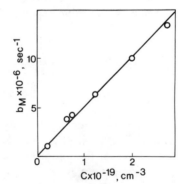

Figure 6 A plot of the phase relaxation rate b_M versus the TMOPO radical concentration for $\langle \sin^2 \theta/2 \rangle \approx 1$.

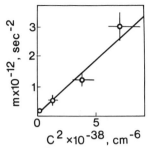

Figure 7 A plot of the spectral diffusion parameter m versus the TMOPO radical concentration.

For this system the contribution to the signal decay from the instantaneous diffusion mechanism is proportional to $\exp(-2b\tau)$ and the contribution due to spectral diffusion is governed by $\exp(-m\tau^2)$. The kinetics of the ESE signal decay for TMOPO radical solutions are in agreement with those calculated using expressions (22) and (27) for the case of comparatively slow spectral diffusion. The observed spin concentration dependence of m given by Eq. (47) indicates that spectral diffusion in this system is caused by modulation of the S-S interaction by the spin-diffusion process.

Recently the phase relaxation kinetics for tri-*tert*-butylphenoxyl radicals (TBP) in frozen glassy solutions of toluene and $\dot{C}H_2OH$ radicals in γ-irradiated methyl alcohol have been examined in detail.[14] In these systems the two-pulse ESE signal decay for various concentrations of the paramagnetic particles ($C \approx 10^{18} - 10^{19}\ cm^{-3}$) at temperatures 77 and 4.2 K is described by Eq. (13).[14] The relaxation rate b was observed to be

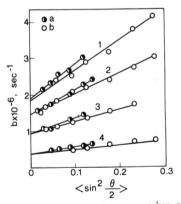

Figure 8 B_1 dependence of the phase relaxation rate of $\dot{C}H_2OH$ radicals at various radical concentrations (cm^{-3}): (1) 3×10^{19}; (2) 2.2×10^{19}; (3) 1.3×10^{19}; (4) 0.4×10^{19}. Points a and b correspond to excited extrema and central components of the ESR spectrum.

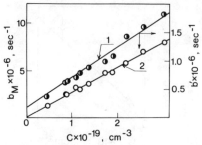

Figure 9 Plots of the phase relaxation rate (1) b' and (2) b_M versus the $\dot{C}H_2OH$ radical concentration.

a linear function of the concentration but temperature independent. Analysis of the dependence of the relaxation rate on the amplitude of the microwave field shows that the relaxation rate can be represented by (see Figs. 8 and 9)

$$b = b' + b_M \left\langle \sin^2 \frac{\theta}{2} \right\rangle_g = b_0 + \alpha_c C + \alpha_M \left\langle \sin^2 \frac{\theta}{2} \right\rangle_g C. \qquad (48)$$

The influence of the concentration of paramagnetic centers on the relaxation rate indicates that the ESE signal decay for these radicals results from S-S interactions. Unlike the case for the hydrogen atoms, for $\dot{C}H_2OH$ and TBP radicals, the contribution of the S-S interaction to the ESE signal decay cannot be associated with the instantaneous diffusion mechanism for which, according to (27), $\alpha_c = 0$ and $\alpha_M = 8.1 \times 10^{13}$ cm^3/s. One can expect that this difference is due to the random modulation of the dipolar S-S interaction by sufficiently fast spin flip-flops. Since the dipolar interaction for $\dot{C}H_2OH$ and TBP, both at 4.2 and 77 K, is manifested by a similar decay of the ESE signal, the spin flip-flop rates at these two temperatures should be considered equal. Therefore mutual flip-flops, but not spin-lattice relaxation, constitute an effective mechanism of spin reorientation. Estimations of the relaxation rate using Eq. (41) show that over the concentration range studied $(C < 10^{19}$ cm$^{-3})$, the flip-flop rate is $W_{max} \approx 10^4$–10^5 s^{-1}. According to the theoretical calculations (see Fig. 2), for such flip-flop rates, the contribution of the instantaneous diffusion to the ESE signal decay must decrease for times $\tau \approx 10^{-6}$ s. Apparently this is the cause of the decrease in the contribution of the instantaneous diffusion mechanism to the phase relaxation of the $\dot{C}H_2OH$ and TBP radicals. For these radicals the relaxation rate does not approach the limiting value $\alpha_M = 8.1 \times 10^{-13}$ cm^3/s, which is characteristic of the instantaneous diffusion mechanism (Table 1).

The contribution of the instantaneous diffusion mechanism to the ESE signal decay is negligible for systems in which the number of echo-generating spins is small (~0.01) in comparison to the total number of spins. Under these conditions the basic contribution to the ESE signal decay is made only by B-type spins, and spectral diffusion becomes the principal mechanism of phase relaxation. The data obtained for VO^{2+} ions in glassy sulfuric acid solutions at 77 K and for frozen alcohol solutions of stable DPPH radicals at 77 K serve as examples of such systems.[22]

The ESR spectrum of VO^{2+} ions at 77 K consists of 16 hyperfine structure components *each* with a width of 10 G. The total splitting of the spectrum is about 1200 G. The kinetics of the two-pulse signal decay has been studied for samples with ion concentrations in the range 10^{18}–6×10^{19} cm^{-3}. Experiments show that for these concentrations the decay rate of the ESE signal is described by a simple exponent and that the phase relaxation rate is a linear function of the spin concentration,

$$b = b_0 + \alpha_c C_i = (0.3 \times 10^6 + 1.2 \times 10^{-13} C_i) \, \text{s}^{-1}. \tag{49}$$

Here C_i is the mean concentration of VO^{2+} ions in cm^{-3} and b_0 is the initial relaxation rate determined by the extrapolation of the linear dependence of $b(C_i)$ to zero concentrations. (See Fig. 10a.) The mechanism of instantaneous diffusion makes a negligible contribution to the ESE signal decay in the case of VO^{2+} ions.

The intensities of the various hyperfine components of the VO^{2+} ESR spectrum differ considerably from each other. Therefore instantaneous diffusion must make a greater contribution to the signal decay for the components with the greater intensity. However, it appears that when the length of the exciting microwave pulses are equal, the signal decay rates for the various hyperfine components are practically the same. An estimation of the spin-lattice relaxation time of the VO^{2+} ion indicates that T_1

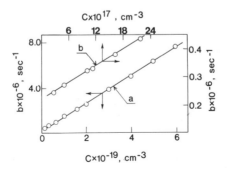

Figure 10 The concentration dependences of the phase relaxation rate for VO^{+2} ions (a) and stable DPPH radicals (b).

lies in the region of 10^{-4}–10^{-6} s. For such spin-reorientation rates one should expect some noticeable decrease in the contribution of the instantaneous diffusion mechanism to the echo signal decay.

Two-pulse ESE signal decay investigations for solutions of the stable DPPH radical in methanol over the range of spin concentrations 3×10^{17}–2×10^{18} cm^{-3} reveal some features analogous with those for VO^{2+} ions. The signal decay over this radical concentration range is described by an exponent with the decay rate increasing linearly with concentration (see Fig. 10b),

$$b = (0.23 \times 10^6 + 0.9 \times 10^{-13} C) \, \text{s}^{-1}. \tag{50}$$

The anisotropically broadened ESR spectrum for frozen DPPH solutions consists of five hyperfine components with intensity ratios $1:4:6:4:1$. The ESE signals were studied for the hyperfine components with relative intensities 4 and 6. The ESE signal decay rate appeared to be independent of which hyperfine component was excited by the microwave pulses. This indicates that the basic contribution to the ESE signal decay for DPPH radicals is made by the spectral diffusion mechanism of phase relaxation.

Summing up the results, it should be clear that the electron phase relaxation in magnetically dilute solids finds its interpretation within the physical models of the mechanisms of instantaneous and spectral diffusion.

5 STUDIES OF THE SPATIAL DISTRIBUTION OF PARAMAGNETIC SPECIES BY THE ESE METHOD

The problem of the spatial distribution of active particles is very important for at least two areas of investigation: (1) for studies of heterogeneous reactions involving paramagnetic particles and (2) for studies of the kinetics and mechanisms of chemical reactions in irradiated substances.

In heterogeneous solid systems the spatial spin distribution is controlled by the phase state of the system, that is, by the composition of the system and the ambient conditions such as temperature and pressure. In irradiated homogeneous solid systems an external factor, the *irradiation*, may serve as a source of inhomogeneity. Irradiated solid matrices have attracted much attention by investigators interested in the problem of spatial distribution using both the stationary ESR method along with its modifications[24] and the ESE technique.[25] Using both techniques, spin spatial distribution data can be obtained from the analysis of the dipolar S-S interactions between the paramagnetic particles.

5.1 The Local Concentration

Intensity measurements of ESR spectra, optical spectra, and other physical methods such as static magnetic susceptibility measurements enable one to determine the total number of paramagnetic particles N in a sample of a volume V, and thus the average concentration of particles $C_{av} = N/V$. For some systems the paramagnetic particles are not distributed uniformly throughout the volume of the sample, and it is perhaps more useful to introduce the local concentration C_{loc}, where $C_{loc} > C_{av}$.

The foregoing theoretical calculations and the interpretation of the experimental results have assumed that the paramagnetic particles are uniformly distributed in the lattice. Formally, the theory can be readily generalized for systems with an arbitrary distribution law. For this purpose we introduce the radical distribution function of paramagnetic particles, $g(\mathbf{r}_k, \mathbf{r}_n)$, which characterizes the conditional probability of finding a particle located at the point with the radius vector $\mathbf{r}_n$, if it is known that some given paramagnetic center is located at the point given by the radius vector $\mathbf{r}_k$. The contribution of the S-S interaction between the particles to the ESE signal decay is then given by

$$V(t) \approx V_{av} \int f(\mathbf{r}_k) \, d^3\mathbf{r}_k \, \exp\left[-C_{av}\int g(\mathbf{r}_k, \mathbf{r}_n)\right.$$

$$\left. \times (1 - \langle\langle \exp(iA_{kn}X(t, \omega))\rangle_t\rangle_g) \, d^3\mathbf{r}_n\right]. \quad (51)$$

Here $A_{kn} = \gamma^2\hbar(1 - 3\cos^2\theta_{kn})r_{kn}^{-3}$, C_{av} is the average concentration of particles for the sample ($C_{av} = N/V$), and $f(\mathbf{r}_k)$ is the probability of finding a particle at the point $\mathbf{r}_k$. For a uniform spatial distribution $g(\mathbf{r}_k, \mathbf{r}_n) = 1$ and Eq. (51) reduces to Eq. (4) and (28).

The data for a uniform particle distribution can be easily generalized for the case of a nonuniform distribution in the following manner. Suppose that the particles occupy only a portion V' of the total volume V and in V' the particles are uniformly distributed. Then $g(\mathbf{r}_k, \mathbf{r}_n)C_{av} = N/V'$ and the ESE signal decay due to S-S interactions will be described by the relation obtained for a uniform distribution. However, instead of the average concentration C_{av} we must use the concentration in the occupied volume (i.e., the local concentration),

$$C_{loc} = C_{av}\frac{V}{V'} \geq C_{av}.$$

According to Eq. (51), for an arbitrary law of particle distribution in the lattice, the ESE signal decay kinetics may have a complicated form.

Unfortunately, this problem has not been thoroughly investigated. In the following discussion when considering track effects, we assume the existence of such inhomogeneous particle distributions whenever the particles are stabilized only in certain volumes and, in addition, whenever each volume and the number of particles in them are considered to be sufficiently large. However, when the number of particles in such volumes is small, more precise calculations can be made using Eq. (51).

Information about the spatial distribution of the paramagnetic particles can be obtained from ESE data in the following manner. A theoretical and experimental analysis of the phase relaxation rate in a system with a uniform spatial particle distribution indicates that the rate of the two-pulse ESE signal decay caused by S-S interactions can be described by the relation

$$b = b_0 + \alpha_p C, \tag{52}$$

to a sufficiently good approximation. Here the quantity b_0 includes all those contributions to the phase relaxation rate that are independent of the concentration of paramagnetic centers and the quantity α_p includes the contributions made by the mechanisms of spectral (α_c) and instantaneous ($\alpha_m \langle \sin^2 \theta/2 \rangle_g$) diffusions. The influence of concentration on the ESE signal decay rate in Eq. (52) reflects, in fact, that the interparticle distances, and thus the magnitude of the S-S interactions, change with concentration. For a uniform particle distribution, $C = C_{av}$ in Eq. (52); however, if the distribution is nonuniform, and the particles occupy only a part of the volume one uses in Eq. (52), $C = C_{loc} \geq C_{av}$.

Thus ESE investigations of the spatial distribution should include the following steps:

1. The decay of a two-pulse ESE signal is studied in samples with various average concentrations of paramagnetic particles. If the value of the average concentration of the paramagnetic particles changes during a certain chemical process under study, then its absolute value must be determined independently (e.g., by ESR, weighing, optical spectroscopy, etc.).

2. The data obtained in Step 1 are analyzed to obtain the concentration dependence of Eq. (52) during the chemical process.

3. If the linear dependence given by Eq. (52) holds over the whole range of average concentrations studied and if $\alpha = \alpha_p$, then the spatial distribution of the paramagnetic centers can be considered uniform during the reaction under study, that is, $C = C_{av}$.

4. Any deviation from a linear concentration dependence of the phase relaxation rate must be treated as evidence for a nonuniform distribution

of the paramagnetic particles. Interpretation of the data obtained depends, however, on the nature of the studied chemical process in which the paramagnetic centers participate.

As was noted, most ESE studies of the spatial distribution of particles are confined to investigations of free radicals, atoms, and ion radicals in irradiated substances.

5.2 Track Effects in Irradiated Solids

The energy of the ionizing particle is spent for ionization and excitation of molecules of irradiated substances. A γ-quantum passing through condensed systems causes the formation of spurs containing two or three pairs of excited and ionized molecules, blobs consisting of several overlapping spurs, and short tracks composed of dozens of overlapping spurs (Fig. 11). An α-particle passing through a substance causes the formation of the main track by secondary electrons with energies $\leq 100\,\text{eV}$, and δ-tracks by those with energies $>100\,\text{eV}$. A heavy particle (e.g., U^{235} fission fragments) causes the formation of a greater number of δ-tracks than an α-particle. As a result, the δ-tracks overlap. The number of secondary electrons and energy losses of the ionizing particle per unit path length (linear energy transfer, LET) is determined by the rate of energy losses for the substance and by the type and energy of the ionizing irradiation.

According to this localized manner of energy transfer from ionizing particles to the substance, the radical products of radiolysis may also be localized in some regions of the solid if there is no effective migration of the charges and excited states, which are the precursors of the radical products, from the track.

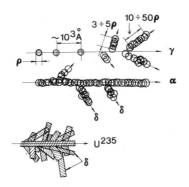

Figure 11 Diagrammatic drawing of the regions of energy losses for particles of various types.

Track effects can be observed most effectively in substances where a portion of the secondary electrons can be stabilized or accepted, since in this case it is possible to expect that the radical products will also be stabilized in certain regions of the solid.

If radical tracks are formed, then to a first approximation, it is possible to consider that the local concentration of radicals inside the tracks is constant and the dependence of the phase relaxation rate on the irradiation dose D can be predicted. The irradiation dose determines the average radical concentration that can be measured (e.g., by the stationary ESR method, $C_{av} = G_R D$); here G_R is the radical yield that corresponds to the number of radicals formed per 100 eV of energy absorbed.

When the radicals are stabilized in the spurs or tracks, the local concentration—and thus the relaxation rate—is independent of the average concentration or dose during the accumulation of the radicals in spurs or tracks that are not *overlapping*. At the onset of overlapping tracks (spurs) the mean distance between the radicals inside the tracks (spurs) is comparable with that between radicals stabilized in the different tracks (spurs). As the radical concentration increases, the relaxation rate becomes a linear function of the average concentration as for the case of a uniform radical distribution. This transition, from the absence of any dependence to the linear dependence with $b \approx \alpha C$, takes place when the average concentration of the radicals C_{av} is approximately equal to the local concentration in the tracks or spurs ($C_{av}^* \approx C_{loc}$).

5.3 Methanol: Radical Tracks, LET Effects

The analysis of the concentration dependence of the decay rate for a two-pulse ESE signal enables one to study in detail the spatial distribution of paramagnetic particles in irradiated solids.

Some examples of such ESE investigations follow. The most detailed studies were carried out for methanol.[26-30] For irradiated methanol frozen at 77 K the mechanism of radical formation has been identified quite reliably. Methanol is a convenient system for ESE investigations of track effects, since, as will be shown, it is possible to change the situation artificially from radicals localized in tracks to those uniformly distributed throughout the volume.

The hydrogen isotope T, sulfur isotope ^{35}S (β-sources), ^{210}Po (α-source), and the nuclear reactions of ^{10}B $(n, \alpha)^7Li$, ^6Li $(n, \alpha)T$, and $^{235}U(n, f)Pr$ have been employed as internal radiation sources. Irradiation with thermal neutrons was carried out in the channel of a nuclear reactor. The mean values of the LET for various sources are listed in Table 2. All

Table 2 Experimental Data on the Radiation-Chemical Yields, Track Radii, and Local Concentrations in Irradiated Methanol at 77 K

Source	LET (eV/Å)	G_R (1/100 eV)	$C_{loc} \times 10^{-18}$ (cm^{-3})	R_{tr} (Å)
$\beta(T)$	1.1	5.5	2.5	90
^{6}Li$(n, \alpha)T$	16	2.6	7	130
^{210}Po(α)Pb	20	3.3	8	155
^{10}B$(n, \alpha)^{7}$Li	43	1.0	7.5	130
^{235}U(n, f)Pr	540	0.2	7.5	180
$\beta(^{35}$S$)$	1	6	1.7	90

samples were transparent glasses at 77 K. Small amounts of the corresponding isotopes were dissolved in methanol before freezing and the phase relaxation rate for the $\dot{C}H_2OH$ radical was investigated by the two-pulse ESE technique. The average concentration of radicals in the various samples under irradiation was determined by the ESR method.

Let us consider some typical experimental results for β-irradiated methanol (Fig. 12). Plots of the relaxation rate b versus the average $\dot{C}H_2OH$ radical concentration are given in Fig. 12a (curve 1).

As is seen from the figure, in the region of small concentrations the relaxation rate b is independent of the radical concentration and equals $b^* = 0.4 \times 10^6$ s^{-1}. Starting from a concentration of about 4×10^{18} cm^{-3}, the relaxation rate becomes a linear function described by the relation

$$b = (0.12 \times 10^6 + 1.1 \times 10^{-13} C_{av}) \, \text{s}^{-1}.$$

The dependence of the average concentration of $\dot{C}H_2OH$ radicals on the irradiation dose that determines G_R is given in Fig. 12b.

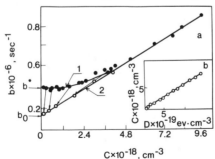

$C \times 10^{-18}$, cm^{-3}

Figure 12 (a) The concentration dependence of the phase relaxation rate of $\dot{C}H_2OH$ radicals under (1) β-irradiation and (2) after photolysis of the irradiated samples. (b) Curve for $\dot{C}H_2OH$ radicals during β-irradiation.

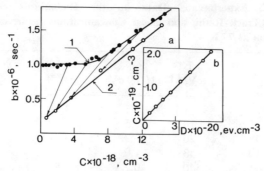

Figure 13 (*a*) Concentration dependence of the phase relaxation rate of ĊH₂OH radicals (1) under α-irradiation and (2) after photolysis of irradiated samples. (*b*) Curve for ĊH₂OH radicals during α-irradiation.

Similar experiments were performed with other types of radiation. The dependence of the phase relaxation rate on the average radical concentration always has the form characteristic of track spatial inhomogeneity. At small concentrations, up to the region of overlapping tracks, the relaxation rate is constant; as the dose increases, the dependence becomes linear, with its slope being typical for a uniform spatial distribution of paramagnetic particles. Figure 13 represents the results obtained for α-irradiated methanol. For samples irradiated by neutrons the relaxation rate shows, in fact, no linear dependence on the average concentration. This effect (Fig. 14) is perhaps associated with both high local concentrations for these samples and comparatively small irradiation doses used in the experiments with neutrons.

The trends found are indicative of an inhomogeneous spatial distribution of radicals. An independent corroboration of the effects of such a

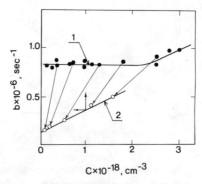

Figure 14 The concentration dependence of the phase relaxation rate of ĊH₂OH radicals in neutron-irradiated samples [reaction ⁶Li (n, α)T], curve 1; the same after UV-photolysis, curve 2.

distribution can be obtained experimentally by studying the phase relaxation of $\dot{C}H_2OH$ radicals after ultraviolet (uv) photolysis of irradiated samples.

It is well known that under uv light methanol radicals decompose into $\dot{C}H_3$, H, and CO.[31] Methyl radicals and hydrogen atoms are unstable in an alcohol matrix at 77 K. Under dark conditions these diffuse and react with each other and with the alcohol molecules. This results in the formation of new $\dot{C}H_2OH$ radicals and a certain decrease in their concentration as compared to the initial $\dot{C}H_2OH$ radical concentration in the irradiated sample. As a result, the initial inhomogeneity of their spatial distribution must decrease. It is thus possible to expect that if irradiated samples are illuminated by uv light and then kept in darkness, the spatial distribution of the $\dot{C}H_2OH$ radicals will be nearly uniform.

Experiments with uv light illumination were performed simultaneously with relaxation ESE measurements. Their results are also represented in Figs. 12–14 (curves 2). The arrows show changes in the concentration and the phase relaxation rate for $\dot{C}H_2OH$ radicals after uv photolysis and storage of the samples in darkness for a long time. As was expected, the phase relaxation rate becomes a linear function of the average concentration over the whole sample region and corresponds to the case of a uniform distribution of paramagnetic centers.

The results obtained allow one to determine the local concentrations of radicals in methanol irradiated by various types of ionizing particles. For a uniform radical distribution the contribution of the S-S interaction to the ESE signal decay is characterized by the parameter $\alpha_p = 1.1 \times 10^{-13}$ cm^3/s. Therefore it is assumed that the spatial distribution of radicals inside the tracks is uniform, then the local concentration (cm^{-3}) is

$$C_{\text{loc}} = 0.9 \times 10^{13}(b^* - b_0).$$

Here b^* is the phase relaxation rate for average concentrations (doses) where track effects are observed, and b_0 is the initial phase relaxation rate. The total concentrations calculated in this way are listed in Table 2. Note that the similar values of the local concentrations could also be obtained from the condition $C_{\text{loc}} = C^*$, where C^* is the radical concentration corresponding to the intermediate part of the curve $b(C_{\text{av}})$. This also verifies qualitatively the track model of spatial distribution.

From the local radical concentration data one can estimate the geometrical dimensions of the regions of their stabilization, that is, the dimensions of the radical tracks. Suppose that the radical tracks are cylindrical with a radius R_{tr} and a length l equal to the path of ionizing particles, then

$$R_{\text{tr}} = \left(\frac{G_R \cdot \text{LET}}{100 \pi C_{\text{loc}}}\right)^{1/2}. \tag{53}$$

Local radical concentrations in methanol irradiated by various types of particles, as seen from Table 2, vary from $1.6 \times 10^{18} \, cm^{-3}$ to $8 \times 10^{18} \, cm^{-3}$. Starting from a LET of about 16 eV/Å, the local concentration remains approximately constant. When the LET changes from 11 to 540 eV/Å, the track radii increase from 90 to 180 Å. These data can be theoretically interpreted in terms of a methanol radiolysis mechanism that includes the intratrack recombination of ions and electrons and the secondary radical reactions.[26]

ESE phase relaxation measurements allow one to elucidate track effects in a number of other polar compounds irradiated by particles with high LET at low temperatures.[30] We shall not consider here these works in detail, since the experimental techniques and the analysis of the data are similar to those described for methanol.

5.4 γ-Irradiated Organic Compounds

The effects of spatial localization of $\dot{C}H_2OH$ radicals in spurs, blobs, and short tracks formed after γ-irradiation of frozen methanol were also observed.[28,30] It appears that in γ-irradiated methanol the local concentration is much lower ($C_{loc} = 1.7 \times 10^{18} \, cm^{-3}$) than that for other types of irradiation. Hence the region of track effects corresponds to very small doses (about 0.2 Mrad).

To date γ-irradiated methanol at 77 K is the only system where the effect of spatial inhomogeneity has been reliably observed. Phase relaxation measurements have been made for a number of other organic compounds such as alcohols, dibasic acids, hydrocarbons, polymers, and others γ-irradiated at 77 K.[22] The structure of the radicals formed after γ-irradiation have been well characterized. Their ESR spectra exhibit hyperfine structure extending over a range of 20–100 G. The amplitude of the pulsed microwave magnetic field of the ESE spectrometer in these experiments did not exceed 2–3 G. These investigations were carried out in the region of average concentrations of 5×10^{17}–$5 \times 10^{19} \, cm^{-3}$ (i.e., in the dose region of about 0.5–50 Mrad). For all compounds studied it was found that the kinetics of the two-pulse ESE signal decay could be explicitly described by the simple exponent $V(2\tau) = V_0 \exp(-2b\tau)$ and that the relaxation rate b grows linearly with the average concentration of radicals or the irradiation dose,

$$b = b_0 + \alpha C_{av}.$$

Values of α lie within $(1-2) \times 10^{-13} \, cm^3/s$, and b_0 varies for different radicals from 0.1×10^6 to $0.8 \times 10^6 \, s^{-1}$.

Both spectral and instantaneous diffusion mechanisms contribute to the two-pulse ESE signal decay of organic free radicals. However, since these spin systems are characterized by inhomogeneously broadened ESR spectra, one assumes on the basis of ESE studies of model systems that for organic free radicals in polyoriented matrices the main contribution to the two-pulse echo decay is made by spectral diffusion resulting from the random modulation of the S-S interaction between the radicals. Variations of the experimental quantity α are perhaps associated with insufficient accuracy of the ESR determinations of the average concentration of radicals C_{av}.

Thus these experiments revealed neither track nor spur inhomogeneity. Moreover, we think that γ-irradiated organic compounds at irradiation doses exceeding 0.5–1 Mrad can be considered good examples of matrices with homogeneously distributed paramagnetic centers.* Nevertheless, it follows from the results for methanol that the region of small doses of γ-irradiation must be studied more carefully.

5.5 Acid and Alkaline Solutions: A Detailed Structure of Radical Tracks

In a great number of organic and inorganic compounds the intermediate products of radiolysis have different precursors. Some products result from reactions of mobile secondary electrons; others are formed due to reactions of heavy cations. Differences in the mobility of these precursors may lead to different spatial distributions of the paramagnetic particles forming radical tracks. Therefore a detailed investigation of the spatial distribution of radicals in irradiated compounds would be extremely useful.

In frozen acid solutions of methanol the $\dot{C}H_2OH$ radicals are the only stable particles at 77 K.[26,27,30] Though local radical concentrations are determined by the ranges of mobile electrons and the migration of positive charges formed in the initial process

$$CH_3OH \rightsquigarrow CH_3OH^+ + e_m^-,$$

so far it is still impossible to study the spatial distribution of radicals that

---

* Note that for nonoverlapping tracks, $C_{loc} =$ constant up to the point where the tracks begin to overlap. Then $C_{loc} = C_{av}$ and the dose or C_{av} dependence of the phase relaxation rate obeys Eq. (52). In addition to this track inhomogeneity effect one can postulate another type of inhomogeneity in the spatial distribution of radicals in irradiated solids that results from radicals being stabilized at sites near crystal defects or at the surface. In this case the linear relation given by Eq. (52) must always hold during irradiation, where $C_{loc} = \eta C_{av}$, $\eta > 1$, and the measured slope is given by $\alpha = \eta \alpha_p$. Results of this type may be found in the literature.[24]

are formed from the electron and ion components of radiolysis. Such experiments were, however, carried out on irradiated frozen aqueous acid and alkali solutions.

In irradiated alkaline glasses at 77 K two paramagnetic particles are stabilized, trapped electrons (e_{tr}^-) and ion radicals (O^-). The spatial distribution of e_{tr}^- must be determined by the path length of the mobile electrons e_m^- formed during the initial ionization process

$$H_2O \rightsquigarrow H_2O^+ + e_m^-,$$

before their trapping. O^- ion radicals are formed by the reactions[32]

$$H_2O^+ + OH^- \rightarrow H_2O + \dot{O}H,$$

$$\dot{O}H \rightarrow O^- + H^+,$$

which may be preceded by a number of positive charge transfers between H_2O^+ and H_2O. These reactions result in the migration of positive charges throughout the medium and, in the end, determine the spatial distribution of the corresponding trapped O^- ion radicals. ESR spectra of O^- and e_{tr}^- overlap negligibly because of the difference in their g-factors. Using the ESE method it is possible to obtain data separately on the phase relaxation of these particles. Thus there appears the possibility of studying e_{tr}^- and O^- spatial distributions. Frozen solutions of $9M$ NaOH containing small amounts of T_2O for β-irradiation as well as γ-irradiated samples were studied.[33] Mean O^- concentrations were measured by the ESR method and e_{tr}^- concentrations were determined by both the ESR technique and spectrophotometrically by the absorption band at $\lambda = 585$ nm. The phase relaxation rate of trapped electrons was determined by the two-pulse ESE method at 77 K. O^- ion radicals at this temperature have a spin-lattice relaxation time $T_1 \approx 10^{-7}$ s and thus its ESE signal cannot be measured. At 4.2 K the O^- spin-lattice relaxation time increases to $T_1 \approx 10^{-3}$ s, which makes it possible to investigate the phase relaxation of this paramagnetic center.

Detailed investigations of the dependences of the e_{tr}^- and O^- phase relaxation rates on the concentration and the amplitude of the pulsed microwave field under β- and γ-irradiation show that instantaneous diffusion is the principal mechanism of phase relaxation for these centers. The influence of temperature on the phase relaxation rate of O^- ion radicals is indicative that this system is a T_1-type sample with a spin reorientation rate $W \leq 10^3$ at 4.2 K. The e_{tr}^- phase relaxation rate is described by

$$V(2\tau) = V_0 \exp(-2b\tau - m\tau^2),$$

and for O^- it is given by

$$V(2\tau) = V_0 \exp(-2b\tau - a\tau^3).$$

The quantities m and a are independent of the concentration of the paramagnetic centers and the difference in the kinetics is perhaps associated with the difference in nuclear local environments surrounding e_{tr}^- and O^-. In both cases b depends on B_1 and the concentration of paramagnetic centers and is determined by the relation typical of instantaneous diffusion,

$$b = \alpha_M \left\langle \sin^2 \frac{\theta}{2} \right\rangle_g \, C = b' \left\langle \sin^2 \frac{\theta}{2} \right\rangle_g. \tag{54}$$

The phase relaxation rates b during irradiation were determined experimentally. Values of b' were calculated by Eq. (54) using a Gaussian function for $g(\omega)$ with a width of 5 G for e_{tr}^-. In the case of O^-, $g(\omega)$ was calculated from the experimental ESR spectrum. Experimental results are given in Figs. 15 and 16.

It follows from these data that the effect of spatial inhomogeneity is obviously observed only in β-irradiated samples for O^- ion radicals (Fig. 16, curve b). In the case of e_{tr}^- the concentration dependences of the relaxation rate coincide for β- and γ-irradiated samples. The slope of the straight line for the e_{tr}^- (Fig. 15) gives the value of b',

$$b' = \alpha_M C_{av} = (6 \pm 2) \times 10^{-13} C_{av},$$

which is close to the theoretically calculated value $\alpha_M = 8.1 \times 10^{-13}$ cm^3/s, for a uniform distribution of particles with $S = \frac{1}{2}$. This is also the case for O^- ion radicals trapped in γ-irradiated samples. Here $\alpha_M = 7.2 \times 10^{13}$ cm^3/s,

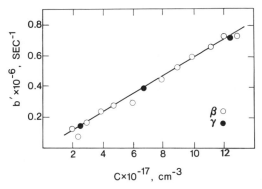

Figure 15 A plot of the phase relaxation rate b' versus the average concentration of trapped electrons in γ- and β-irradiated samples.

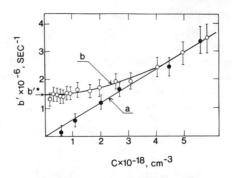

Figure 16 Plots of the phase relaxation rate versus the average concentration of O^- ion radicals for (*a*) γ-irradiated and (*b*) β-irradiated samples.

For O^- ion radicals in β-irradiated samples it is possible to determine the local concentration in a track. It is $C_{loc} = b'^*/\alpha_M = 2 \times 10^{18} \text{ cm}^{-3}$, where b'^* is the phase relaxation rate in the region where b' is independent of the average O^- concentration. Such a value of the local concentration corresponds to $R_{tr} = 75$ Å.

As for e_{tr}^- in β- and γ-irradiated samples, starting from at least a concentration of about $2 \times 10^{17} \text{ cm}^{-3}$, these particles are uniformly distributed. It is possible to estimate the value of C_{loc} for e_{tr}^- in the region of very small doses utilizing the sensitivity of the ESE spectrometer (i.e., to consider that $C_{loc} < 2 \times 10^{17} \text{ cm}^{-3}$). This gives $R_{tr} > 200$ Å for the e_{tr}^-. Thus it follows that trapped electrons and O^- ion radicals are distributed non-uniformly with respect to each other.[33]

Since the mean distance between O^- ion radicals in a track is about 70 Å, it is possible to conclude that before the formation of O^- from H_2O^+ there occurs at least 15 charge transfer events.

Frozen glassy solutions of sulfuric and phosphoric acids are also systems in which information on the structure of radical tracks can be obtained by the ESE method. In these systems mobile electrons e_m^- become trapped hydrogen atoms because of the reaction with H^+, and positively charged H_2O^+ ions, after charge migration, react with the acid ions (e.g., with SO_4^{-2}) to give the corresponding ion radicals (SO_4^-).

We have shown that trends in the phase relaxation for γ-irradiated samples of these acids for both the hydrogen atoms and the corresponding ion radicals could be adequately described in terms of a uniform spatial distribution of these particles. Therefore the decay rate of the two-pulse ESE signal was studied for α- and β-irradiated solutions of $8M$ H_2SO_4 and $12M$ H_3PO_4.[34]

We do not consider these investigations in detail, since the phase

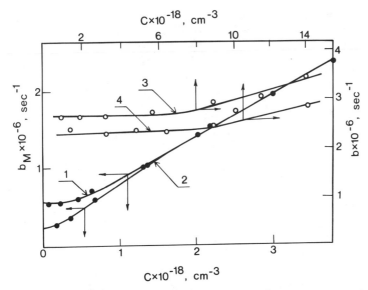

Figure 17 Plots of the phase relaxation rate versus the average concentration of paramagnetic particles in α-irradiated frozen aqueous acid solutions. Hydrogen atoms (1) in H_2SO_4 and (2) in H_3PO_4; (3) SO_4^- ion radicals and (4) PO_4^{-2} ion radicals.

relaxation of the hydrogen atoms and the ion radicals in α- and β-irradiated samples is similar to that in γ-irradiated samples except for the concentration dependence of the relaxation rate under irradiation for hydrogen atoms, PO_4^- and SO_4^-. These dependences show the presence of strong track effects for anions. Nevertheless, one can still estimate C_{loc} for the hydrogen atoms in α-irradiated samples.

The data obtained for glassy aqueous acid and alkaline solutions are listed in Table 3. Note that the local hydrogen atom concentrations in α- and β-irradiated H_2SO_4 solutions were reliably estimated by the ESE method so that the dependence of the spin-lattice relaxation rate at 77 and 4.2 K on the hydrogen atom concentration could be established. It was found that $C_{loc} = (0.75 - 3) \times 10^{18}$ cm^{-3} in α-irradiated samples and $C_{loc} = (0.5 - 2) \times 10^{16}$ cm^{-3} in β-irradiated samples.

It follows from the preceding data that the anion radicals resulting from the reaction of positively charged H_2O^+ ions are localized in a region with a radius of about 70 Å and the trapped electrons or hydrogen atoms, whose precursors are mobile electrons, are localized in much larger regions with radii exceeding 200–300 Å. By ESE investigations of dipolar interactions it is thus possible not only to determine local concentrations and other parameters characterizing track inhomogeneity but also to achieve a better understanding of the distribution of various paramagnetic

Table 3 Radiation-Chemical Yields, Local Concentrations, and Track Radii in Irradiated Frozen Aqueous Acid and Alkaline Solutions at 77 K[33,34]

Solution	Radical	Irradiation	G_R (1/100 eV)	$C_{loc} \times 10^{17}$ (cm^{-3})	R_{tr} (Å)
$8M \ H_2SO_4$	H	$\beta(T)$	2.0	≤ 2	≥ 200
	H	$\alpha(^{210}Po)$	0.85	~ 10	~ 250
	SO_4^-	$\beta(T)$	2.0	15	65
	SO_4^-	$\alpha(^{210}Po)$	1.7	120	100
$12M \ H_3PO_4$	H	$\alpha(^{210}Po)$	0.46	~ 4	~ 300
	PO_4^{-2}	$\alpha(^{210}Po)$	1.1	110	85
$8M \ NaOH$	e_{tr}^-	$\beta(T)$	2.7 ± 0.3	≤ 2	≥ 200
	O^-	$\beta(T)$	3.7 ± 1	20	75

centers in radical tracks. The results obtained may be considered a direct verification of the model representations of the structure of ionizing particle tracks in compounds that were put forward as far back as 1951 by Lea, Gray, and Platzman.[36]

6 STUDIES OF THE DISTRIBUTION OF RADICAL PAIRS IN SOLID MATRICES

Track spatial inhomogeneity in radical distributions results from the local character of energy losses of high-energy radiation in the substance. Another type of spatial inhomogeneity may appear in uv-irradiated solid matrices containing molecules or ions sensitive to the spectral region of applied uv light. For example, if photolysis results in photoionization, cation-electron pairs are the initial particles. Then the electrons either are stabilized in the matrix or react with acceptors and form another paramagnetic particle. In any case, particle pairs are formed. These pairs are uniformly distributed in the matrix; however, the distances separating the pairs may be different. It is thus possible to speak about an intrapair distance distribution: parent ion (cation)—electron (anion, ion radical, radical). Recently the ESE method was employed for the first time to study the pair distribution with respect to the distance between the partners and the variation of this distribution resulting from diffusion.[37] This approach was further developed in work by Dzuba et al.[38]

The main idea of the method is as follows. Let us consider pairs of r-distant paramagnetic particles A and B. We will observe the ESE signal from A-type particles whose precessional phases are randomized by their partners in the pairs, the B spins. The contribution of the B spins to the ESE signal decay depends on the rate of B spin reorientation and on the distance between A and B in the pairs. If this distance is sufficiently short, then the ESE spectrometer fails to detect the echo signal of such pairs. In fact, only those A spins whose B partners are at a distance exceeding some critical value can be observed. A certain region of effective relaxation exists around each B spin in which A spins make no contribution to the experimentally observed ESE signal. The volume of this region can be found in the simplest way when the B particles have sufficiently short paramagnetic relaxation times, T_{1B} and T_{2B}. Then the contribution of the B spins to the phase relaxation rate of the A spins is

$$b_{AB} = \langle \Delta\omega^2 \rangle T_{1B} \equiv K T_{1b} r^{-6}, \tag{55a}$$

where

$$K = \frac{16\pi}{15} S_B (S_B + 1) \gamma_A^2 \gamma_B^2 \hbar^2, \tag{55b}$$

and $\langle \Delta\omega^2 \rangle$ is the mean-square shift of the resonant A spins resulting from the dipolar interaction with the B particles. Let τ_{min} be the shortest time interval between the microwave pulses. Those A spins, for which $b_{AB}\tau_{min} > 1$, are assumed to be in the region of fast relaxation. The boundary of this region is determined by $b_{AB}\tau_{min} \approx 1$, hence, using Eq. (55a, b), we obtain for the volume of the rigion of fast relaxation,

$$v = \tfrac{4}{3}\pi \sqrt{K T_{1B}\tau_{min}}.$$

On the basis of the foregoing considerations, the distribution of hydrogen atoms with respect to Fe^{3+} was investigated using the ESE technique.[37] These pairs are formed at 77 K after photolysis of frozen aqueous H_2SO_4 solutions. Figure 18 represents the dependence of the logarithm of the amplitude of the two-pulse ESE signal from hydrogen atoms. Curve 1 is obtained immediately after photolysis and curve 2 after thermal annealing for an hour at 94 K. As seen from the figure, after annealing, the kinetics of the ESE signal decay does not appreciably change; however, the amplitude of the ESE signal increases noticeably. Note that there are no significant changes in the form and in the intensity of the ESR spectrum for the hydrogen (i.e., the total number of hydrogen atoms in the sample does not change.

The increase of the ESE signal amplitude can be attributed to the fact that a considerable portion of the hydrogen atoms are in close proximity to the Fe^{3+} ions so that the phase relaxation time T_2 of the hydrogen

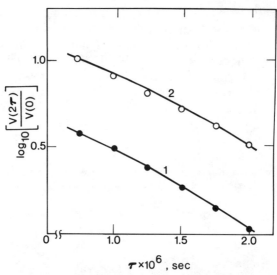

Figure 18 Plots of the two-pulse ESE signal amplitude for hydrogen atoms versus the time duration between pulses: (1) immediately after photolysis and (2) 90 min after annealing at 90 K.

atoms in this region is much shorter than the minimum time interval τ_{min} between microwave pulses.

Those hydrogen atoms that are in the region of fast relaxation make no contribution to the ESE signal decay. However, during annealing these diffuse, leaving this region and thus enhancing the ESE signal. The experimental dependence of the ratio of the echo amplitude V to the initial amplitude V_0 on the annealing duration at various temperatures is represented in Fig. 19. A threefold increase of the observed echo signal amplitude is indicative of the fact that initially about 65% of the hydrogen atoms are in the region of fast relaxation.

To determine the contribution of Fe^{3+} ions to the kinetics of the hydrogen atom phase relaxation, the experimental dependence of the logarithm of the ESE signal amplitude $V_0(2\tau)$ for samples containing no Fe^{3+} was subtracted from the analogous dependence for samples containing Fe^{3+}, $V_F(2\tau)$. Within the limits of experimental error, the influence of Fe^{3+} on the kinetics of the ESE signal decay for hydrogen atoms can be described by a dependence of the form of Eq. (9)

$$V_F(2\tau) = V_0(2\tau) \exp(-f\sqrt{2\tau}), \qquad (56)$$

where

$$f = kn + f_0, \qquad (57)$$

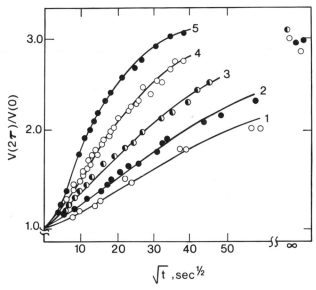

Figure 19 Plots of the two-pulse ESE signal amplitude for hydrogen atoms versus the duration of annealing at temperatures: (1) 90.1 K; (2) 91 K; (3) 93 K; (4) 95 K, and (5) 97 K.

and n is the Fe^{3+} concentration. The presence of the constant term f_0 in Eq. (57) is probably associated with the presence of some paramagnetic impurities in the solutions. Comparing Eq. (9) with the experimental value of $k = 6.7 \times 10^{-16}$ cm^3 s$^{1/2}$ and taking into account that $S_B = \frac{5}{2}$, we find that the spin-lattice relaxation time of hydrated Fe^{3+} ions is $T_1 = 4 \times 10^{-18}$ s. In reference 37 τ_{min} is set equal to 10^{-6} s from which the volume of fast relaxation was found to be equal to 3.5×10^{-19} cm^3 and the radius of this region R equal to 6.2×10^{-7} cm.

The experimental data (Fig. 19) on the dependence of the hydrogen atom ESE amplitude on annealing duration at various temperatures provides an estimate of the activation energy of hydrogen atom diffusion in frozen H_2SO_4 solutions. According to the model proposed, an increase in the echo signal amplitude during annealing is determined by the product of the diffusion coefficient for the hydrogen atoms at the temperature of annealing and the annealing duration. Setting the time at which the half limiting value is reached in the curves of Fig. 19 to be the characteristic time of the increase and assuming an Arrhenius law for the temperature dependence of the diffusion coefficient, we have

$$t_{1/2} \cong \frac{1}{D} = \frac{1}{D_0} \exp\left(\frac{E}{kT}\right). \tag{58}$$

Here E is the activation energy of diffusion and T is the absolute temperature of annealing. One finds from the temperature dependence of $t_{1/2}$ that $E = 5.5 \pm 0.5$ kcal/mole and $D_0 \approx 7 \times 10^{-4}$ cm^2/s. Therefore the T_{1B} dependence of the ESE signal amplitude can yield information about the pair distribution. In a recent study this method was used to investigate the distribution of $\dot{C}H_2OH - Fe^{2+}$ pairs that appear after photolysis of $FeCl_3$ in frozen methanol solutions at 77 K.[38] On freezing, T_1 for the Fe^{2+} ions increases and thus the characteristics of the $\dot{C}H_2OH$ radical relaxation change. The temperature was varied from 8 to 77 K and it was shown that 70% of the radicals in this system are ≤ 14 Å-distant from Fe^{2+}.[38] Over the region of 14–25 Å, the probability $n(r)$ for finding a radical in the volume element d^3r at a distance r from the parent ion can be approximated by $n(r) = 1.29 \times 10^{21} \exp(-0.34r)$ cm^3, where r is in Å.

REFERENCES

1. W. B. Mims, in S. Geschwind, Ed., *Electron Paramagnetic Resonance*, Plenum, New York, 1972, pp. 263–351.
2. K. M. Salikhov, A. G. Semenov, and Yu. D. Tsvetkov, *Electron Spin Echoes and Their Applications*, Science, Novosibirsk, 1976, p. 342.
3. B. Herzog and E. L. Hahn, *Phys. Rev.*, **103**, 148 (1956).
4. W. B. Mims, K. Nassau, and J. L. McGee, *Phys. Rev.*, **123**, 2059 (1961).
5. J. R. Klauder and P. W. Anderson, *Phys. Rev.*, **125**, 912 (1962).
6. E. L. Hahn, *Phys. Rev.*, **80**, 580 (1950).
7. W. B. Mims, *Phys. Rev.*, **168**, 370 (1968).
8. G. M. Zhidomirov and K. M. Salikhov, *J. Exptl. Theoret. Phys.*, **56**, 1933 (1969).
9. A. D. Milov, K. M. Salikhov, and Yu. D. Tsvetkov, *J. Exptl. Theoret. Phys.*, **63**, 2329 (1972).
10. A. D. Milov, K. M. Salikhov, and Yu. D. Tsvetkov, *Fiz. Tverd. Tela*, **15**, 1187 (1973).
11. A. M. Raitsimring, K. M. Salikhov, B. A. Umansky, and Yu. D. Tsvetkov, *Fiz. Tverd. Tela*, **16**, 756 (1974).
12. S. R. Hartmann and P. Hu, *J. Mag. Resonance*, **15**, 226 (1974).
13. A. M. Raitsimring, K. M. Salikhov, S. A. Dikanov, and Yu. D. Tsvetkov, *Fiz. Tverd. Tela*, **17**, 3174 (1975).
14. A. M. Raitsimring, K. M. Salikhov, S. F. Bychkov, and Yu. D. Tsvetkov, *Fiz. Tverd. Tela*, **17**, 484 (1975).
15. N. Blombergen, S. Shapiro, P. S. Pershan, and J. O. Artman, *Phys. Rev.*, **114**, 445 (1959).
16. P. W. Anderson, *Phys. Rev.*, **109**, 1492 (1958).
17. J. P. Gordon and K. D. Bowers, *Phys. Rev. Lett.*, **1**, 368 (1958).
18. S. Chandrasekhar, *Rev. Mod. Phys.*, **15**, 1 (1943).
19. A. D. Milov, K. M. Salikhov, and Yu. D. Tsvetkov, *Fiz. Tverd. Tela*, **14**, 2259 (1972).
20. G. R. Khutsishvili, *Usp. Fiz. Nauk.*, **87**, 211 (1965).

21. A. D. Milov, K. M. Salikhov, and Yu. D. Tsvetkov, *Fiz. Tverd. Tela*, **14,** 2211 (1972).

22. A. M. Raitsimring, Candidate Thesis, Novosibirsk, 1971.

23. P. B. Ayscough, H. E. Evans, and A. P. McGann, *Nature (Lond.)*, **203,** 226 (1964).

24. Ya. S. Lebedev and V. I. Muromtsev, *ESR and Relaxation of Trapped Radicals*, Khimiya, Moscow, 1972, p. 255.

25. Yu. D. Tsvetkov, *Izv. Sib. Otd. Akad. Nauk SSSR, Ser. Kim. Nauk.*, **4,** 90 (1976).

26. A. M. Raitsimring, Yu. D. Tsvetkov, and V. M. Moralev, *Int. J. Rad. Phys. Chem.*, **5,** 249 (1973).

27. A. M. Raitsimring, V. M. Moralev, and Yu. D. Tsvetkov, *Khim. Vys. Energ.*, **4,** 180 (1970).

28. A. M. Raitsimring, A. P. Pomytkin, and Yu. D. Tsvetkov, *Khim. Vys. Energ.*, **4,** 369 (1970).

29. Yu. D. Tsvetkov and A. M. Raitsimring, *Rad. Effects*, **3,** 61 (1970).

30. A. M. Raitsimring, V. M. Moralev, and Yu. D. Tsvetkov, *Khim. Vys. Energ.*, **7,** 125 (1973).

31. B. N. Shelimov, N. B. Fok, and V. V. Voevodsky, *Kinet. Katal.*, **4,** 539 (1963).

32. A. K. Pikaev, *Solvated Electron in Radiational Chemistry*, Nauka, Moscow, 1969, p. 289.

33. A. M. Raitsimring, P. I. Samoilova, V. M. Moralev, and Yu. D. Tsvetkov, preprint, *Instit. Chem. Kinet. Combust., Novosibirsk*, **1,** 4 (1976).

34. A. M. Raitsimring, V. M. Moralev, and Yu. D. Tsvetkov, *Khim. Vys. Energ.*, **9,** 517 (1975).

35. A. M. Raitsimring and Yu. D. Tsvetkov, *Khim. Vys. Energ.*, in press.

36. D. E. Lea, *Actions of Radiations on Living Cells*, London, 1955; L. H. Gray, *J. Chim. Phys.*, **48,** 172 (1961); H. Fröhlich and R. L. Platzman, *Phys. Rev.*, **92,** 1152 (1953).

37. P. P. Borbat, V. M. Berdnikov, A. D. Milov, and Yu. D. Tsvetkov, *Fiz. Tverd. Tela*, **19,** 1080 (1977).

38. S. A. Dzuba, A. M. Raitsimring, and Yu. D. Tsvetkov, *Theor. Exp. Chem.*, in press.

8 MODULATION OF ELECTRON SPIN-ECHO DECAY IN SOLIDS

Larry Kevan

Department of Chemistry
Wayne State University
Detroit, Michigan

1 ORIGIN OF ELECTRON SPIN-ECHO MODULATION

Spin echoes can be produced in response to suitable resonant pulse
sequences in magnetic resonance experiments. The pulse sequences reo-
rient the magnetic dipoles such that they dephase and then rephase in
response to all time-dependent magnetic interactions in the system; the
rephasing of the magnetic dipoles to reform the macroscopic magnetic
moment constitutes the echo. The echo thus measures the net magnetization of the spin system at a given time. In general, the echo intensity
decays as the time scale of the pulse sequences increases, which gives
information about various relaxation times in the spin system. This is the
topic of Chapters 5, 6, and 7 in this volume.

In the present chapter we are concerned with deriving *structural* information from the echo signal. This is possible when the pulse sequences
cause precessing magnetic nuclei to produce a regular time dependence of
the magnetic field at the unpaired electron. This results in a modulation
of the echo signal as the time scale of the pulse sequences is changed.
Basically this modulation is due to weak hyperfine interactions and hence
gives structural information akin to that obtained from electron nuclear
double resonance (ENDOR).[1]

Spin echoes were discovered in nuclear magnetic resonance by Hahn,
and his original paper[2] reports the observation of modulation in nuclear
spin echoes reflecting spin-spin coupling between different nuclei. This
nuclear spin-echo modulation phenomenon was later studied in more
detail.[3]

Modulation phenomena in electron spin echoes were first reported in 1961 and the basic theory of the modulation effect was given by Rowan, Hahn, and Mims,[4] who described how to calculate the expected modulation pattern from a given geometrical model.

1.1 Two-Pulse Echo Modulation

The physical origin of electron spin-echo modulation can be qualitatively described in both classical and quantum-mechanical terms.[4–6] Figure 1 shows a 90–180° two-pulse spin-echo sequence in which the time between the pulses is τ. As τ is varied the echo traces out an envelope that may show modulation. In order to gain some physical insight into the origin of the modulation we will describe classical and quantum-mechanical pictorial models. It must be remembered that these models are incomplete and do not give a clear, quantitative picture of the modulation frequencies associated with the electron magnetization observed at various times in the pulsing sequence. However, they do serve to illustrate several relevant physical principles. We will later describe a quantum-mechanical density matrix treatment of echo modulation that will give quantitative results on the expected modulation pattern.

The classical picture of spin-echo envelope modulation is illustrated by Fig. 2. Initially we assume an electron spin system with an isotropic g-factor characterized by a spin-angular momentum vector S in an external magnetic field H_0. We define the H_0 direction as the z-direction. The z-component S_z will be aligned along H_0 as shown in Fig. 2a. Typically H_0 is ~3000 G. A nearby nucleus with spin I located a distance r from S is subject to H_0 and the dipolar magnetic field from the electron spin H_e. since $H_e \sim \mu_e/r^3$ where μ_e is the electron magnetic moment, H_e is a significant fraction of H_0 for $r = 2$–5 Å. For example, $H_e = 343$ G for $r = 3$ Å. The H_0 and H_e fields combine vectorially to produce an effective field H_{eff} about which the nuclear spin vector I precesses. The precession cone is shown in Fig. 2a. The nuclei also produce a dipolar field H_n back at the electron, but H_n is typically so much smaller than H_0 that the precession of S around H_0 is little perturbed. For example, $H_n \sim \mu_n/r^3$

Figure 1 Illustration of modulation of a two-pulse electron spin-echo decay envelope. Microwave pulses 1 and 2 separated by time τ produce the echo signal at time τ after pulse 2. As τ is increased the echo amplitude changes and traces out an echo envelope that may be modulated.

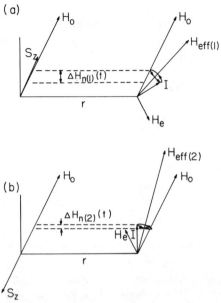

Figure 2   Diagrams showing origin of electorn spin-echo modulation. (*a*) The z-component of the electron spin S_z is oriented along the external magnetic field H_0. A nucleus of spin I at distance r from the electron precesses around an effective magnetic field $H_{\text{eff}(1)}$ due to the vector sum of H_0 and the electron dipolar field at the nucleus H_e. The precession of I produces a time-dependent magnetic field of amplitude $\Delta H_{n(1)}(t)$ at the electron, which modulates the electron Larmor frequency at the nuclear frequency $\gamma_n H_{\text{eff}(1)}$ where γ_n is the nuclear gyromagnetic ratio. (*b*) Now the electron spin is suddenly reversed by a microwave pulse so the direction of H_e changes and consequently so does $H_{\text{eff}(2)}$. *Some I* nuclei at the right precession position in (*a*) can follow the sudden field change and begin to precess about $H_{\text{eff}(2)}$ at frequency $\gamma_n H_{\text{eff}(2)}$ to produce a different amplitude magnetic field $\Delta H_{n(2)}(t)$ at the electron. The interference between the modulation of the electron motion at both $\gamma_n H_{\text{eff}(1)}$ and $\gamma_n H_{\text{eff}(2)}$ produces modulation beats in the electron precession at the nuclear frequencies. After a second microwave pulse this modulation appears in the electron spin-echo signal.

and is 658 times smaller than H_e for protons and 4288 times smaller the H_e for deuterons. Thus H_n is less than 1 G for $r > 2.5$ Å. However, the precession of I about H_{eff} causes a time-dependent variation in H_n at S. This is shown as $\Delta H_{n(1)}(t)$ in Fig. 2*a* where one sees that the magnitude of $\Delta H_{n(1)}(t)$ reflects the changing magnitude of the I_z component and depends on the orientation of H_{eff}. Thus $\Delta H_{n(1)}(t)$ modulates the electron Larmor frequency by its frequency, which is the nuclear Larmor frequency $\omega_{n(1)}$.

When the electron spins are flipped by a resonant microwave pulse, the direction of S_z instantaneously changes to $-S_z$ as shown in Fig. 2*b*. This instantaneously reverses the direction of H_e at the nuclei and alters the direction of H_{eff}. If this microwave pulse occurs in a time short compared to the nuclear Larmor period (~500 ns for deuterons and ~80 ns for protons in a 3000 G field) all the nuclei cannot adiabatically follow the new $H_{eff(2)}$ direction. Those nuclei that cannot follow continue to precess around $H_{eff(1)}$ as in Fig. 2a and to produce $\Delta H_{n(1)}(t)$ at frequency $\omega_{n(1)}$. Those nuclei that are near H_0 in the precession path in Fig. 2a are able to follow adiabatically the new $H_{eff(2)}$ field and begin to precess about $H_{eff(2)}$ to produce $\Delta H_{n(2)}(t)$ at frequency $\omega_{n(2)}$. The net result of the microwave pulse is to produce a *branching* of the precessing nuclei into two sets that in turn produce two different nuclear modulation fields at the electrons. The electron Larmor frequency is thus modulated by magnitudes $\gamma_e |\Delta H_{n(1)}|$ and $\gamma_e |(\Delta H_{n(2)}|$ at frequencies $\omega_{n(1)} = \gamma_n |H_{eff(1)}|$ and $\omega_{n(2)} = \gamma_n |H_{eff(2)}|$ where γ_e and γ_n are the electron and nuclear magnetogyric ratios. Interference between these two nuclear frequencies produces beats in the electron free precession signal at the nuclear frequencies.

The free precession signal can be observed in a pulsed ESR spectrometer with a bimodal cavity.[7] However, the experiments with this spectrometer were only carried out on liquids where the electron-nuclear dipolar coupling is motionally averaged and no modulation would be expected. Thus the nuclear modulation effect on the free precession signal has not yet been observed.

The modulation of the electron Larmor frequency by the nuclear frequencies also appears after a second resonant microwave pulse is applied to produce an electron spin echo. The echo measures the magnitude of the electron magnetization which is proportional to S_z and which is modulated by the nuclear fields. A quantitative calculation shows that the two-pulse echo is modulated by $\omega_{n(1)}$, $\omega_{n(2)}$, $\omega_{n(1)} + \omega_{n(2)}$ and $\omega_{n(1)} - \omega_{n(2)}$. These nuclear frequencies serve to identify the interacting nuclei. From Fig. 2 one can directly see that the amplitude of the echo modulation $\Delta H_n(t)$ depends on the tilt of the nuclear precession cone with respect to H_0, hence on the magnitude of H_{eff}, and thus on the electron-nuclear distance r. H_{eff} also depends on the number of equivalent nuclei at r. Thus we see qualitatively that an analysis of the echo modulation pattern should give information on the identity, number, and distance of weakly interacting nuclei.

The quantum-mechanical picture of spin-echo envelope modulation is best described in terms of a spin energy level diagram. The simplest such diagram is that for a $S = \frac{1}{2}$, $I = \frac{1}{2}$ spin system, which is shown in Fig. 3. The electron and nuclear spin quantum numbers M_s and M_I are shown as labels for each energy level but each energy level is, in fact, a mixed state whose wavefunction depends on both nuclear spin orientations. This

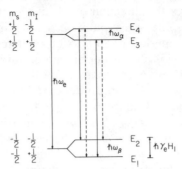

Figure 3 Energy level diagram for an $S = \frac{1}{2}$, $I = \frac{1}{2}$ spin system with positive dipolar coupling. The energy levels are denoted by their electron (m_s) and nuclear (m_I) spin quantum numbers and H_1 is the microwave magnetic field. The solid and dashed lines correspond to allowed and forbidden transitions, respectively.

means that M_I is not a "good" quantum number and is not strictly conserved in a spin transition. Consequently, both allowed ESR transitions ($M_I = 0$) and forbidden ESR transitions ($M_I = \pm 1$) can occur. These are shown by solid and dashed lines, respectively, in Fig. 3.

In a spin-echo experiment each pulse, whether it is 90°, 180°, or some other "classical" turning angle, excites spins from the lower states (E_1 and E_2) to the upper states (E_3 and E_4). Furthermore, if the pulse magnitude $\hbar \gamma H_1$ is greater than $E_2 - E_1$, both allowed and forbidden ESR transitions will be excited with the same pulse. Thus electron spins in both E_1 and E_2 will be excited to E_3. This constitutes an analogous kind of *branching* transition to that we discussed classically. The electrons from E_1 will be precessing at a frequency different from that of the electrons from E_2 by the nuclear hyperfine frequency ω_β. In other words, the first microwave pulse produces two different sets of spins in energy level E_3 that can interfere with each other's motion to generate a modulation of the electron magnetization at the nuclear frequency ω_β. The same situation obtains for spins excited to E_4. Likewise, the spins that are excited from both levels E_3 and E_4 into levels E_1 and E_2 introduce modulation of the electron magnetization at the nuclear frequency ω_α. If a second microwave pulse is applied before these induced modulations have decayed away, the magnetization will show the additional modulation frequencies $\omega_\alpha + \omega_\beta$ and $\omega_\alpha - \omega_\beta$. Thus the two-pulse spin echo is modulated by ω_α, ω_β, $\omega_\alpha + \omega_\beta$, and $\omega_\alpha - \omega_\beta$. In general, for more complex spin systems (e.g., $I > \frac{1}{2}$) the two-pulse spin echo will be modulated by the nuclear frequencies and all their sums and differences. This sounds like a very complex modulation pattern to unravel. However, for *weak* nuclear coupling to the electron ($r > 3 \text{ Å}$), $\omega_\alpha \sim \omega_\beta$ and the spin echo will only show modulation at the nuclear frequency ω_α and at double the nuclear frequency $2\omega_\alpha$.

1.2   Three-Pulse Echo Modulation

The physical origin of modulation of the three-pulse electron spin echo is the same as for the two-pulse echo.[8] It essentially depends on transient excitation of branched ESR transitions involving two nuclear spin orientations. Figure 4 shows a 90°–90°–90° three-pulse spin-echo sequence. The time between the first two pulses is τ, as in the two-pulse echo, and, in fact, a weak two-pulse echo is generated. However, τ is normally held fixed at a value corresponding to a *minimum* in the two-pulse echo modulation pattern. The time between the second and third pulses is T and it is this time that is varied to generate the echo envelope. The echo occurs at time τ after the third pulse.

One important feature of the three-pulse echo modulation is that the modulation frequencies include only the nuclear frequencies ω_α and ω_β without also including their sums and differences as in the two-pulse case. A qualitative argument for this simplification of the three-pulse echo modulation can be seen by referring to Fig. 3. In the presence of branching transitions the first pulse generates modulation effects at the nuclear frequencies; the second pulse additionally incorporates the sums and differences of the nuclear frequencies. A third pulse causes these

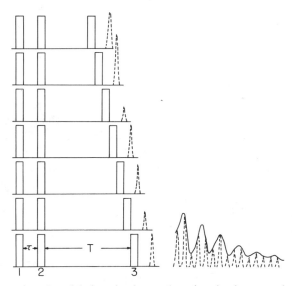

Figure 4   Illustration of modulation of a three-pulse spin-echo decay envelope. The time τ between microwave pulses 1 and 2 is kept fixed while the time T between pulses 2 and 3 is varied. The echo appears at time τ after pulse 3. As T is increased the echo amplitude varies as shown from top to bottom and traces out the echo envelope shown at the bottom, which may be modulated.

sums and differences to interfere and leave only the nuclear frequencies in the modulation pattern.

The three-pulse echo modulation appears to have the following advantages over two-pulse echo modulation. Fewer nuclear frequencies are involved and the modulation occurs over a longer time scale in the three-pulse case. Furthermore, from our discussion thus far it would seem that the two-pulse and three-pulse echo modulation carry duplicate information about the nuclear coupling. Our later discussion will show that, in practice, the two-pulse and three-pulse echo modulation patterns give *complementary* information. Because of the difference in time scales two-pulse echo experiments are more sensitive to coupling with closer nuclei, whereas three-pulse echo experiments are more sensitive to farther nuclei.

1.3   Conditions Required to Observe Modulation

Three conditions are required to observe electron spin-echo modulation. First, there must be branching transitions in the energy level diagram with the weaker transition intensity $\geqslant 10^{-2}$ of the stronger transition intensity. In other words, there must be magnetic nuclei in the vicinity of the unpaired electron that couple to it. Second, the nuclear hyperfine coupling and/or the free nuclear frequency must be strong enough so that the nuclear frequencies ω_α and ω_β are larger than the frequency width of the spin packets that make up the inhomogeneously broadened ESR line. This ensures that several modulation periods occur within the echo decay time. In the classical diagram in Fig. 2 condition 1 is equivalent to saying that the electron dipolar field at nucleus H_e must be large enough so that $H_{\text{eff}(1)}$ and $H_{\text{eff}(2)}$ are oriented significantly differently. Thus sufficiently distant nuclei do not contribute to the modulation. The third condition is instrumental and requires that the branching transitions in the energy level diagram must be excited. In other words, the microwave magnetic field H_1 must be large enough that its frequency $\gamma_e H_1$, where γ_e is the electron magnetogyric ratio, is larger than the nuclear frequencies ω_α and ω_β. (See Fig. 3). If H_1 is too small the branching transitions will not be excited. This fact can sometimes be used to advantage when different kinds of nuclei of quite different nuclear frequencies are coupled to the electrons. It may then be possible to vary H_1 so as to excite branched transitions for only one nuclear type. The most common case is for hydrogen and deuterium nuclei. Figure 5 shows an example for a solvated electron localized in a partially deuterated organic glass, 2-methyltetrahydrofuran. For $H_1 \sim 3$ G branched transitions associated only

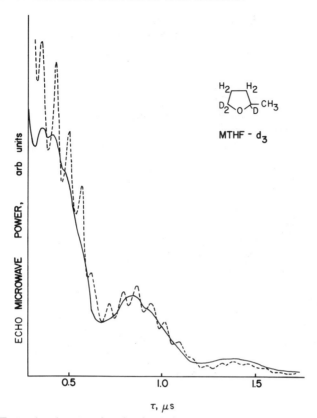

Figure 5 Two-pulse electron spin echo signals from a solvated electron in MTHF-d_3 glass at 77 K. The solid curve is for $H_1 = 3.0$ G and shows only deuteron modulation. The dashed curve is for $H_1 = 9.5$ G and shows both deuteron and proton modulation. The microwave power was detected in this experiment instead of the microwave magnetic field. [From L. Kevan et al., *J. Chem. Phys.*, **63**, 409 (1975).]

with deuterium nuclei are excited and the modulation shows a characteristic period of $\sim 2\ \mu$s in a ~ 3000 G field. However, when H_1 is increased to ~ 9 G both deuterium and proton branched transitions are excited and the faster proton modulation is seen superimposed on the deuterium modulation. The classical analog of this third condition for echo modulation is that the microwave pulse length should be shorter than the shortest modulation period to be observed. This point was already made when Fig. 2 was discussed.

We have portrayed the two-pulse sequence in Fig. 1 as consisting of 90 and 180° pulses where the turning angle is given by $\gamma_e H_1 t_p$ and t_p is the pulse width. This idealized pulse sequence makes it easy to explain the origin of spin echoes based on a classical model of precessing electron spin vectors. However, most other two-pulse sequences (e.g., 90–90°, 90–120°, etc.) also give an echo, but of reduced intensity. The optimum 90–180° pulse sequence can be obtained by varying H_1 or t_p. To vary H_1 requires two microwave sources with different attenuations. To vary t_p on successive pulses requires a complex pulse control system. Both options are cumbersome, and experimentally it is much easier to generate echos with pulses of the same intensity and width. Most current electron spin-echo spectrometers described in the literature operate this way. For equal-length pulses the relative intensity at the center of the echo I_{rel} is given by

$$I_{rel} = |-0.5 \sin (\gamma_e H_1 t_p) + 0.25 \sin (2\gamma_e H_1 t_p)|, \qquad (1)$$

where $I_{rel} = 1$ for a 90–180° pulse sequence.[9] The echo signal is then maximized for $\gamma_e H_1 t_p = 120°$. Equation (1) assumes the absence of relaxation effects and is derived for H_1 larger than the ESR linewidth. However, even when H_1 is less than the ESR linewidth, Eq. (1) applies at the center of the echo signal.

2 ANALYSIS OF TWO-PULSE SPIN-ECHO MODULATION

To extract explicit hyperfine information from spin-echo modulation patterns it is necessary to calculate the expected modulation pattern for a certain assumed hyperfine interaction or electron-nuclear geometry. Then by varying the assumed hyperfine interaction for a simulated pattern until it fits the observed pattern the weak hyperfine interaction parameters may be determined.

Rowan, Hahn, and Mims[4] first presented a density matrix formalism for simulating echo modulation patterns and obtained explicit expressions for the $S = \frac{1}{2}$, $I = \frac{1}{2}$ case. This was extended by Zhidomirov and Salikhov[10] to the $S = \frac{1}{2}$, $I = 1$ case. Later Mims[11,12] rewrote the formalism and carefully discussed the approximations and limitations involved for both two- and three-pulse echo modulation. Other workers have simulated patterns for $S = \frac{3}{2}$, $I = \frac{5}{2}$[6] and $S = \frac{1}{2}$, $I = \frac{7}{2}$[5] cases to compare with experiment but they have not given explicit expressions for these cases and have presumably used numerical methods for evaluation.

2.1 Simulation of Modulation Patterns for one Nucleus

Essentially one wants to calculate the echo intensity as a function of τ due to the excitation of branching transitions between two electron spin states in the absence of relaxation effects. Relaxation processes cause the echo to decay monotonically as discussed in Chapters 5–7. The total time dependence of the spin-echo intensity (V) can be written as

$$V_{echo} = V_{decay} V_{mod},\qquad(1)$$

and we are currently concerned with the calculation of V_{mod}. Instead of resorting to time-dependent perturbation theory to calculate V_{mod}, it is easier for this problem to use its variant, the density matrix formalism.[13] The density matrix takes the place of the wavefunction in the calculation of the expectation value of a physical observable.

The relationship of a density matrix to a wavefunction is as follows. Any wavefunction for a system can be expanded in terms of a complete set of orthonormal functions U_n

$$\psi(q, t) = \sum_n C_n(q, t) U_n(q),\qquad(2)$$

where the time dependence is all in the coefficients C_n. To calculate a property such as energy E of a real system consisting of many spins we must find the ensemble average of the energy operator $\mathcal{H}$, which can be written as

$$E = \sum_{n,m} \overline{C_m^* C_n} \langle U_m | \mathcal{H} | U_n \rangle,\qquad(3)$$

where we have used (2) to replace the wavefunction ψ and the bar denotes an ensemble average. The density matrix ρ is now defined as the matrix of the coefficient products $\overline{C_m^* C_n}$. An important result of this formalism is that a physical observable such as energy is given by

$$E = \text{Trace}(\rho \mathcal{H}).\qquad(4)$$

Furthermore, the matrix elements of ρ are calculable from the equation of motion for the density matrix, which is

$$\frac{d\rho}{dt} = \left(\frac{i}{\hbar}\right)[\rho, \mathcal{H}],\qquad(5)$$

where the brackets represent the commutator of ρ and $\mathcal{H}$. Equation (4) shows formally that physical properties can be calculated by using the density matrix instead of the wavefunction of a system.

To apply Eq. (4) we must first write the appropriate Hamiltonian and then obtain an expression for ρ. The amplitude of the echo is just the magnetization determined from a spin Hamiltonian. We do not need to include any relaxation terms in the spin Hamiltonian because these will affect V_{decay} but not V_{mod}. Thus the appropriate Hamiltonian for echo modulation alone is

$$\mathcal{H} = \mathcal{H}_0 + \mathcal{H}_t, \tag{6}$$

where $\mathcal{H}_0$ is the ordinary static Hamiltonian involving interaction with one nucleus in the absence of a microwave pulse and $\mathcal{H}_t$ is the time-dependent part associated with a pulse. These are given by Eq. (7) and (8) where

$$\mathcal{H}_0 = \beta H_0 \cdot g \cdot S - g_n \beta_n H_0 \cdot I + h S \cdot A \cdot I, \tag{7}$$

$$\mathcal{H}_t = 2g\beta H_1 S_y \cos \omega_e t, \tag{8}$$

H_0 is the external magnetic field in G, g is the electron g-tensor, A is the hyperfine tensor with elements in Gz, H_1 is the rotating component of the microwave magnetic field of frequency ω_e, and the other symbols have their usual meanings in magnetic resonance. The terms in $\mathcal{H}_0$ represent the electron Zeeman, nuclear Zeeman, and electron-nuclear hyperfine interactions, respectively. At this point we have neglected any quadrupole interaction; we will include this later. We can simplify $\mathcal{H}_0$ for the typical experiment at magnetic fields greater than about 3 kG (i.e., X-band ESR) by choosing H_0 to be along the z-axis, by choosing the nucleus to lie in the xz plane so there is no I_y term, and by assuming the high field approximation that makes the S_x and S_y terms negligible. The high-field approximation is justified above 3 kG for the weak hyperfine interactions that contribute to the echo modulation. In addition, we choose g to be isotropic as justified for a typical free radical; when g is not isotropic, as for many paramagnetic metal ions, $\mathcal{H}_0$ may be treated as in any continuous wave (cw) electron spin resonance (ESR) spectral analysis.[4] In addition, we assume a point dipole approximation for the anisotropic hyperfine interactions. With these simplifications we have Eq. (9), where

$$\mathcal{H}_0 = g\beta H_0 S_z - g_n \beta_n H_0 I_z + h\mathbf{a} I_z S_z$$

$$+ \frac{g g_n \beta \beta_n}{r^3} [(3 \cos^2 \theta - 1) I_z + 3 \sin \theta \cos \theta I_x] S_z, \tag{9}$$

a is the isotropic hyperfine coupling in Hz, r is the distance between the electron and nucleus, and θ is the angle between H_0 and the position vector **r** joining the electron and nucleus.

Now let us return to the application of Eq. (4) to calculate the echo modulation. The density matrix $\rho(2\tau)$ is time dependent so Trace $(\rho \mathcal{H})$ will depend only on the time-dependent part of $\mathcal{H}$ given by (8). The normalized echo modulation signal $V_{mod}(2\tau)$ is then given by

$$V_{mod}(2\tau) = \frac{\mathrm{Tr}\,(\rho(2\tau)S_y)}{\mathrm{Tr}\,(\rho(0)S_y)}, \tag{10}$$

where the denominator is the normalization factor and $\rho(0)$ is the time-independent equilibrium density matrix in the high-temperature approximation that exists prior to the first pulse.

The electron Zeeman energy dominates $\mathcal{H}_0$ so $\rho(0)$ is given by[4]

$$\rho(0) = \hbar\omega_e \sum_{j=1}^{N} \frac{S_{zj}}{kT\,\mathrm{Tr}\,1} = \frac{\hbar\omega_e \hat{S}_z}{kT\,\mathrm{Tr}\,1}, \tag{11}$$

where the sum is over N electrons per cm^{-3} and $\hat{S}$ represents that the summation has been taken.

An expression for $\rho(2\tau)$ can be obtained from $\rho(0)$ by applying operators appropriate to the pulse and free precession periods in the two-pulse spin-echo sequence.

The thermal equilibrium density matrix is given by

$$\rho(0) = \frac{1}{Z} \exp\left(\frac{-\mathcal{H}_0}{kT}\right), \tag{12a}$$

where $\mathcal{H}_0$ is time independent and Z is the partition function.[13] Since the electron Zeeman energy dominates $\mathcal{H}_0$, it will determine the thermal equilibrium properties. In the high-temperature approximation $Z = 2S + 1$, and the exponential is expanded keeping only the first two terms. Then

$$\rho(0) = \frac{1}{2S+1} \left(1 - \frac{g\beta H_0 S_z}{kT}\right), \tag{12b}$$

where we must remember that S_z is an ensemble average over all electron spins.

Now from Eq. (5) we may obtain a formal solution when $\mathcal{H}$ is independent of time, which is

$$\rho(t) = \exp\left(\frac{-i\mathcal{H}t}{\hbar}\right)\rho(0)\exp\left(\frac{i\mathcal{H}t}{\hbar}\right). \tag{13}$$

To apply (13) we eliminate the time dependence in $\mathcal{H}$ that is contained in $\mathcal{H}_t$ by transforming our Hamiltonian to the electron rotating frame. This is equivalent to subtracting the electron Zeeman term so we have

$$\mathcal{H}_r = \mathcal{H} - g\beta H_0 S_z, \tag{14}$$

$$\mathcal{H}_r(\text{pulse off}) = -g_n\beta_n H_0 I_z + h a I_z S_z + \frac{g g_n \beta \beta_n}{r^3}$$

$$\times [(3\cos^2\theta - 1)I_z + 3\sin\theta\cos\theta I_x]S_z, \tag{15}$$

$$\mathcal{H}_r(\text{pulse on}) = H_r(\text{pulse off}) + 2g\beta H_1 S_y \approx 2g\beta H_1 S_y, \tag{16}$$

where H_r (pulse off) is neglected during the microwave pulses since the pulse energy $g\beta H_1$ is larger than the nuclear Zeeman or hyperfine energies for small hyperfine couplings such as we are considering. Although this assumption seems reasonable it may be borderline for longer pulses (hence smaller H_1) and for nuclei closer than ≈ 3 Å. This assumption is crucial to obtain simple equations for the echo modulation and some discrepancies between simulations and experiment may be associated with the partial failure of this assumption.

The density matrix $\rho(2\tau)$ is now obtained from Eq. (13) by successively applying the appropriate time-independent Hamiltonians during the first pulse (90°), the free precession time after the first pulse, the second pulse (180°), and the free precession time after the second pulse. Let $R_i = \exp(-i\mathcal{H}_i t/\hbar)$, then

$$R_1 = \exp\left(\frac{-i2g\beta H_1 S_y \pi}{2\hbar}\right), \tag{17}$$

$$R_2 = \exp\left(\frac{-i\mathcal{H}_r(\text{pulse off})\tau}{\hbar}\right), \tag{18}$$

$$R_3 = \exp\left(\frac{-i2g\beta H_1 S_y \pi}{\hbar}\right), \tag{19}$$

and

$$\rho(2\tau) = R_2 R_3 R_2 R_1 \rho(0) R_1{}^* R_2{}^* R_3{}^* R_2{}^*. \tag{20}$$

Using (20) $V_{\text{mod}}(2\tau)$ is finally evaluated from Eq. (10). Details of this evaluation of the trace are given in the appendix of reference 4 and in more general form in reference 11.

2.1.1 CASE FOR $S=\frac{1}{2}$ AND $I=\frac{1}{2}$

For a $S=\frac{1}{2}$, $I=\frac{1}{2}$ system the result is

$$V_{\text{mod}}=(2\tau, I=\tfrac{1}{2})=1-2k\sin^2\left(\frac{\omega_\alpha\tau}{2}\right)\sin^2\left(\frac{\omega_\beta\tau}{2}\right), \qquad (21)$$

where

$$k=\left(\frac{\omega_I B}{\omega_\alpha\omega_\beta}\right)^2, \qquad \omega_I=\frac{g_n\beta_n H_0}{\hbar},$$

$$\omega_\alpha=\left[\left(\frac{A}{2}+\omega_I\right)^2+\left(\frac{B}{2}\right)^2\right]^{1/2},$$

$$\omega_\beta=\left[\left(\frac{A}{2}-\omega_I\right)^2+\left(\frac{B}{2}\right)^2\right]^{1/2}, \qquad (22)$$

$$A=\frac{1}{\hbar}\frac{gg_n\beta\beta_n}{r^3}(3\cos^2\theta-1)+2\pi\mathbf{a},$$

and

$$B=\frac{1}{\hbar}\frac{gg_n\beta\beta_n}{r^3}(3\cos\theta\sin\theta).$$

In the preceding expressions ω_I is the free nuclear frequency in rad/s at the applied field, θ is angle between H_0 and the electron-nuclear axis, and ω_α and ω_β are the nuclear frequencies in the upper and lower electron spin states as shown in Fig. 3.

Equation (21) can also be written in the equivalent form[10]

$$V_{\text{mod}}(2\tau, I=\tfrac{1}{2})=1-\frac{k}{4}[2-2\cos\omega_\alpha\tau-2\cos\omega_\beta\tau+$$

$$\cos(\omega_\alpha-\omega_\beta)\tau+\cos(\omega_\alpha+\omega_\beta)\tau], \quad (23)$$

which is not as compact as (22) but more clearly shows that the nuclear frequencies and their sums and differences occur in the modulation pattern.

2.1.2 CASE FOR $S=\frac{1}{2}$ AND $I=1$

For a $S=\frac{1}{2}$, $I=1$ system with negligible quadrupole interaction one finds

$$V_{\text{mod}}(2\tau, I=1)=1-\frac{16}{3}k\sin^2\left(\frac{\omega_\alpha\tau}{2}\right)\sin^2\left(\frac{\omega_\beta\tau}{2}\right)$$

$$+\frac{16}{3}k^2\sin^4\left(\frac{\omega_\alpha\tau}{2}\right)\sin^4\left(\frac{\omega_\beta\tau}{2}\right), \quad (24)$$

where $\omega_\alpha = \omega_{ab} = \omega_{bc}$ and $\omega_\beta = \omega_{de} = \omega_{ef}$ as defined in Fig. 6. If the electron-nuclear distance is large (typically >4 Å), B will be small and the k^2 term can be neglected. Then the form of Eq. (24) for $I = 1$ is the same as Eq. (21) for $I = \frac{1}{2}$.

The parameter k can be regarded as a modulation depth parameter. For the $I = \frac{1}{2}$ case we note that the modulation depth is the same for any $I = \frac{1}{2}$ nucleus, since all the factors in k scale by the same amount.[14] For the $I = 1$ case in Eq. (24) when the k^2 term can be neglected, it is seen that the modulation depth is greater by $\frac{8}{3}$ than for the $I = \frac{1}{2}$ case. Thus we expect deeper modulation for deuterons than for protons. This can, in fact, be seen in Fig. 5. In general for $S = \frac{1}{2}$ spin systems when the electron dipolar field at the nucleus is less than the nuclear Zeeman field (i.e., B is small and k^n terms for $n \geq 2$ can be neglected) and quadrupole interaction can be neglected, the modulation pattern is given by Eq. (21) with k replaced by[14]

$$k'(I) = \tfrac{4}{3}I(I+1)k. \tag{25}$$

Thus in this limiting case of very weak electron-nuclear dipolar coupling and negligible quadrupole interaction the modulation depth is proportional to $I(I+1)$.

When the preceding approximations are not valid and $I > 1$, the analytical expressions are too cumbersome to tabulate and the simulations are best performed by the general matrix formulation described by Mims.[11,12]

2.1.3 CASE FOR g-ANISOTROPY

In the case of axial g-anisotropy, Eq. (21) and (24) can still be used with

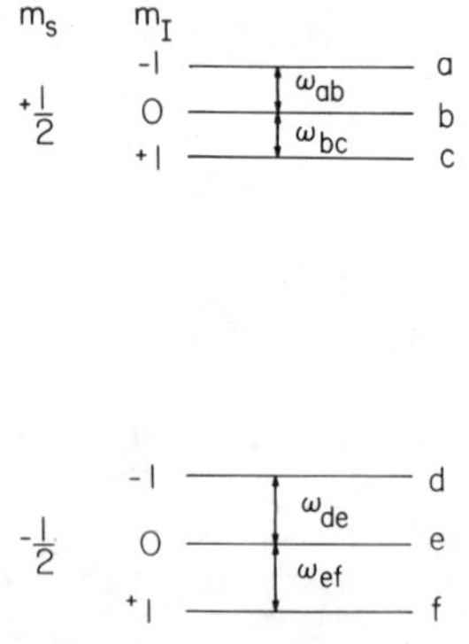

Figure 6 Energy level diagram for an $S = \frac{1}{2}$, $I = 1$ spin system with positive dipolar coupling and zero quadrupole coupling. The energy levels are denoted by their electron (m_s) and nuclear (m_I) spin quantum numbers. The nuclear transitions are shown.

the following redefinitions for A and B:[4]

$$g^2 = g_\parallel^2 \cos^2 \theta_0 + g_\perp^2 \sin^2 \theta_0, \tag{26}$$

$$A = \left(\frac{gg_n\beta\beta_n}{\hbar r^3}\right)\left[\left(\frac{3}{g^2}\right)(g_\parallel^2 \cos \theta_0 \cos \theta_I + g_\perp^2 \sin \theta_I \sin \phi_I)\right.$$

$$\left. \times (\cos \theta_0 \cos \theta_I + \sin \theta_0 \sin \theta_I \cos \phi_I) + \left(\frac{ar^3}{g_n\beta_n\beta}\right) - 1\right], \tag{27}$$

$$B = (B'^2 + C'^2)^{1/2}, \tag{28}$$

$$B' = \left(\frac{gg_n\beta\beta_n}{\hbar r^3}\right)\left[\left(\frac{3}{g^2}\right)(g_\parallel^2 \cos \theta_0 \cos \theta_I + g_\perp^2 \sin \theta_0 \sin \theta_I \sin \phi_I)\right.$$

$$\times (\cos \theta_0 \sin \theta_I \cos \phi_I - \sin \theta_0 \cos \phi_I)$$

$$\left. + \left(\frac{ar^3}{g_n\beta_n\beta} - 1\right)\left(\frac{g_\perp^2 - g_\parallel^2}{g^2}\right) \sin \theta_0 \cos \theta_0\right], \tag{29}$$

and

$$C' = \left(\frac{gg_n\beta\beta_n}{\hbar r^3}\right)\left[\left(\frac{3}{g^2}\right)(g_\parallel^2 \cos \theta_0 \cos \theta_I + g_\perp^2 \sin \theta_0 \sin \theta_I \cos \phi_I)\right.$$

$$\left. \times (\sin \theta_I \sin \phi_I)\right]. \tag{30}$$

In Eq. (26)–(30), θ_0 is the angle between H_0 and the z-axis, θ_I is the polar angle between the z-axis and the electron-nuclear axis, and ϕ_I is the azimuthal angle about the z-axis between the xz plane that contains H_0 and the electron-nuclear axis.

2.1.4 SUMMARY OF APPROXIMATIONS

It will be well to summarize the approximations made in the derivation of the basic simulation equations (21) and (24) so that comparison with experimental modulation patterns may be critically assessed.

1. The high-field approximation has been made so that S_x and S_y terms are dropped from the static Hamiltonian. This should generally be valid even for ESR spectra having large hyperfine interactions, since only a single ESR transition is being observed in each echo experiment.
2. An isotropic g-factor has been assumed. This is not a limiting assumption because g-anisotropy can be readily incorporated, as shown previously.
3. The point dipole approximation has been made for the anisotropic hyperfine interaction. The validity of this approximation depends on the localization of the unpaired electron wavefunction and the

electron-nuclear distance. Its validity must be assessed for each specific system, but it should often be valid for the weak hyperfine interactions typically investigated by echo modulation experiments.

4. Quadrupole interaction has been neglected. Of course, this is only a limitation for $I \geq 1$. This can be a serious approximation and will be discussed in more detail in Section 2.4. However, for the important case of deuterons it appears that errors introduced by neglect of the quadrupole interaction are small.[14,15] For nitrogen nuclei the effect of the quadrupole interaction appears more serious.[15]

5. The high-temperature approximation has been used to simplify the expression for the equilibrium density matrix [Eq. (12)]. This should be valid above 1 K.

6. To obtain the final density matrix the nuclear Zeeman and hyperfine energies have been assumed to be negligible compared to the pulse energy [Eq. (16)]. This assumption may be borderline in many cases but it is crucial to obtain Eq. (21) and (24).

2.2   Simulation for Several Nuclei

If two nuclei couple to the unpaired electron the Hamiltonian contains additional terms and may be written in abbreviated form as

$$\mathcal{H}_0 = \mathcal{H}_S + \mathcal{H}_{I_1} + \mathcal{H}_{SI_1} + \mathcal{H}_{I_2} + \mathcal{H}_{SI_2} + \mathcal{H}_{I_1 I_2}, \tag{31}$$

where the subscripts show the spin operators in each term, $\mathcal{H}_s$ is the electron Zeeman term, and so on. From Eq. (18), (19), and (20) it can be seen that this Hamiltonian will introduce additional exponential factors into R_2 and hence into $\rho(2\tau)$. If the $\mathcal{H}_{I_1 I_2}$ term is neglected, as is justified, the net result is

$$V_{mod}^{I_1 I_2} = V_{mod}^{I_1} V_{mod}^{I_2} \tag{32}$$

and for n nuclei[4,11]

$$V_{mod}^n = \prod_{i=1}^{n} V_{mod}^i. \tag{33}$$

This product rule holds for both identical and nonidentical nuclei. For equivalent, identical nuclei it is seen that the modulation depth increases rapidly as the number of interacting nuclei increases.

2.3   Simulation in Disordered Systems

In disordered systems we must average over all orientations to find the average echo modulation pattern. For interaction of the unpaired electron with one nucleus, this is given by

$$\langle V_{mod} \rangle_\theta = \frac{1}{4\pi} \int\int V_{mod}(\theta) \sin\theta \, d\theta \, d\phi. \tag{34}$$

For four nuclei arranged tetrahedrally around the electron we can explicitly define the angular correlation between the nuclei. For example, for the first nucleus let $\cos\theta_1 = \cos\theta$, for a second nucleus $\cos\theta_2 = \sin\theta \cos\phi$ $\sin\alpha + \cos\alpha \cos\theta$ where α is the tetrahedral angle, and so on. The average modulation pattern is then explicitly given by

$$\langle V_{mod}^{tetra} \rangle_\theta = \frac{1}{4\pi} \int_{\phi=0}^{2\pi} d\phi \int_{\theta=0}^{\pi} \prod_{i=1}^{4} V_{mod}(\cos\theta_i) \sin\theta \, d\theta, \tag{35}$$

where we have used Eq. (33). Figure 7 shows simulated modulation patterns for a tetrahedral configuration for $I = 1$ with $r = 3$ and 4 Å.

For the typical disordered experimental system the explicit nuclear geometry around the electron is not known, so approximations to the explicit formulation of (35) are necessary. Kevan et al.[14] first suggested and justified a spherical model in which an effective interaction distance is used for the closest shell of nuclei and the expression for the echo modulation is given by

$$\overline{V_{mod}^n} = [\langle V_{mod}(r, \mathbf{a}) \rangle_\theta]^n, \tag{36}$$

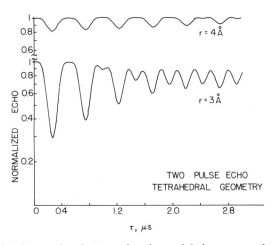

Figure 7   Calculated two-pulse electron spin-echo modulation patterns for an $S = \frac{1}{2}$, $I = 1$ spin system with four nuclei arranged tetrahedrally around the electron spin at $r = 3$ and 4 Å. The applied magnetic field is 3150 G and the isotropic hyperfine coupling is zero.

which is a function of r, $\mathbf{a}$, and n. The formulation in Eq. (36) is equivalent to an average modulation for nuclei *randomly* distributed on a sphere of radius r. In other words, no allowance is made for correlation between nuclear positions and the formulation includes configurations with two nuclei located at the same position. This is not a serious problem because for a random distribution the probability for such a configuration is small. Formally, Eq. (36) corresponds to an inversion of the product and integration operations in Eq. (35). The justification for Eq. (36) is given by a comparison of Fig. 8, which shows simulations for $n = 4$ at 3 and 4 Å using Eq. (36), and Fig. 7, which shows simulations for an explicit tetrahedral geometry for $I = 1$ nuclei. At $r = 4$ Å there is no difference between the tetrahedral geometry and the spherical model. At $r = 3$ Å the modulation depth is somewhat greater for the tetrahedral geometry, but the spherical model is still a good approximation. Furthermore, the definition of the spherical model implies that it becomes a better approximation as n increases.

An alternate approximation for Eq. (35) is[14]

$$\overline{V^n_{\text{mod}}} = \langle V^n_{\text{mod}} \rangle_\theta, \tag{37}$$

which is equivalent to taking all the n nuclei at the same θ before averaging over θ. This approximation corresponds to complete correlation between the nuclei. However, it generally predicts shallower modulation

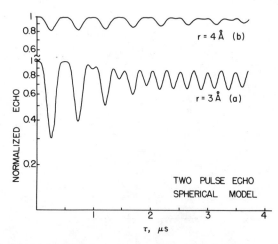

Figure 8 Calculated two-pulse electron spin-echo modulation patterns for an $S = \frac{1}{2}$, $I = 1$ spin system with four nuclei randomly distributed on a sphere of radius $r = 3$ and 4 Å (spherical model). This model corresponds to Eq. (36) in the text. The applied magnetic field is 3150 G and the isotropic hyperfine coupling is zero.

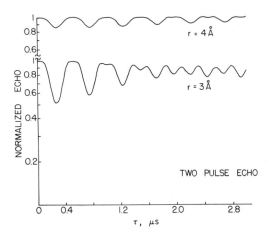

Figure 9   Calculated two-pulse electron spin-echo modulation patterns for an $S = \frac{1}{2}$, $I = 1$ spin system with four nuclei completely correlated on a sphere of radius $r = 3$ and 4 Å. This model corresponds to Eq. (37) in the text. The applied magnetic field is 3150 G and the isotropic hyperfine coupling is zero.

than the spherical model or the explicit configuration model. This is shown by Fig. 9 for $r = 3$ Å.

In the limit of large r and large n all three models represented by Eq. (35), (36), and (37) must become equivalent. In practice for $I = 1$ and $I = \frac{1}{2}$ the required values of "large r" and "large n" are ~ 4 Å and 4; compare the simulations in Figs. 7–9 for $r = 4$ Å. Since the modulation depth per nucleus increases as $I(I+1)$ and the equivalence of the spherical and explicit configuration models is approached for shallow modulation, we expect that the radius beyond which this equivalence holds may also increase like $I(I+1)$.

If we consider two spherical distributions of nuclei at different radii around an unpaired electron, the modulation pattern should be well approximated by

$$V_{\text{mod}} = [\langle V_{\text{mod}}(r_1, \mathbf{a}_1)_\theta \rangle]^{n_1} [\langle V_{\text{mod}}(r_2, \mathbf{a}_2) \rangle]^{n_2}, \qquad (38)$$

where $r_1 > r_2 > 3$ Å and $n_1 \geq n_2 \geq 4$. If r_2 is less than 3 Å, it is probably best to assume an explicit configuration for this contribution and average that over θ.

Since the dipolar interaction decreases as r^{-6} more distant nuclei make little contribution to the modulation pattern and probably contributions from two or three discrete shells should adequately represent it. However, it may be desirable in some cases to consider a continuous distribution of nuclei extending from some minimum value r_{min}. This can be done by considering a series of contiguous shells in equal steps of r^{-3} with the number of nuclei in the ith shell given by

$$n_i = \left(\frac{4\pi\rho_n}{3}\right)(r_{i+1}^3 - r_i^3), \tag{39}$$

where ρ_n is the average nuclear density.

2.4 Typical Modulation Patterns

Figure 8 shows typical modulation patterns for the spherical model for $I = 1$ at two different distances. The nuclear coupling is weak so that $\omega_\alpha \sim \omega_\beta \sim \omega_I$ and modulation frequencies at $\sim\omega_I$ and $\sim2\omega_I$ are expected. The intensity of the $2\omega_I$ component decreases with increasing distance and it is not apparent in Fig. 8b at 4 Å over the time range simulated. However, in Fig. 8a at 3 Å the $2\omega_I$ component is clearly seen at longer times. This suggests that observation of a $2\omega_I$ component gives a qualitative indication of the dipolar interaction distance.

Similar results are seen for $I = \frac{1}{2}$ systems at two distances as shown in Fig. 10. At the greater distance of 4 Å the ω_I frequency dominates the modulation but at the latest modulation periods shown some $2\omega_I$ contribution is apparent. At 3 Å the $2\omega_I$ contribution is apparent in the first modulation period and totally dominates the pattern after the third modulation period.

The dominance of the $2\omega_I$ frequency at longer times can be understood from the Fourier transform of the time domain modulation pattern.[15] The Fourier transform of Fig. 10a to give the frequency distribution is shown in Fig. 11. The broad peak at ~13 MHz (i.e., $\sim\omega_1$) is due to the positive $\cos\omega_\alpha\tau$ and $\cos\omega_\beta\tau$ terms in Eq. (23), the narrow negative peak at

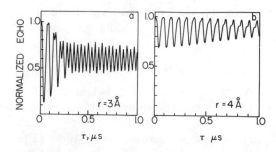

Figure 10 Calculated two-pulse electron spin-echo modulation patterns for an $S = \frac{1}{2}$, $I = 1$ spin system with (a) 18 protons at $r = 3$ Å using the spherical model of Eq. (36), and (b) a continuous density of protons based on the density of water extending from $r = 4$ Å to infinity. The magnetic field is 3200 G and the isotropic hyperfine coupling is zero. [Adapted from W. B. Mims et al., *J. Chem. Phys.*, **66**, 5536 (1977).]

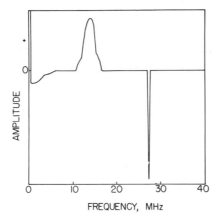

Figure 11 The Fourier transform of Fig. 10a showing the spectral distribution of the frequency amplitude for protons. [From W. B. Mims et al., *J. Chem. Phys.*, **66**, 5536 (1977).]

$\sim$26 MHz arises from the negative cos $(\omega_\alpha + \omega_\beta)\tau$ term, the broad negative peak near 0 MHz arises from the negative cos $(\omega_\alpha - \omega_\beta)\tau$ term and the positive zero frequency peak, which is off scale in Fig. 11, comes from the $1-(k/2)$ term in Eq. (23). The narrowness of the $\omega_\alpha + \omega_\beta$ frequency peak arises from the fact that the electron dipolar field at the nucleus is in opposite directions for the ω_α and ω_β energy levels since the electron spin quantum number is $+\frac{1}{2}$ and $-\frac{1}{2}$ for the two sets of levels, and thus the broadening from the electron dipolar field tends to cancel for $\omega_\alpha + \omega_\beta \approx 2\omega_I$. The narrow, negative amplitude of the $2\omega_I$ frequency first appears as an indentation in the peaks of period ω_I and ultimately dominates the modulation pattern since it is narrower than the ω_I frequency. It is interesting that for a transition between $M_s = \frac{1}{2}$ and $\frac{3}{2}$ electron spin states for a $S = \frac{3}{2}$ spin system the electron dipolar field does not cancel for $\omega_\alpha + \omega_\beta$. Thus one expects the $2\omega_I$ frequency component to be broader than the ω_I frequency component and not to dominate the modulation pattern at longer times. This appears to be the case for $FeCl_3 \cdot 6H_2O$ in a water-ethanol matrix where the proton modulation pattern shows only the ω_I component.[15]

Modulation expressions (21) and (24) indicate that the theoretical ratio of modulation depths from the coefficients of the k terms for a single deuteron environment to a single proton environment is $\frac{8}{3} = 2.7$. The k^2 term lowers this ratio by less than 2% at $r = 3$ Å. This ratio has been tested for Yb^{3+} in frozen aqueous solutions where it has been deduced that there are 18 equivalent protons or deuterons in the first solvation shell.[15] The experimental modulation patterns are shown in Fig. 12. The eighteenth roots of the observed modulation depths, measured as unity

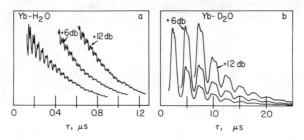

Figure 12 Two-pulse electron spin-echo signals for 10 mM YbCl$_3$ in (a) 20 volume % C$_2$H$_5$OH in H$_2$O glass and (b) 20 volume % C$_2$D$_5$OD in D$_2$O glass at 1.8 K and $H_0 = 6000$ G. [From W. B. Mims et al., *J. Chem. Phys.*, **66**, 5536 (1977).]

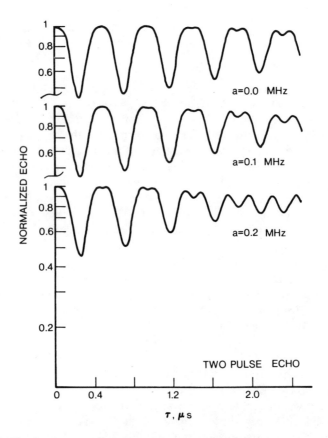

Figure 13 Simulated two-pulse electron spin-echo modulation patterns for an $S = \frac{1}{2}$, $I = 1$ system at $H_0 = 3320$ G with 16 deuterons at 3.8 Å averaged by the spherical model of Eq. (36) and showing the effect of an isotropic hyperfine coupling, **a**.

302

minus the peak-to-trough ratios, are obtained to find the equivalent modulation depth for a single interacting nucleus. The deuteron-to-proton depth ratio is 2.7 averaged over the first five modulation periods, which is in excellent agreement with theory. Since quadrupole interactions were not included in expression (24) for deuteron modulation, this suggests that such interactions are not too important for Yb^{3+}. However, a similar analysis of the D/H modulation depth ratio for Cu^{2+} gave a ratio of only 1.4 when eight equivalent protons were assumed for the first solvation shell.[15] Either quadrupole interactions are more important for hydrated Cu^{2+} than for hydrated Yb^{3+} or the number of first solvation shell nuclei used is incorrect. We consider how effects of quadrupole interaction can be incorporated into the expressions for the modulation pattern in the next subsection. The effect of an isotropic hyperfine coupling on the modulation pattern is shown in Fig. 13 for $I = 1$. As the isotropic coupling increases, the modulation amplitude decays more rapidly and the $2\omega_I$ component becomes more prominent.

For the case of g-anisotropy in single crystals, explicit account of the anisotropic g-factors must be taken to simulate the modulation patterns. (See Eq. 26–30.) However, in disordered systems it has been empirically found that the isotropic g-value expressions [i.e., Eq. (21) and (24)] produce a satisfactory modulation pattern provided that the experimental g-value corresponding to the H_0 value for the echo is used.[17]

2.5 Nuclear Quadrupole Effects

We only consider the case for $I = 1$ when quadrupole corrections to the energy levels are smaller than the electron dipolar corrections. Then Mims[11] has given the following expression for the two-pulse modulation:

$$
\begin{aligned}
V_{\text{mod}}(2\tau, I = 1) = {} & (1 - \tfrac{4}{3}k + \tfrac{3}{4}k^2) + (\tfrac{2}{3}k - \tfrac{1}{2}k^2)(\cos \omega_{ab}\tau + \cos \omega_{bc}\tau \\
& + \cos \omega_{de}\tau + \cos \omega_{ef}\tau) - [\tfrac{1}{6}k - \tfrac{1}{6}k^2 + \tfrac{1}{6}k(1 - k^2)^{1/2}] \\
& \times [\cos (\omega_{ab} + \omega_{de})\tau + \cos (\omega_{ab} - \omega_{de})\tau + \cos (\omega_{bc} + \omega_{ef})\tau \\
& + \cos (\omega_{bc} - \omega_{ef})\tau] - [\tfrac{1}{6}k - \tfrac{1}{6}k^2 - \tfrac{1}{6}k(1 - k^2)^{1/2}][\cos (\omega_{bc} \\
& + \omega_{de})\tau + \cos (\omega_{bc} - \omega_{de})\tau + \cos (\omega_{ab} + \omega_{ef})\tau \\
& \qquad\qquad\qquad\qquad\qquad\qquad\qquad + \cos (\omega_{ab} - \omega_{ef})\tau] \\
& + \tfrac{1}{4}k^2(\cos \omega_{ac}\tau + \cos \omega_{df}\tau) + \tfrac{1}{24}k^2[\cos (\omega_{ac} + \omega_{df})\tau \\
& + \cos (\omega_{ac} - \omega_{df})\tau] \\
& \qquad\qquad\qquad - \tfrac{1}{12}k^2[\cos (\omega_{de} + \omega_{ac})\tau + \cos (\omega_{de} - \omega_{ac})\tau \\
& + \cos (\omega_{df} + \omega_{ab})\tau + \cos (\omega_{df} - \omega_{ab})\tau \\
& + \cos (\omega_{df} + \omega_{bc})\tau + \cos (\omega_{df} - \omega_{bc})\tau \\
& + \cos (\omega_{ef} + \omega_{ac})\tau + \cos (\omega_{ef} - \omega_{ac})\tau], \quad (40)
\end{aligned}
$$

where the frequencies ω_{ab}, ω_{bc}, and so on, refer to the energy level diagram in Fig. 6 corrected for a quadrupole correction Δ as follows:

$$\omega_{ab} = \omega_\alpha + \Delta, \qquad \omega_{bc} = \omega_\alpha - \Delta, \qquad \omega_{de} = \omega_\beta + \Delta, \qquad \omega_{ef} = \omega_\beta - \Delta,$$

where ω_α and ω_β are the nuclear splittings in the $m_s = +\frac{1}{2}$ and $-\frac{1}{2}$ states. The angular dependence of the quadrupole correction is given by

$$\Delta = \Delta_0(3 \cos^2 \theta - 1), \tag{41}$$

where $2\Delta_0$ denotes the quadrupole correction in the parallel direction.

Typical simulations based on Eq. (40) averaged by Eq. (36) for a disordered system are given in Figs. 14 and 15.[18] A small quadrupole coupling $\Delta_0 = 0.2$ MHz has about the same effect as an isotropic hyperfine coupling of the same magnitude (compare Fig. 14 top with Fig. 13 bottom); it causes the ω_I component to decay faster and accentuates the $2\omega_I$ component. For a larger quadrupole coupling of $\Delta_0 = 0.4$ MHz a characteristically different pattern appears (Fig. 14). The modulation consists of two components denoted as a and b in Fig. 14. At shorter

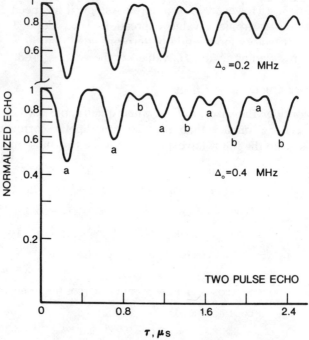

Figure 14 Simulated two-pulse electron spin-echo modulation patterns for an $S = \frac{1}{2}$, $I = 1$ system at $H_0 = 3320$ G with 16 deuterons at 3.8 Å showing the effect of a quadrupole interaction Δ_0 based on Eq. (40) averaged by the spherical model. The isotropic hyperfine coupling is zero. The a and b designations are discussed in the text.

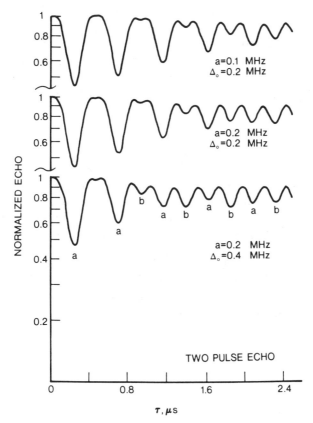

Figure 15 Simulated two-pulse electron spin-echo modulation patterns for an $S = \frac{1}{2}$, $I = 1$ system at $H_0 = 3320\,\text{G}$ with 16 deuterons at 3.8 Å based on Eq. (40) averaged by the spherical model. The effect of both isotropic hyperfine **a** and quadrupole coupling Δ_0 is shown.

times component a is stronger than b but beyond a third-period component b is stronger. This produces a pattern where the ω_I frequency still dominates but with a 180° phase shift. This pattern is quite different from dominance of the $2\omega_I$ frequency at longer times.

Figure 15 shows the effect of both an isotropic hyperfine constant and a quadrupole coupling. The addition of the isotropic hyperfine increases the decay of the modulation pattern, but at larger values of the quadrupole coupling the dominance of the b components is again seen.

Overall it appears that small quadrupole couplings will not be distinguishable from small isotropic hyperfine couplings. But if the quadrupole coupling is large enough it will produce a characteristic pattern.

2.6 Analysis to Compensate for Echo Decay

The experimentally observed echo modulation signal is the product of V_{mod} and an echo decay function. [See Eq. (1).] Sometimes the echo decay function can be taken to be an exponential, but in general the form is complex and unknown.[19] The decay function can be estimated from the envelope of the maxima of the echo modulation, but inaccuracies obtain because the first few modulation periods are not observed experimentally. One method to compensate for echo decay is to work with ratios of the maxima and minima of the modulation pattern. Then the ratios from the experimental V_{obs} and the ratios from the calculated V_{mod} should be directly comparable. The simplest method is to plot the ratio from adjacent minima and maxima versus τ and, when the $2\omega_I$ component becomes significant, to use the mean of adjacent minima or maxima.[17] A possibly more exact method is to trace envelopes of the minima and maxima and to plot the ratio at a given τ versus τ.[20]

The procedure to obtain the calculated ratios may be sketched as follows.[20] It is analogous to the data analysis suggested for three-pulse echoes.[21] It should be cautioned that inclusion of an isotropic coupling in the Hamiltonian is essential for accuracy unless other considerations suggest that $\mathbf{a} = 0$.

Assuming $\omega_I \gg A$ and B, $\omega_\alpha \sim \omega_I + (A/2)$ and $\omega_\beta \sim \omega_I - (A/2)$, we can write Eq. (24) for $I = 1$ as

$$V_{mod}(2\tau, I = 1) = 3^{-1}[(2 - k(\cos \omega_I \tau - \cos (A/2)\tau)^2)^2 - 1]. \tag{42}$$

By setting the time derivative to zero it is found that

$$V_{max}(2\tau, I = 1) = 3^{-1}[(2 - k(1 - \cos (A/2)\tau)^2)^2 - 1], \tag{43}$$

$$V_{min}(2\tau, I = 1) = 3^{-1}[(2 - k(1 + \cos (A/2)\tau)^2)^2 - 1], \tag{44}$$

and we define

$$\overline{(R_\theta)^n} = \left[\frac{\displaystyle\int_0^\pi V_{min} \sin \theta \, d\theta}{\displaystyle\int_0^\pi V_{max} \sin \theta \, d\theta} \right]^n, \tag{45}$$

where we have assumed the spherical model with n equivalent interacting nuclei. The experimental ratio R and the calculated ratio $(R_\theta)^n$ are then compared as a function of τ to determine the best set of values for n, r, and $\mathbf{a}$. This comparison can be conveniently carried out by minimizing the least squares difference between the experimental and calculated ratios. Analogous equations may be derived for the $I = \frac{1}{2}$ case.

3   ANALYSIS OF THREE-PULSE SPIN-ECHO MODULATION

3.1 Simulation for One or Several Nuclei

As shown in Fig. 4, three-pulse echo modulation usually occurs over a longer time scale than two-pulse modulation because of the extra time interval T between the second and third pulses. Thus it may be sensitive to weaker hyperfine coupling than can be seen by two-pulse echo modulation.

The simulation of three-pulse echo modulation is a straightforward extension of the formalism used for the two-pulse case.[11,12] The normalized modulation is given by

$$V_{\text{mod}}(2\tau + T) = \frac{\text{Tr}\,(\rho(2\tau + T)S_y)}{\text{Tr}\,(\rho(0)S_y)}, \tag{46}$$

which is analogous to Eq. (10). The density matrix is given by

$$\rho(2\tau + T) = R_2 R_1 R_5 R_1 R_2 R_1 \rho(0) R_1^* R_2^* R_1^* R_5^* R_1^* R_2^*, \tag{47}$$

where R_1 and R_2 are defined by Eq. (17) and (18) and R_5 is given by

$$R_5 = \exp\left(\frac{-i\mathcal{H}_r(\text{pulse off})T}{\hbar}\right). \tag{48}$$

Using the Hamiltonian in Eq. (9) with an isotropic g-factor, the normalized three-pulse echo modulation for $S = \frac{1}{2}$ and $I = \frac{1}{2}$ is given by[11]

$$V_{\text{mod}}(2\tau + T, I = \tfrac{1}{2}) = 1 - (k/2) + (k/4)[\cos \omega_\alpha \tau + \cos \omega_\beta \tau$$
$$+ (1 - \cos \omega_\beta \tau) \cos \omega_\alpha (\tau + T) + (1 - \cos \omega_\alpha \tau) \cos \omega_\beta (\tau + T)], \tag{49}$$

which may be written more compactly as

$$V_{\text{mod}}(2\tau + T, I = \tfrac{1}{2}) = 1 - k\left[\sin^2\left(\frac{\omega_\alpha \tau}{2}\right)\sin^2\left(\frac{\omega_\beta(\tau + T)}{2}\right)\right.$$
$$\left. + \sin^2\left(\frac{\omega_\beta \tau}{2}\right)\sin^2\left(\frac{\omega_\alpha(\tau + T)}{2}\right)\right]. \tag{50}$$

Similarly, the three-pulse echo modulation for $S = \frac{1}{2}$ and $I = 1$ is compactly given as

$$V_{\text{mod}}(2\tau + T, I = 1) = 1 - \left(\frac{8}{3}\right)k\left[\sin^2\left(\frac{\omega_\alpha \tau}{2}\right)\sin^2\left(\frac{\omega_\beta(\tau \pm T)}{2}\right)\right.$$
$$\left. + \sin^2\left(\frac{\omega_\alpha(\tau + T)}{2}\right)\sin^2\left(\frac{\omega_\beta \tau}{2}\right)\right]$$
$$+ \left(\frac{8}{3}\right)k^2\left[\sin^4\left(\frac{\omega_\alpha \tau}{2}\right)\sin^2\left(\frac{\omega_\beta(\tau + T)}{2}\right) + \sin^4\left(\frac{\omega_\alpha(\tau + T)}{2}\right)\sin^2\left(\frac{\omega_\beta \tau}{2}\right)\right]. \tag{51}$$

Note that these expressions contain only the hyperfine frequencies ω_α and ω_β *without* their sums and differences in contrast to the two-pulse case; compare with Eq. (21). In addition, the amplitude of the ω_α frequency component is dependent on the ω_β frequency and vice versa. For example, if $\omega_\beta \tau = 2n\pi$ the second sin term in Eq. (50) goes to zero and the ω_α component disappears from the modulation pattern. And if $\omega_\alpha \tau = 2n\pi$ also no modulation would be seen at all. Recall that τ is held constant in the normal three-pulse experiment and is selected by the experimenter. For weak hyperfine interaction $\omega_\alpha \sim \omega_\beta \sim \omega_I$ so $\tau = n\pi/\omega_I$ is required to observe maximum modulation. This means that τ must be selected as the value giving the maximum two-pulse modulation depth in order to see the maximum three-pulse modulation effect.

Mims[12] has shown that the three-pulse modulation expression can be obtained very simply from the two-pulse modulation expression without using the density matrix and Eq. (46). The procedure is as follows. Start with the two-pulse expression in the form such that the sum and difference frequencies do not appear explicitly as the argument of a sin or cos function. Equation (21) is already in that form. In Eq. (23) $\cos (\omega_\alpha + \omega_\beta)\tau + \cos (\omega_\alpha - \omega_\beta)\tau$ must be replaced by $2 \cos \omega_\alpha \tau \cos \omega_\beta \tau$ to obtain this form.

1. Then write the two-pulse expression with $\cos \omega_\alpha \tau$ or $\sin \omega_\alpha \tau$ replaced by $\cos \omega_\alpha (\tau + T)$ or $\sin \omega_\alpha (\tau + T)$.

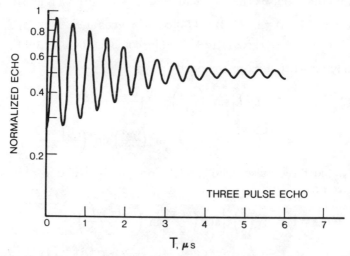

Figure 16 Simulated three-pulse electron spin-echo modulation pattern for an $S = \frac{1}{2}$, $I = 1$ system with 16 deuterons at 3.9 Å averaged by the spherical model with the following parameters: $H_0 = 3610 \, \text{G}$, $\tau = 0.63 \, \mu\text{s}$, and $\mathbf{a} = 0.04 \, \text{MHz}$.

2. Write the two-pulse expression again with $\cos \omega_\beta \tau$ or $\sin \omega_\beta \tau$ replaced by $\cos \omega_\beta(\tau + T)$ or $\sin \omega_\beta(\tau + T)$.
3. Add the expressions of (1) and (2) and divide by 2 to renormalize.

Simulation of three-pulse modulation patterns for several nuclei proceeds the same as for the two-pulse case and is carried out by the product relation in Eq. (33). Also, the simulation procedure in disordered systems may be done by Eq. (36) in the framework of the spherical model. A typical three-pulse simulation for $I = 1$ using Eq. (51) averaged by Eq. (36) is given in Fig. 16. It is seen that the modulation only exhibits the $\sim \omega_I$ frequency component and that the decay of the modulation is relatively slow compared to two-pulse modulation.

3.2 Phase Changes in the Three-Pulse Echo Modulation Pattern

In certain cases it has been observed experimentally[22,23] and theoretically[24] that the three-pulse modulation exhibits a 180° phase change. A typical experimental result is shown in Fig. 17 for trapped electrons in deuterated aqueous matrices. The vertical dashes are equispaced and it

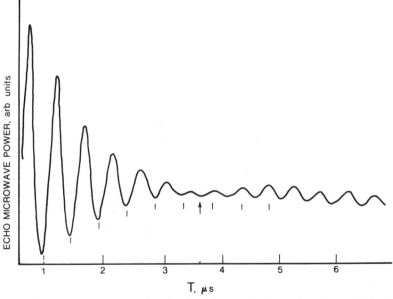

Figure 17 Three-pulse electron spin-echo signal from radiation-produced trapped electons in $10M$ NaOD/D$_2$O at 77 K. The value of τ is 0.7 μs. The arrow indicates the position of a change in the modulation phase. The microwave power was detected instead of the microwave magnetic field. [From P. A. Narayana et al., *J. Chem. Phys.*, **63**, 3365 (1975).]

can be seen that the arrow denotes the position of a phase change.

Let us define the delay time corresponding to the phase change as T_p. Then if the isotropic hyperfine interaction is zero and the dipolar hyperfine interaction is weak, a simple expression relating T_p and r can be derived.[19] For weak interactions ($r \geqslant 3$ Å), $k^2 \approx 0$, $\omega_\alpha = \omega_I + A/2$, and $\omega_\beta = \omega_I - A/2$. With these approximations Eq. (46) reduces to

$$V_{mod}(2\tau + T, I = 1) = 1 - (\tfrac{8}{3})k \sin^2 \frac{\omega_\alpha \tau}{2}[1 - \cos \omega_I(\tau + T) \cos (A(\tau + T)/2)].$$

$$(52)$$

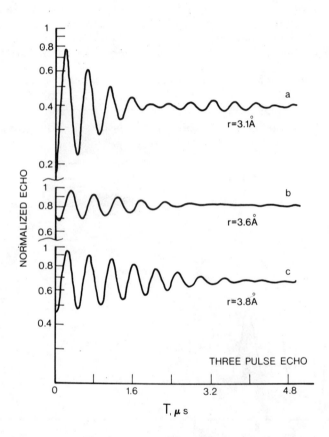

Figure 18 Simulated three-pulse electron spin-echo modulation patterns for an $S = \frac{1}{2}$, $I = 1$ spin system averaged by the spherical model showing the variation in the time of phase change with the electron-nuclear dipolar distance r. The other parameters are $H_0 = 3320$ G, $a = 0$ MHz and $\tau = 0.7$ μs. The phase reversal times are (a) 2.2 μs, (b) 3.6 μs and (c) 4.5 μs; they are independent of the number of interacting deuterons.

Thus the modulation is made up of two time-varying terms for fixed τ. The more rapidly varying term $\cos \omega_I(\tau + T)$ gives the modulation and the more slowly varying term $\cos A(\tau + T)/2$ introduces a phase change. In fact, the modulation changes its phase when $\cos A(\tau + T)/2$ becomes zero. But for a disordered system one has to average over all orientations before the condition for phase reversal is obtained. Thus the function

$$F(r) = \int_0^{\pi} \cos^2 \theta \sin^2 \theta \cos \left[\frac{g g_n \beta \beta_n}{2hr^3}(\tau + T)(3 \cos^2 \theta - 1) \right] \sin \theta \, d\theta \quad (53)$$

must be zero for phase reversal. The $\cos^2 \theta \sin^2 \theta$ in the integrand comes from the term k and the expression for A in the argument of the cosine factor has been substituted from Eq. (51). When $F(r)$ is numerically evaluated, it becomes zero when $g g_n \beta \beta_n (\tau + T)/hr^3$ is equal to 7.25 and thus

$$r(\text{Å}) = [10.7(T_p + \tau)_{\mu s}]^{1/3}. \quad (54)$$

So if T_p is measured experimentally for a known τ, the distance r can be calculated.

The conditions under which the approximate analysis of (54) is valid can be investigated by carrying out complete simulations of the three-pulse echo modulation pattern.[24] Figure 18 shows simulations for $r = 3.1$, 3.6, and 3.8 Å with $\mathbf{a} = 0$. The phase reversal times from Eq. (49) for these distances are 2.2, 3.6, and 4.5 μs, respectively, which agree well with the simulations. However, this is not the case when a small isotropic hyperfine coupling is incorporated into the simulation. Figure 19 shows simulations for $r = 3.1$ Å with $\mathbf{a} = -0.1$, 0.1, 0.2 MHz, respectively. A small negative isotropic coupling moves T_p to slightly shorter times (compare Figs. 18a and 19a) while a small positive isotropic coupling moves T_p to significantly longer times (compare Figs. 19b and 19c with 18a). However, at larger distances (3.6–3.8 Å) a small positive isotropic coupling moves T_p to much shorter times than in the absence of an isotropic coupling. These results clearly demonstrate that the approximate phase reversal analysis of Eq. (54) breaks down if even a small isotropic hyperfine interaction ($\geqslant 0.1$ MHz) exists for the nuclei contributing to the three-pulse modulation.

3.3 Nuclear Quadrupole Effects

As for the two-pulse case we will only consider the case for $I = 1$ when

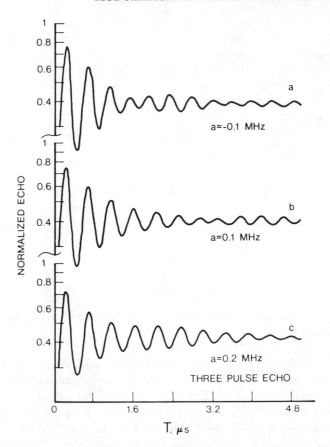

Figure 19 Simulated three-pulse electron spin-echo modulation patterns for an $S = \frac{1}{2}$, $I = 1$ spin system averaged by the spherical model showing the variation in the time of phase change with the isotropic coupling **a**. The other parameters are $H_0 = 3320$ G, $r = 3.1$ Å, and $\tau = 0.7$ μs. The phase reversal times are $(a) \sim 2.4$ μs, (b) 3.2 μs and $(c) > 4.5$ μs; they are independent of the number of interacting deuterons, which is 12 in the simulations.

the quadrupole corrections are smaller than the electron dipolar corrections. The modulation is then given by[11]

$$
\begin{aligned}
V_{\text{mod}}(2\tau + T, I = 1) = {}& (1 - \tfrac{4}{3}k + \tfrac{3}{4}k^2) + (\tfrac{1}{3}k - \tfrac{1}{4}k^2)[\cos \omega_{ab}\tau + \cos \omega_{bc}\tau + \cos \omega_{de}\tau \\
& + \cos \omega_{ef}\tau + \cos \omega_{ab}(\tau + T) + \cos \omega_{bc}(\tau + T) \\
& \hspace{6cm} + \cos \omega_{de}(\tau + T) \\
& + \cos \omega_{ef}(\tau + T)] - [\tfrac{1}{6}k - \tfrac{1}{6}k^2 + \tfrac{1}{6}k(1 - k^2)^{1/2}] \\
& \hspace{3.5cm} \times [\cos \omega_{ab}\tau \cos \omega_{de}(\tau + T) \\
& + \cos \omega_{bc}\tau \cos \omega_{ef}(\tau + T) + \cos \omega_{de}\tau \cos \omega_{ab}(\tau + T)
\end{aligned}
$$

$$+\cos \omega_{ef}\tau \cos \omega_{bc}(\tau+T)]-[\tfrac{1}{6}k-\tfrac{1}{6}k^2-\tfrac{1}{6}k(1-k^2)^{1/2}]$$
$$\times[\cos \omega_{ab}\tau \cos \omega_{ef}(\tau+T)+\cos \omega_{bc}\tau \cos \omega_{de}(\tau+T)$$
$$+\cos \omega_{de}\tau \cos \omega_{bc}$$
$$(\tau+T)+\cos \omega_{ef}\tau \cos \omega_{ab}(\tau+T)]$$
$$+\tfrac{1}{8}k^2[\cos \omega_{ac}\tau+\cos \omega_{df}\tau+\cos \omega_{ac}$$
$$(\tau+T)+\cos \omega_{df}(\tau+T)]$$
$$+\tfrac{1}{24}k^2[\cos \omega_{ac}\tau \cos \omega_{df}(\tau+T)$$
$$+\cos \omega_{df}\tau \cos \omega_{ac}(\tau+T)]$$
$$-\tfrac{1}{12}k^2[\cos \omega_{ab}\tau \cos \omega_{df}(\tau+T)$$
$$+\cos \omega_{df}\tau \cos \omega_{ab}(\tau+T)$$
$$+\cos \omega_{bc}\tau \cos \omega_{df}(\tau+T)+\cos \omega_{df}\tau \cos \omega_{bc}(\tau+T)$$
$$+\cos \omega_{de}\tau \cos \omega_{ac}(\tau+T)+\cos \omega_{ac}\tau \cos \omega_{de}(\tau+T)$$
$$+\cos \omega_{ef}\tau \cos \omega_{ac}(\tau+T)+\cos \omega_{ac}\tau \cos \omega_{ef}(\tau+T)],$$

$$(55)$$

where the frequencies ω_{ab}, ω_{bc}, and so on, are defined as in Section 2.5. The angular dependence of Δ is given by Eq. (41).

Typical simulations based on Eq. (55) averaged by Eq. (36) for a disordered system are given in Fig. 20. Systematic changes due to the quadrupole coupling are less clear than in the two-pulse case. Nevertheless, it can be seen that a quadrupole coupling can produce a phase reversal (Fig. 20a, vs. 20b) and generally decreases the modulation depth. These effects also occur in the presence of isotropic hyperfine coupling and may damp the modulation almost completely (Fig. 20c).

3.4 Analysis to Compensate for Echo Decay

As in the two-pulse case (Section 2.6) one may compensate for the echo decay in the experimental signal by comparing calculated and experimental ratios of the maxima and minima of the modulation pattern. For $I=1$ the expressions for the maxima and minima are

$$V_{max}(2\tau+T, I=1)=3^{-1}+(\tfrac{2}{3})k^2[k^{-1}+(1-\cos (A/2)\tau)$$
$$\times(\cos (A/2)(\tau+T)-1)]^2$$
$$-3^{-1}k^2(\cos \omega_I\tau-\cos (A/2)\tau)^2(\cos (A/2)(\tau+T)-1)^2 \quad (56)$$
$$V_{min}(2\tau+T, I=1)=3^{-1}+(\tfrac{2}{3})k^2[k^{-1}+(1-\cos \omega_I\tau \cos (A/2)\tau)$$
$$\times(-\cos (A/2)(\tau+T)-1)]^2$$
$$-3^{-1}k^2(\cos \omega_I\tau-\cos (A/2)\tau)^2(-\cos(A/2)(\tau+T)-1)^2 \quad (57)$$

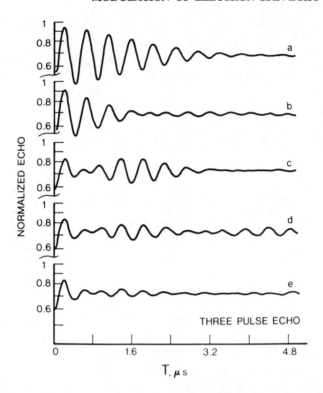

Figure 20 Simulated three-pulse electron spin-echo modulation patterns for an $S = \frac{1}{2}$, $I = 1$ spin system based on Eq. (55) averaged by the spherical model showing the effects of varying isotropic hyperfine **a** and quadrupole coupling Δ_0. The constant parameters are $H_0 = 3320$ G, 16 deuterons at 3.8 Å, and $\tau = 0.7$ μs. The variable parameters are (a) **a** = 0, $\Delta_0 = 0$; (b) **a** = 0, $\Delta_0 = 0.2$ MHz; (c) **a** = 0, $\Delta_0 = 0.4$ MHz; (d) **a** = −0.1 MHz, $\Delta_0 = 0.4$ MHz; and (e) **a** = 0.2 MHz, $\Delta_0 = 0.4$ MHz.

and the calculated ratio is still given by Eq. (45). The preceding expressions reduce to those implied by Yudanov et al. when **a** = 0 is assumed.[21]

3.5 Consistency of Parameters from Simulation of Two- and Three-Pulse Modulation

The experimental parameters $(n, r, \mathbf{a})$ defining the radical environment that one wishes to deduce from the echo modulation patterns are expected to be overdetermined when both two- and three-pulse echo modulation data are available. In other words, the parameters determined from a fit of the simulated and experimental three-pulse echo modulation should satisfactorily reproduce the two-pulse echo modulation. However, this will only be valid for more distant nuclei ($\geqslant 3.5$ Å), since it is these nuclei that give the main contribution to the three-pulse echo modulation at the longer times. Closer nuclei also contribute to the three-pulse echo modulation, but their contribution seems significant only over the first portion of the observable time scale for the modulation pattern. In contrast, over the shorter observable time scale for two-pulse modulation both close and distant nuclei make significant contributions over the entire time scale.

There have been relatively few systems where parameters from two- and three-pulse echo modulation analysis have been compared. But where $a = 0$ seems to hold, the comparison has shown self-consistency as for nitroxide radicals in toluene glass.[21] If self-consistency between two- and three-pulse modulation analysis is not found, one should suspect $a \neq 0$ and interacting nuclei at two or more distances to be present. Further analysis in this area is needed.

4 APPLICATIONS

4.1 Environment of Ions and Radicals in Single Crystals

Although two-pulse electron spin-echo modulation was first observed for the $S = \frac{1}{2}$ ion, Ce^{3+}, in a $CaWO_4$[25a] and lanthanum magnesium nitrate[25b] single crystals, relatively few single crystal studies have been attempted. The first analysis of the two-pulse modulation was done for 0.01 weight % Ce^{3+} in $CaWO_4$ single crystal[4] at 4 K where the modulation is due to the 14% abundant ^{183}W with $I = \frac{1}{2}$. By assuming the known crystal structure of pure $CaWO_4$ and that Ce^{3+} was substituted for calcium it was found that the experimental modulation at various angles could be reasonably well simulated by assuming zero isotropic coupling and includ- ing dipolar interactions with the 10 nearest tungsten nuclei. To obtain agreement with experiment it was necessary to enhance the dipolar interaction for all nuclei by a factor $\rho = 4$. If this enhancement is assumed due to lattice distortion by the foreign ion the effective lattice distance is $r_{eff} = r_{lattice}\, \rho^{-1/3}$ and r_{eff} varies from 2.30 to 3.48 Å for the nearest 10 tungsten nuclei. A less detailed analysis was made for Ce^{3+} in cubic CaF_2 where it was found that $\rho = 8$ and $r_{eff} = 2.36$ Å for eight equivalent nearest fluorine nuclei.

The $S = 1$ system Ti^{2+} in CdS has also been studied at 27 K with some success.[26] The Ti^{2+} is substituted for Cd^{2+} and is modulated by the ^{111}Cd and ^{113}Cd isotopes, both with $I = \frac{1}{2}$. The dipolar interaction is relatively weak and gives a *regular* modulation pattern characterized by a single frequency that changes with angle. This indicates that the modulation frequency corresponds to a nuclear frequency in the ESR spectrum. If this frequency is not resolved in the ESR spectrum due to inhomogeneous line broadening, the electron spin-echo pattern may reveal it as in this case. Here the detailed modulation pattern is not analyzed. The modulation frequency is simply assigned to a particular transition and its angular dependence is fit to the spin Hamiltonian. For Ti^{2+} in CdS the modulation frequency was assigned to the frequency difference between allowed electron flip transitions and forbidden transitions involving both electron and cadmium nuclear spin flips.

A similar study was earlier made for $CH(COOH)_2$ radical in γ-irradiated malonic acid single crystal at room temperature.[27] The occurrence of two forbidden and two allowed transitions gives two sets of doublets. The observed spin-echo modulation was observed only over restricted angular ranges where the forbidden transition probability was high and was successfully assigned to the frequency difference between allowed and forbidden transitions.

Three-pulse echo experiments offer some advantage for determining the angular dependence of nuclear modulation frequencies in single crystals. The longer time scale of the three-pulse echo increases the accuracy of the modulation frequency measurement. This has been nicely demonstrated for $CH_2(CH_2)_2COOH$ radicals in γ-irradiated glutaric acid crystal at 77 K.[28] The nuclear frequencies associated with the two β-protons were determined as a function of orientation around three mutually perpendicular axes and were used to determine the hyperfine tensors.

In planes where there are two or more sets of nonequivalent radicals the modulation frequencies from both sets of radicals interfere and produce a complex modulation pattern as shown in Fig. 21a. However, as discussed in connection with Eq. (51), specific modulation frequencies can be supressed in the three-pulse echo pattern at appropriate values of time τ. The effect of varying τ is shown in Fig. 21b, c, where the individual modulation frequencies are separated, thus making the hyperfine tensor analysis feasible. The regularity of the modulation pattern is also sensitive to crystal orientation; interference beats distort the modulation pattern for misalignments as small as $\sim 2°$. This fact can be used to determine accurately the directions of the crystal axes.[28]

The main advantage of hyperfine tensor determination in single crystals by spin-echo modulation is its greater accuracy, by perhaps fivefold, for protons with low anisotropy over ordinary ESR spectra. However, single-crystal electron-nuclear double resonance (ENDOR) is probably superior to the echo method for most applications.[1]

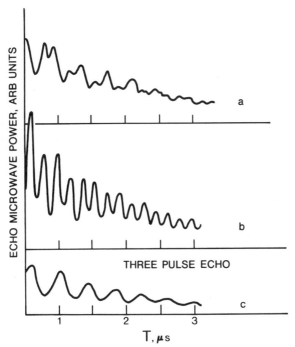

ECHO MICROWAVE POWER, ARB UNITS

THREE PULSE ECHO

a

b

c

$T, \mu s$

Figure 21 Three-pulse electron spin-echo modulation of $\cdot CH_2(CH_2)_2COOH$ radicals in γ-irradiated single crystal glutaric acid at 77 K. The modulation is due to β-proton interaction and the orientation is in a plane where there are two nonequivalent sets of radicals: (a) modulation pattern at arbitrary time τ where the modulation frequencies from two nonequivalent radicals interfere, (b) and (c) modulation patterns at specific τ values where one or the other modulation frequency is suppressed. [Adapted from V. F. Yudanov et al., *J. Struct. Chem.*, **15**, 510 (1974).

When $I > \frac{1}{2}$, the single crystal results have been less encouraging. The $S = \frac{1}{2}$ ion, V^{4+}, has been studied in TiO_2^5 at 4 K where the complex modulation pattern is due to 7.75% abundant ^{47}Ti with $I = \frac{5}{2}$ and 5.5% abundant ^{49}Ti with $I = \frac{7}{2}$. Simulations of the two-pulse modulation were made by assuming a substitutional V^{4+} interacting with the 10 nearest titanium nuclei with $I = \frac{7}{2}$ in the known TiO_2 structure and $\mathbf{a} = 2.4$–2.6 MHz. As in $CaWO_4$, comparison with experiment suggests that the electron-nuclear dipolar interaction is considerably enhanced with $\rho = 7$–8 being preferred. If this were due to lattice distortion, $r_{eff} = 1.5$–1.9 Å is implied, which seems quite short and probably presages breakdown of the point dipole approximation assumed in the analysis. In general, the agreement between experiment and simulation at different angles is not very good. This probably reflects the neglect of quadrupole interactions, the approximation of the $I = \frac{5}{2}$ nuclei by $I = \frac{7}{2}$, and the breakdown of the point dipole approximation. The experimental data may also have been distorted because of the incomplete excitation of some of the hyperfine intervals.

The great sensitivity of the two-pulse echo modulation pattern to the hyperfine parameters in single crystals is illustrated by the results on the $S = \frac{3}{2}$ ion, Cr^{3+}, in Al_2O_3 (i.e., ruby) at $4\,K^{6,29-31}$ where the complex modulation pattern is from aluminum nuclei with $I = \frac{5}{2}$. Instead of beginning with the pure lattice geometry the modulation simulation was based on known ENDOR hyperfine and quadrupole parameters for the nearest 13 aluminum nuclei to the Cr^{3+}. This gave a simulated modulation pattern in fairly good agreement with experiment for pulse separations of 0.8–2.5 μs but the agreement was poor at larger pulse separations.[6] (See Fig. 22a, b.) Subsequent electron spin-echo ENDOR experiments, in which the nuclear rf pulse is applied between the second and third pulses

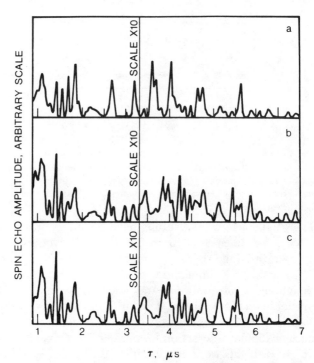

τ, μs

Figure 22 Two-pulse electron spin -echo modulation of single-crystal ruby (Cr^{3+}-doped Al_2O_3) at $H_0 = 3300\,G$ and $\sim 2\,K$. (a) Simulated echo modulation using ENDOR parameters of N. Laurance et al., *J. Phys. Chem. Solids*, **23**, 515 (1962). (b) Experimental echo modulation. (c) Simulated echo modulation using a more general quadrupole interaction in the Hamiltonian and spin-echo ENDOR parameters. [Adapted from P. F. Liao and S. R. Hartmann, *Phys. Rev. B*, **8**, 69 (1973).

of a three-pulse echo sequence, showed that the quadrupole interaction must be written more generally in terms of an asymmetry of the quadrupole components and a tilt of the quadrupole axes from the hyperfine axes.[31] These corrections to the Hamiltonian only amounted to about a 3% change in the aluminium nuclear frequencies, but the effect on the two-pulse echo modulation pattern at longer pulse separations is profound. (See Fig. 22b, c.)

For single-crystal studies it appears that electron spin-echo modulation can best be analyzed to give new structural information for $I = \frac{1}{2}$ and probably for some $I = 1$ systems. For larger I, especially with a significant quadrupole interaction, the modulation pattern is complex and very sensitive to all the various magnetic parameters.

4.2 Biradical Systems

Biradical systems can be composed of two adjacent radicals that are magnetically coupled to form an $S = 1$ state; this is commonly called a radical pair. Also, two relatively isolated unpaired electrons on the same molecular species can constitute a biradical. Biradicals introduce electron dipole-electron dipole and exchange interactions that may lead to electron spin-echo modulation effects just as the electron-nuclear dipole interactions we have previously discussed. Yudanov et al. first demonstrated the echo modulation from such interactions.[32]

Ultraviolet irradiation of potassium persulfate ($K_2S_2O_4$) single crystals at 77 K produces SO_4^- radical pairs separated by 15.8 Å. This distance is deduced from the electron dipolar splitting observed in the ESR spectrum, which is given by

$$\Delta\nu = \left(\frac{3g^2\beta^2}{2hr^3}\right)(1 - 3\cos^2\theta), \qquad (57)$$

where θ is the angle between the electron-electron dipolar axis and the magnetic field. The two-pulse electron spin echo of the radical pair lines exhibits a regular modulation pattern with a frequency $\nu_{mod} = \Delta\nu$. The modulation frequency also varies with angle, as does $\Delta\nu$ between the observable limits of 3–20 MHz for the modulation. This coincidence indicates that the modulation is due to the electron dipole-dipole interaction. It is interesting that these modulation effects are absent in a polycrystalline sample containing SO_4^- pairs even though the two radical pair lines can still be observed.[32]

The following biradical A

(A)

exhibits a 4 MHz modulation of the two-pulse spin-echo signal in both protiated and deuterated toluene glass at 77 K.[32] It was suggested that the modulation is not due to electron-nuclear dipolar interaction and is associated with the exchange interaction J in the biradical. The exchange interaction will only lead to echo modulation if the interacting magnetic moments have different Larmor precession frequencies measured by $\Delta\omega = \omega_1 - \omega_2$. For $J \ll \Delta\omega$ the modulation frequency is identified as J. For $J \gg \Delta\omega$ the modulation frequency is identified as $(\Delta\omega)^2/2J$. A selection between these two interpretations cannot be made without additional information.

In contrast to biradical A, the butyl bisverdazyl biradical exhibits modulation in toluene at 77 K that is mainly assigned to electron-nuclear dipolar interaction with nitrogen nuclei.[33] An analogous assignment for biradical A does not seem to have been excluded.

Too little work on echo modulation associated with biradicals has yet been done to ascertain whether exchange interactions can be generally studied by this method or not.

4.3 Radical Solvation Structure in Disordered Systems

Probably the most useful application of electron spin-echo modulation thus far has been to reveal radical solvation structure in disordered, mainly glassy systems. In particular the solvation structure of localized electrons in various glassy matrices has been delineated for the first time.

The first observation that two-pulse electron spin-echo modulation is retained in a disordered system was reported by Zhidomirov et al. for trapped hydrogen atoms in deuterated sulfuric acid.[34] In this and related work,[35,36] two-pulse spin-echo modulation was reported at 77 K for hydrogen atoms in $2M$ H_2SO_4 showing proton modulation, deuterium atoms in $2M$ D_2SO_4 showing deuterium modulation, diphenylpicrylhydrazyl (DPPH) in CH_3OH showing proton[34] and nitrogen[35] modulation, DPPH in CD_3OD showing deuterium modulation, CH_2OH in NaA zeolite showing aluminum modulation, FOO· in γ-irradiated teflon showing fluorine modulation, and CO_2^- in γ-irradiated Na_2CO_3 showing sodium modulation. These results showed the generality of observing spin-echo modulation for a variety of radiation-produced and stable radicals involving different types of matrix nuclei. Attempts were made to analyze these initial results by relating the normalized modulation depth λ to an "effective" interaction distance $r_{\text{eff}} = (n/r^6)^{-1/6}$ by relation (58)

$$\lambda = \frac{16}{5} \frac{g^2 \beta^2}{H_0^2 r_{\text{eff}}^6} I(I+1). \tag{58}$$

This relation is derived from Eq. (23) and (25) under the limiting assumption that $\omega_\alpha = \omega_\beta = \omega_I$ followed by averaging over angle (i.e., set $\theta = 45°$). The results generally gave $r_{\text{eff}} \sim 2.3$–2.7 Å, whereas the very weak hyperfine assumptions leading to (58) are only valid for $r > 3.6$ Å.[37]

The first attempt at a more complete analysis of radical solvation structure in glassy systems by full simulation of the echo modulation pattern was made by Bowman, Kevan, and Brown, who studied radiation-produced localized electrons in alkaline aqueous glasses.[38] Subsequent work established that quantitative conclusions on radical solvation structure are best accomplished by full simulation of the modulation patterns.[14,15,17,22,39] Two examples may be cited as test systems.

The deuterium modulation pattern of CH_2OD radicals produced in γ-irradiated CH_3OD glass at 77 K may be expected to be dominated by the deuterium atom *on* the radical.[39] The normalized two-pulse echo modulation pattern is fit quite well by a simulation with parameters $n = 1$, $r = 2.11$ Å and $\mathbf{a} = 0.76$ MHz. These values compare well with $r = 2.0$ Å, which is the C–D distance in the CH_3OD molecule[40] and $\mathbf{a} = 0.74$ MHz determined from $\mathbf{a} = 4.9$ MHz for OH protons from the low-temperature (223 K) liquid phase ESR spectrum.[41] A contribution to the echo modulation from the matrix deuterons is also expected, although it was not included in the preceding analysis; it should be more easily detectable by three-pulse echo modulation.

The two-pulse spin-echo proton modulation pattern of Nd^{3+} in a 4:1 water:ethanol glass at 4 K is simulated fairly well by $n = 18$, $r = 3.0 \pm 0.1$ Å and $\mathbf{a} = 0$.[17] No attempt was made to divide out the echo decay function so the comparison of parameters with experiment is only semiquantitative. However, the parameters found agree well with a solvation shell of nine water molecules as found in crystals of neodymium ethyl sulfate and neodymium bromate and with the sum of the Nd^{3+} ionic radius plus the bonding distance in H_2O assuming that the negative ends of the water dipoles are oriented toward Nd^{3+}.

The most dramatic new chemical information from electron spin-echo modulation studies derives from investigations of radiation-produced localized electrons in glassy matrices. Localized electrons exist as transient species in the radiolysis or photoionization of water, alcohols, and a number of other organic liquids, and they may also be stably trapped in aqueous and organic glasses at sufficiently low temperature.[42] From a variety of both optical and ESR studies it appears that the injected electrons are initially localized in shallow potential wells, which are then deepened by a solvation process in which molecular orientation occurs to produce an equilibrium solvated electron species. Thus it is important to try to establish the solvation structure of these localized electrons. The ESR spectrum alone is a single broad line that hides the dipolar interactions to near neighbor magnetic nuclei and gives limited information about the possible solvation structures. However, electron spin-echo modulation has been able to be observed for these species and makes possible the detection of weak dipolar hyperfine interaction between the localized electron spin and nearby nuclear spins to probe the proton arrangement and hence the molecular arrangement around the localized electron.

The first system to be discussed will be solvated electrons in 2-methyltetrahydrofuran (MTHF) organic glass. In order to determine not only the number and distances of near-neighbor magnetic nuclei to the localized electron but also the solvent molecular orientation with respect to the localized electron, specifically deuterated MTHF molecules were used.[14]

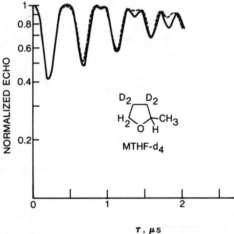

τ, μs

Figure 23 Simulated (solid curve) and experimental (dashed curve) two-pulse electron spin-echo modulation from deuterons around solvated electrons in MTHF-d_4 glass at 77 K. The parameters of the simulated curve are $r = 3.9$ Å, $a = 0.2$ MHz, $H_0 = 3220$ G, and $n = 16$ deuterons, which corresponds to four MTHF molecules. [Corrected from L. Kevan et al., *J. Chem. Phys.*, **63**, 409 (1975).]

Typical two-pulse electron spin-echo signals for solvated electrons in MTHF-d_3 have already been shown in Fig. 5 along with the structure of MTHF-d_3. In addition, Fig. 23 shows the normalized two-pulse echo amplitude along with a simulation for electrons in MTHF-d_4 and the structure of that partially deuterated molecule. The analysis of the echo modulation was carried out for the deuterium interactions only. The prominent modulation frequency is ~2 MHz, which is near the free deuteron frequency, and the double frequency only comes in slightly after several modulation cycles. This alone suggests that the dominant electron-deuteron interaction distance is >3 Å because closer distances give more prominent contributions from the double frequency. The modulation depths in the MTHF-d_3 and MTHF-d_4 are quite similar, with the depth in MTHF-d_4 being only slightly larger, because of the additional deuteron. This suggests that the localized electron interacts about equally with the d_3 and d_4 deuterons and roughly places the electron above or below the plane of the four carbons in the MTHF ring. If the localized electron were located on the side of the MTHF molecule opposite the oxygen, for example, it would interact more strongly with deuterons in MTHF-d_4 than in MTHF-d_3 and give a much larger modultion depth in MTHF-d_4. A similar analysis of the modulation depth in fully deuterated MTHF-d_{10} shows that the MTHF-d_{10} depth is only slightly more than the product of the MTHF-d_3 and MTHF-d_4 depths. This indicates that the localized electron interacts considerably more weakly with the CD_3 deuterons than with the ring deuterons and places the localized electron on the side of the ring opposite CD_3. The quantitative simulations of the modulation patterns fully bear out this analysis and suggest a solvation structure, as shown in Fig. 24.

In the simulations we have used an average electron-deuteron interaction distance for all the ring deuterons in each matrix. Since some of the deuterons are above the ring and below the ring, this effective interaction distance corresponds to the distance from the electron to the plane of the ring carbons. We can find a consistent set of parameters for all three deuterated molecules that correspond to the same number n equidistant MTHF molecules. The best common set of parameters for these simulations is $n = 3$, $r = 3.7$ Å and $\mathbf{a} = 0.2$ MHz (4 deuterons) or $n = 4$, $r = 3.9$ Å, and $\mathbf{a} = 0.2$ MHz. A small positive isotropic hyperfine interaction is indicated by the simulations. For zero isotropic coupling or for negative isotropic coupling the fits are not satisfactory. The small positive isotropic hyperfine interaction is also consistent with the absence of a phase change in the three-pulse modulation pattern.[24] The preceding parameters are slightly revised from those published originally[14] because of a computer programming error. As can be seen from the two possible sets of

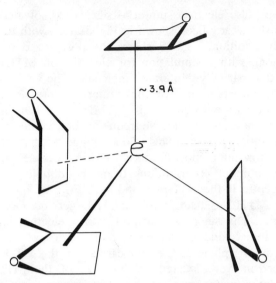

Figure 24 Schematic model for solvated electron structure in 2-methyltetrahydrofuran (MTHF) glass at 77 K based on electron spin and second-moment analyses. The four MTHF molecules are equivalent and are assumed to be arranged tetrahedrally around the electron; the indicated distance is from the electron to the plane of the ring carbons. An alternate model with three MTHF molecules arranged trigonally at 3.7 Å from the electron also fits the data. [Corrected from L. Kevan et al., *J. Chem. Phys.*, **63**, 409 (1975).]

parameters given earlier there is some correlation between the number and the distance of interacting nuclei that often prevents a totally unambiguous structural determination to be made. This ambiguity can probably be reduced by finding the best set of parameters that simultaneously satisfy two- and three-pulse echo modulation patterns. Work of this type is in progress.[20]

Although the ESR spectrum alone of localized electrons does not give any direct information about the dipolar interactions, a second-moment analysis of the ESR linewidths of the localized electron in the variously deuterated MTHF molecules does give sets of acceptable numbers and distances for interacting deuterons.[14] Such second-moment results independently support the structural conclusions of Fig. 24 but again do not distinguish between the acceptable combinations of n and r. The second-moment analyses also confirm the fact that the methyl group on the MTHF molecules is farther away from the localized electron than are the ring deuterons. It can be seen that when selective deuteration, or other appropriate isotopic substitution, is used in conjunction with the analysis of electron spin-echo modulation patterns a quite detailed structure of a radical's environment in a disordered matrix can be deduced.

The structure of solvated electrons in a typical alkane, 3-methylpentane glass at 77 K, has also been studied by the analysis of electron spin-echo modulation patterns.[43] Only the fully deuterated 3-methylpentane glass was studied so it was not possible to determine uniquely the orientation of the surrounding matrix molecules oriented by the localized electron. The best comparison of the simulated modulation pattern to the experimental one gave $n = 18$–21, $r = 3.0$–3.2 Å and $\mathbf{a} = \sim 0$. From chemical and theoretical considerations it was concluded that the most probable orientation of the solvating alkane molecules is probably that with one methyl group from each molecule oriented toward the electron. Thus 18–21 interacting deuterons implies 6–7 first solvation shell molecules. It would be interesting to carry out more detailed experiments with selectively deuterated 3-methylpentane to try to pin down the orientation of the alkane molecule in this solvated electron species.

Solvated electrons in aqueous matrices have also been studied by electron spin echo. Here the analysis is much more complicated because there are interacting nuclei closer than 3 Å. In the preceding cases of solvated electrons in MTHF and 3-methylpentane glasses the nearest interacting nuclei were >3 Å away and the modulation pattern could be successfully simulated with one shell of nuclei, the shells of nuclei at greater distances making a negligible contribution. In the earliest analysis of solvated electrons in aqueous glasses the pattern was interpreted mainly as due to surrounding deuterons at >3 Å,[38] but closer analysis shows that there are distinctive sharp features in the first several periods of the modulation pattern that can only be explained as contributions from nuclei much closer than 3 Å. Thus to analyze the echo modulation pattern for solvated electrons in aqueous glasses it is necessary to consider contributions from both the nearest and next nearest magnetic nuclei.

The system that was studied in detail was solvated electrons in $10M$ NaOD aqueous glass at 77 K. To sort out the nearest to next nearest deuteron interaction the three-pulse echo pattern was examined and found to exhibit a phase change (see Fig. 17), which indicated an interaction with deuterons at $r = 3.6$ Å. These deuterons were interpreted as being the next nearest neighbors to the solvated electron and the number of deuterons at this distance was deduced from the deuteron density in the aqueous glass. The total number of deuterons within 3.6 Å ± 0.5 Å radius was found to be 12, which must be divided between those in the nearest and next nearest neighbor shells. By using the constraints of the second moment of the ESR line of the solvated electron on the values of the isotropic and anisotropic constants the range of possibilities for the parameters n, $\mathbf{a}$, and r_1 for the nearest-neighbor

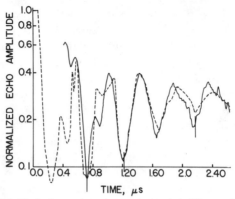

Figure 25 Simulated (dashed curve) and experimental (solid curve) two-pulse electron spin-echo modulation from deuterons around solvated electrons in $1OM$ NaOD/D_2O aqueous glass at 77 K. The simulated curve corresponds to nearest neighbor deuterons with parameters $r_1 = 2.1$ Å, $n_1 = 6$, and $a = 0.9$ MHz and next nearest deuterons with parameters $r_1 = 3.6$ Å, $n_2 = 6$, and $a_2 = 0$. [From P. A. Narayana et al., *J. Chem. Phys.*, **63**, 3365 (1975).]

deuterons was limited. The best final results were obtained for a model where the nearest-neighbor deuterons were characterized by $n_1 = 6$, $r_1 = 2.1$ Å, and $a_1 = 0.9$ MHz and the next nearest-neighbor deuterons by $r_2 = 3.6$ Å, $n_2 = 6$, and $a = 0$. The product of the simulated modulation from the two shells of interacting deuterons gave the final simulated modulation pattern shown in Fig. 25 which agrees reasonably well with the experimental pattern. In particular the sharp features in the first two modulation periods are reproduced. It is to be noted that the experimental modulation pattern could not be simulated by only considering deuterons at a single distance, close or far. Together with subsequent data based on ^{17}O enriched matrices the solvated electron structure shown in Fig. 26[44] was deduced. It is particularly interesting that the OH bonds of

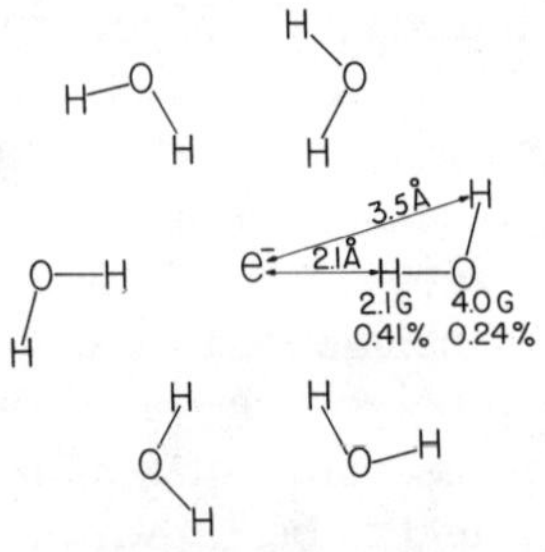

Figure 26 Geometrical structure of solvated electrons in $1OM$ NaOH/H_2O aqueous glass at 77 K deduced from electron spin-echo and second-moment analyses. The six waters are equivalent and are assumed to be arranged octahedrally. The %'s refer to the percent of solvated electron unpaired spin density per hydrogen or oxygen nucleus. [From S. Schlick et al., *J. Chem. Phys.*, **64**, 3153 (1976).]

the solvating water molecules rather than the water molecule dipoles are oriented toward the electrons. This structure is also fully compatible with the second moment and the satellite line intensities observed.[45] Determination of this detailed geometrical picture for solvated electrons in aqueous glasses has only been possible through the analysis of electron spin-echo modulation patterns.

Subsequent studies of the three-pulse echo modulation for solvated electrons in five different aqueous glasses formed with different types of solutes at 77 K have indicated that $r \sim 3.5$ Å and $n = 6$–9 for all systems.[23] This suggests that the solvated electron geometry is similar in different aqueous matrices. Three-pulse analysis was carried out using Eqs. 45, 56, and 57 with the assumption that the isotropic coupling was zero. This study really only shows that the next-nearest-neighbor deuteron interactions are similar in the different aqueous matrices, but based on the model in Fig. 26 this implies that the nearest-neighbor interacting deuteron must also have similar parameters in the different aqueous matrices. In addition, an important feature of this more complete three-pulse analysis is that it confirms the number of interacting nuclei assigned to the next-nearest-neighbor interaction.

Another system that has been studied is the solvation of silver atoms in aqueous matrices.[46] Here it has been possible to form initially the silver atom in a desolvated state at 4 K and then thermally to induce solvation by brief warming to 77 K. Both the desolvated and solvated states have been studied by electron spin-echo techniques. The two-pulse electron spin-echo modulation pattern has been analyzed to show that the initial desolvated silver atom produced by reduction of a silver ion in an aqueous matrix with electrons at 4 K is surrounded by eight equivalent deuterons at 3.1 ± 0.1 Å, which suggests a tetrahedral model for four waters with their dipoles pointed away from the silver atom (i.e., a solvated silver ion geometry). In other words, it appears that the solvated Ag^+ geometry is retained on initial reduction of Ag^+ to Ag^0 by electrons at 4.2 K. Brief warming to 77 K causes a decrease in the two-pulse electron spin-echo modulation depth and suggests that there is one less deuteron interacting at 3.1 Å. This is interpreted to be due to solvation of the silver atom by hydrogen bonding forces via rotation of one water molecule around one of its OD bonds to put one deuteron very close at approximately 1.8 Å from Ag^0 and leaving seven deuterons at the 3.1 Å distance. The close deuteron does not contribute to the echo modulation pattern observed at low pulse powers because the nuclear splitting is larger than the microwave magnetic field. However, this close deuteron has been seen by spin echo at high pulse powers.[47] This picture of solvation is shown in Fig. 27. It has been confirmed and supported by a

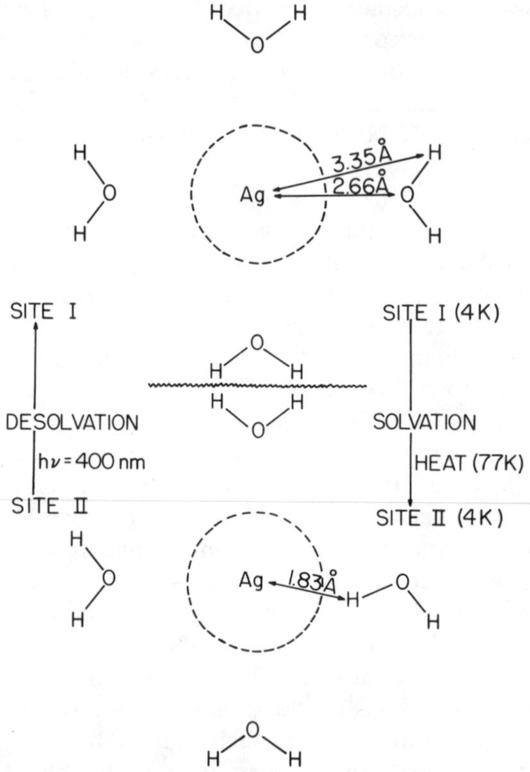

Figure 27 Schematic of geometry changes in silver atom solvation and desolvation in aqueous matrices.

detailed analysis of the forbidden proton spin-flip satellite transitions, which seem to be due to eight protons at a distance of about 3.2 Å in the case of the desolvated Ag^0 and to one very close proton at ~ 1.8 Å in the solvated Ag^{0}.[48] The preceding electron spin-echo results were obtained by complete simulation of the two-pulse modulation pattern.

Irradiation of oxyanion acids at 77 K produces trapped hydrogen atoms with magnetic resonance parameters that are almost identical to those of the free atom.[49] It has long been wondered exactly what the nature of the atomic hydrogen traps is.[50] For trapped hydrogen atoms in sulfuric acid glass it is possible to observe directly satellite lines resulting from forbidden transitions associated with weak dipolar coupling to matrix protons. From the relative intensities of these satellite lines and the main allowed line transitions and their energy separation it is possible to determine both the distance to and the number of matrix nuclei that contribute to a given satellite transition.[51] For hydrogen atoms trapped in gamma-irradiated sulfuric acid glass at 77 K this type of analysis indicates that interaction occurs with four matrix protons at a distance of 2.15 Å.

The two-pulse electron spin-echo modulation pattern of hydrogen atoms in sulfuric acid glass at 77 K has also been analyzed by the approximate method shown in Eq. (45) and the assumption that $a = 0$.[37] The analysis was made for deuterium modulation associated with deuterium atoms trapped in $8M$ D_2SO_4 glass and gave $n = 24$ and $r = 3.65$ Å. This result does not seem consistent with the satellite line analysis. Although the results have not been published in detail, a later analysis involving a full simulation of the two-pulse and three-pulse echo modulation pattern for $\tau > 2.5$ μsec is consistent with the parameters $n = 16$, $r = 3.6$ Å, and $a = 0$.[52] These parameters suggest a model for the hydrogen atom formed by eight water molecules situated at the apices of a cube oriented with the oxygen of each water molecule closest to the trapped hydrogen atom. This is an appealing model but the discrepancy between the spin-echo results and the analysis of the satellite lines remains to be resolved.

Trapped hydrogen atoms have also been studied by electron spin echo in water-alcohol mixtures.[52] The analysis of the deuterium modulation pattern for deuterium atoms in methanol-water mixtures with a mole fraction of 0.3 methanol corresponding to the maximum deuterium atom yield gave the following results. For CH_3OD/D_2O mixtures $n = 22$, $r = 3.6$ Å, and $a = 0$, whereas for CD_3OH/H_2O mixtures $n = 20$, $r = 5.3$ Å, and $a = 0$. The big difference in distances between these two mixtures suggests that the alcohol molecules are oriented with their hydroxyl groups toward the deuterium atoms in the trapping site.

Another species that is trapped in alkaline aqueous glasses is the O^- radical anion. This species is a chemically converted hole, that is trapped in the same type of matrices that localize electrons. The two-pulse echo was observed for O^- at 4.2 K in $4.5M$ LiOH, $10M$ KOH, and $5-19M$ NaOH.[53] Modulation was observed from the ^7Li and ^{23}Na alkali cations in these aqueous glasses. No modulation was seen from ^{39}K because its expected modulation period is greater than the echo decay time. The modulation depth was small so the analysis was carried out in terms of r_{eff} based on the approximate analysis given by Eq. (58). This analysis gives $r_{eff} = 3.87$ Å for both the lithium and sodium nuclei; this distance is large enough that the analysis on which Eq. (58) is based is expected to be valid. One surprising feature is that the cation modulation depth was independent of the NaOH concentration in the range from $5-19M$. Thus it was suggested that there is a specific O^--cation stabilization distance indicating an interaction not subject to the statistical distribution of

Table 1 Electron Spin-Echo Analysis of the Deuteron Environment of Nitroxide Radicals in Specifically Deuterated Solvents (Data from References 21 and 52)

Radical	Solvent	Three-Pulse Echo		Two-Pulse Echo	
		n	$r(\text{Å})$	n	$r(\text{Å})$
H—C=NOH ... H$_3$C, C=N, CH$_3$ ... H$_3$C, N, CH$_3$, O	C$_6$H$_5$CD$_3$	10.9	4.76	12	4.6
	C$_6$D$_5$CH$_3$	5.3	3.55	5	3.5
	CH$_3$OD	1.5	3.7		
	CD$_3$OH	20.4	4.6		
H—C=NOH ... H$_3$C, C=N→O, CH$_3$ ... H$_3$C, N, CH$_3$, O	C$_6$H$_5$CD$_3$	11.5	4.91	12	4.6
	C$_6$D$_5$CH$_3$	4.5	3.55	5	3.6
	CH$_3$OD	2.1	3.8		
	CD$_3$OH	4.1	4.4		
Ph, H$_3$C, C=N, CH$_3$, H ... H$_3$C, N, C=NOH, O	C$_6$H$_5$CD$_3$	4.1	3.92	6	4.1
	C$_6$D$_5$CH$_3$	3.2	3.42	5	3.40
	CH$_3$OD	1.6	4.1		
	CD$_3$OH	20.1	4.5		
Ph, H$_3$C, C=N→O, CH$_3$ ... H$_3$C, N, CH$_3$, O	C$_6$H$_5$CD$_3$	6.2	4.16	6	4.0
	C$_6$D$_5$CH$_3$	4.5	3.57	5	3.40
	CH$_3$OD	1.7	3.9		
	CD$_3$OH	20.7	4.3		
(H$_3$C)$_3$C, C(CH$_3$)$_3$, N, O	C$_6$H$_5$CD$_3$	7.9	4.86	9	4.6
	C$_6$D$_5$CH$_3$	3.2	3.92	5	3.6
	CH$_3$OD	2.2	4.1		
	CD$_3$OH	24.6	4.8		

cations. It would be interesting to make a full calculation of the modulation pattern for O^- to investigate the solvation structure of O^- more fully.

In addition to the environment of small unstable radicals in frozen solutions, electron spin-echo techniques have also been used to probe the environment of stable nitroxide radicals in specifically deuterated solvents.[21,52] The results are shown in Table 1. An approximate analysis for both the three-pulse and two-pulse results was carried out on the basis of Eq. (45) with the assumption that $\mathbf{a} = 0$. The results for a number of interacting deuterons and their average distance are summarized in Table 1. Where both two- and three-pulse echo results were obtained it is seen that there is good self-consistency. Although it is not possible to deduce a detailed picture of the solvation of the nitroxide group in these radicals, the data do suggest that in toluene matrices at least one molecule has its benzene ring situated close to the NO site of the predominant unpaired electron density, whereas in methanol matrices it appears that two methanol molecules are oriented with their hydroxyl groups toward the NO group on the nitroxide. The data also suggest that the unpaired electron on the nitroxide interacts with two or four molecules of toluene and seven or eight molecules of methanol oriented with their methyl groups toward the nitroxide radical center.

The preceding summary of electron spin-echo modulation results on the environment of trapped radicals in disordered systems indicates that a great deal of new and interesting chemical information can be obtained by the detailed analysis of the electron spin-echo modulation pattern that is not readily available from other physical techniques.

4.4 Radicals on Surfaces

In addition to solvation shell studies it appears that electron spin-echo modulation may contribute to the location and perhaps the orientation of radicals on surfaces. One recent study has been reported where CH_2OH radicals produced by radiolysis at 77 K are trapped in A-type zeolites.[54] Zeolites are composed of AlO_4 and SiO_4 tetrahedrons bonded together to form a three-dimensional network composed of sodalite or β-cages and supercages or α-cages. In A-type zeolites the sodalite cage opening is 2.2 Å and the supercage opening is controlled by the nature of the charge-compensating cation in the zeolite. Zeolites designated $3A$, $4A$, and $5A$ correspond to K^+ zeolite with ~3 Å supercage openings, Na^+ zeolite with ~4 Å openings, and Ca^{++} zeolite with ~5 Å openings. The surface in zeolites is mostly interior in the cages.

Both two- and three-pulse spin-echo results were obtained for CH_2OH and CD_2OH radicals produced in $3A$, $4A$, and $5A$ zeolites. The CH_2OH

radical showed modulation due to aluminum ($I = \frac{5}{2}$) nuclei in the zeolite lattice, but no modulation due to sodium or potassium nuclei was observed. Only the two-pulse data were analyzed and the approximate relation in Eq. (58) was used. The results gave $r_{\text{eff}} = 4$ Å independent of cation type in the zeolite.

The CH_2OH radicals showed modulation from deuterons in neighbouring CD_3OH molecules in the same supercage. The two-pulse results are a complex superposition of aluminum and deuterium modulation. However, in the three-pulse results the modulation of aluminum can be suppressed relative to that from deuterium and vice versa because $\omega_I(^{27}Al)/\omega_I(^{2}D) = 1.7$. Recall that the three-pulse echo modulation signal is maximized at τ corresponding to a maximum depth in the two-pulse echo modulation. See Eq. (51) and discussion following. Since $\omega_I(^{27}Al)/\omega_I(^{2}D) \sim 2$, the three-pulse modulation signal from ^{2}D is maximized when that from ^{27}Al is minimized. Furthermore, no three-pulse echo modulation is seen from the two α-deuterons on CD_2OH, since their hyperfine coupling is too large for simultaneous excitation of their allowed and forbidden transitions. The three-pulse echo results were analyzed on the basis of Eq. (45) with the assumption that $\mathbf{a} = 0$. For $r = 4$ Å, $n = 18$, 6, and 3 for CD_2OH in irradiated pure CD_3OH, in irradiated NaA zeolite saturated with methanol vapor, and in irradiated NaA zeolite only 50–70% saturated with methanol vapor. These results do suggest that CD_3OH molecules neighboring the CD_2OH radical are being detected in the supercage. Again these results were independent of the cation type in the zeolite.

From the preceding results a model for the CD_2OH radical in A-type zeolite supercages has been proposed.[54] From the supercage volume and the saturation pressure of methanol over zeolite A it is deduced that there are about eight molecules per supercage. If each of these is assumed to be situated in an octant of a spherical supercage with their methyl groups oriented toward the center of the cage the geometrical results from the spin-echo analysis can be accounted for reasonably well.

It appears that there is considerable potential for studies of radicals on surfaces by spin-echo methods. Full simulations of both two- and three-pulse modulation patterns should lead to even more detailed information about the radical orientation relative to the surface.

4.5 Biological Systems

Many biological systems contain paramagnetic metal ions that can serve as probes to determine the coordination or binding to certain types of ligands. These systems are typically disordered and the problem is an analogous to the determination of solvation structure, which we discussed in Section 4.3.

The first series of studies in this area has recently been reported by Mims and co-workers on several copper proteins.[55-57] Copper proteins contain two types of mononuclear copper. Type 1 copper sites are characterized by a strong optical absorption near 600 nm and a small value of the $a_{\parallel}$ copper coupling and type II sites have a weak visible absorption and larger $a_{\parallel}$ values. EPR studies have not been able to identify the ligands of type I copper, although ENDOR studies of stellacyanin have indicated nitrogen-containing ligands.[58] However, electron spin-echo modulation patterns for stellacyanin and other copper proteins indicate that one or more imidazole groups are bound to the copper in these proteins.

The following copper proteins have been studied by spin echo: stellacyanin, which contains type I copper[55]; galactose oxidase, which contains type II copper[57]; laccase from *Rhus vernicifera*, which contains both type I and II copper[56]; and porcine ceruloplasmin, which also contains both type I and II copper. A typical two-pulse modulation pattern at 4 K is shown in Fig. 28a for type I copper in laccase from which the type II copper has been removed.[56] The rapid modulation frequencies with periods of ~90 ns and ~45 ns are due to proton modulation (see Fig. 12a), but the interesting feature is the low-frequency modulation with a

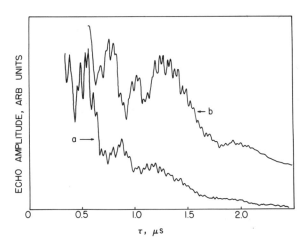

Figure 28 Two-pulse electron spin echo at 4 K for (a) type I copper in 3 mM decuprolaccase in 0.05M phosphate, pH 7 at $H_0 = 2970$ G, and (b) Cu^{2+}-diethylene triamineimidazole complex in 1:1 glycerol-water, pH 8.3 at $H_0 = 3195$ G. The vertical scales are offset for (a) and (b) to avoid overlap. [Adapted from B. Mondovi et al., *Biochemistry*, **16**, 4198 (1977).]

period of ~0.7 μs. This modulation is attributed to nitrogen and specifically to the distal nitrogen in ligated imidazole by comparison to model compounds. Figure 28b shows the two-pulse spin-echo pattern of Cu^{2+}-diethylenetriamine-imidazole, which is similar to copper protein. The ~0.7 μs modulation is assigned to the distal nitrogen, which is not directly bonded to Cu^{2+} because only the proton modulation is seen for Cu^{2+}-glycylglycine in which a nitrogen is directly bonded to copper.[55] Directly bonded nitrogen to copper has a relatively large hyperfine coupling, typically 30 MHz, so that its modulation is not seen because both allowed and forbidden transitions cannot be excited by a single microwave pulse as is required to see modulation effects. The depth of the modulation is indicative of the number of imidazole groups coordinated to the copper. For example, the modulation pattern of Cu^{2+}(imidazole)$_4$ was also obtained and it shows much deeper modulation. A qualitative comparison of Fig. 28a and 28b suggests that only one imidazole is coordinated to Cu^{2+} in laccase.

The environment of Cu^{2+} type II was also studied in laccase and in ceruloplasmin by setting the external magnetic field to the low-field end of the Cu^{2+} EPR spectrum where only type II copper contributes. The resulting modulation pattern is similar to Fig. 28a, so it was concluded that both type I and type II copper are coordinated to one imidazole in these copper proteins.

Stellacyanin contains only type I copper and also shows an echo modulation pattern similar to Fig. 28a.[55] The same interpretation of one coordinated imidazole is further supported by finding a similar echo modulation pattern for Cu^{2+}-bovine serum albumin that is known to be coordinated to one amino nitrogen, two peptide nitrogens, and one imidazole.[55] This also reinforces the conclusion that nitrogen directly bonded to Cu^{2+} does not contribute to the modulation pattern.

Galactose oxidase contains only type II copper; its two-pulse echo modulation pattern also shows the 0.7 μs modulation frequency attributed to a nonbonded imidazole nitrogen, but the modulation depth is deeper than that found for any of the previously mentioned copper proteins or for Cu^{2+}-diethylenetriamine-imidazole.[57] On the other hand, the nitrogen modulation depth is shallower than that in Cu^{2+}-(imidazole)$_4$, so the Cu^{2+} galactose oxidase is presumably coordinated to two or three imidazoles. Two coordinated imidazoles are consistent with previous EPR data, which implies two directly bonded nitrogen ligands.[59]

The effect of F^- and CN^- on the two-pulse echo modulation of galactose oxidase was also studied,[57] since these ions are known to coordinate to the protein by EPR studies. The F^- ion had no effect, as might be expected, since the F^- hyperfine coupling is large and branching transitions involving F^- are not excited by the available microwave pulse power. The CN^- ion decreased the imidazole nitrogen modulation depth slightly. This could be due to interference between different nitrogen modulation frequencies associated with imidazole and CN^- or to a weakening of the imidazole nitrogen coupling due to back π-bonding of CN^- to Cu^{2+}.

Although the comparisons between the imidazole nitrogen modulation in copper proteins and model compounds have been made for two-pulse echo responses, it is noteworthy that the 0.7 μs nitrogen frequency shows up much more clearly in the three-pulse echo modulation pattern. This occurs because only ω_α and ω_β contribute to the three-pulse pattern, whereas ω_α, ω_β, and their sums and differences contribute to the two-pulse pattern.

A quantitative analysis by simulation of the copper protein modulation patterns was not undertaken because the ^{14}N quadrupolar interaction is comparable to the nuclear Zeeman energy at 3000 G fields and greatly complicates the modulation patterns. No simple analytic equations like Eq. (21) and (24) exist for this case; the quadrupole effects discussed in Section 3.3 apply only to quadrupole interactions smaller than the electron dipolar coupling. However, because of the common occurrence of nitrogen coupling in biological systems it is important to investigate model systems of this type where the geometry is known. This has been investigated recently for Cu^{2+}-ethylenetriamine-imidazole and Cu^{2+}-tetraimidazole complexes in a glycerol-water glassy matrix.[60] The analysis was made feasible by first studying ^{15}N-substituted complexes where $I = \frac{1}{2}$ so there is no quadrupole interaction. Three-pulse echo experiments give ω_α and ω_β frequencies of 2.9 and 0.25 MHz so the hyperfine coupling almost cancels the nuclear Zeeman splitting in one of the electron spin manifolds. When the isotropic hyperfine coupling is large the unpaired electron wavefunction is spread out over the interacting nuclei and the point dipole approximation for these nuclei must fail. However, the point dipole formalism can be artificially retained if the electron-nuclear distance r is replaced by r_{eff} where $r_{eff} < r$. This r_{eff} then characterizes a pseudodipolar interaction where the unpaired electron wave function is assumed to have cylindrical symmetry. This approximation was used with Eq. (21) averaged by the spherical model [Eq. (36)] to obtain a satisfactory simulation of the two-pulse echo modulation for the Cu^{2+}-imidazole complexes with $\mathbf{a}(^{15}N) = 2.5$ MHz and $r_{eff} = 2.9$ Å in place of the actual Cu^{2+}-distal N distance of 4.6 Å. The two-pulse echo modulation for the ^{14}N complexes was then simulated with $r_{eff} = 2.9$ Å, $\mathbf{a}(^{14}N) = 1.75$ MHz (scaled from the ^{15}N value) and the previously measured zero field quadrupolar frequencies for the protonated site in imidazole,[61] $Q_{zz} = 0.35$ MHz and $Q_{xx} = Q_{yy} = 0.36$ MHz. The resulting simulation is in semi-quantitative agreement with the experimental modulation pattern and supports the earlier qualitative assignments made for the copper proteins.[55-57] Note that because of the complexities introduced by large hyperfine and quadrupolar couplings, the simulations of the ^{14}N modulation patterns cannot yet be used to determine the magnetic and geometric parameters. Other information about these parameters must first be obtained and ^{15}N-substitution provides an approach to this.

Nitrogen modulation has also been observed in the two-pulse echo from chlorophyll-*a* cations at 77 K prepared by oxidation with I_2 in a 25 volume % solution of methylene chloride in methanol.[62] The ^{14}N pattern was too complex to be analyzed, but substitution by ^{15}N enabled an analysis to be made that placed limits on the isotropic and anisotropic coupling constants and the number of interacting nitrogens. Together with second-moment data and comparison with ESR data for the bacteriochlorophyll cation it was deduced that there are four interacting nitrogens with $\mathbf{a} \sim 2.8$ MHz and an anisotropic coupling of ~ 0.66 MHz.

Two pulse electron spin echo modulation patterns have also been used to study metal ion complexation with various biological phosphate ligands.[63] The depth of the ^{31}P modulation is used to determine the concentration of ligands required to form saturated complexes. Complexation of the paramagnetic ions Ce^{3+}, Nd^{3+}, Er^{3+}, Yb^{3+}, Cu^{2+}, Co^{2+}, Mn^{2+}, and Cr^{3+} with adenine triphosphate (ATP), adenine diphosphate (ADP), adenosine, and hexametaphosphate ligands has been studied at 4 K in glycerol/D_2O glasses. Qualitative interpretations based on the ^{31}P modulation depth versus ligand concentration lead to the following conclusions: ATP complexes more strongly than ADP with rare earth ions, Ce^{3+} complexes more strongly with ATP than does Nd^{3+}, Co^{2+} can bind to adenosine, the interaction between rare earth ions and phosphate groups is strongest for ions with the smallest radius, some coordination sites in rare earth ion–ATP complexes are occupied by water, and more phosphate groups can be bound to rare earth ions in hexametaphosphate complexes than in ATP complexes.

Iron-sulfur proteins have been studied by two-pulse electron spin echo to determine whether exchangeable protons are close to the paramagnetic iron-sulfur centers.[64] Observation of proton modulation for 2- and 4-iron ferredoxins and for high potential iron proteins demonstrate that protons are close to the paramagnetic centers. After standing in D_2O the high potential iron proteins show no change in the modulation pattern whereas the 2- and 4-iron ferredoxins show prominent deuteron modulation. Thus the modulation pattern can be used as a simple diagnostic test for exchangeable protons near the paramagnetic center.

It seems clear that both qualitative and quantitative use of spin-echo modulation patterns will be increasingly valuable for the study of paramagnetic interactions, ligand coordination, and, eventually, active site geometry in a variety of biological systems.

5 OTHER ASPECTS

5.1 Fourier Transform of Echo Envelope Modulation

In principle the echo envelope modulation can be Fourier transformed to obtain a frequency spectrum showing the various nuclear hyperfine frequencies (e.g., see Fig. 11). In practice this method often leads to difficulties and ambiguities.[8] The basic problem seems to be that the echo envelope can only be detected beginning some time after the echo envelope commences at $\tau = 0$. This is due to cavity ringing as well as to detection electronics response times. The omission of the initial part of the function to be Fourier transformed distorts the line shapes and intensities in the frequency spectrum and sometimes introduces spurious peaks. When two or more nonequivalent nuclei are coupled to the unpaired electron, multiple convolutions are required to obtain the Fourier transform. Additional features involving beat frequencies and harmonics may arise in the frequency spectrum, which may be difficult to assign correctly.

Because of these difficulties there have not been any detailed studies of echo modulation patterns by Fourier transform methods. Brown and Kreilick[33] have obtained the Fourier transforms of two- and three-pulse echoes for the butyl bisverdazyl biradical in both polystyrene and toluene matrices at 77 K. They were able to assign the lines in the frequency spectrum to two nonequivalent nitrogens, but some expected combination frequencies are not observed. Bowman[16] has recently studied Fourier transforms of two-pulse echoes associated with chlorophyll cations and suggests that nitrogen hyperfine components can be obtained. Clearly more work needs to be done to assess the utility of Fourier transform methods for analysis of electron spin-echo modulation patterns.

5.2 Spin-Echo ENDOR

ENDOR may be done by using three-pulse spin echo as the detection method. The radio-frequency (rf) pulse is applied during time interval T (see Fig. 4) between the second and third microwave pulses to a coil surrounding the sample in the microwave cavity. As the rf is swept the echo intensity is monitored and shows a decrease when the rf corresponds to a nuclear transition. Mims[65] first demonstrated spin-echo ENDOR in

Ce^{3+} doped $CaWO_4$ at 4.2 K. Ce^{3+} substitutes for Ca^{2+} and should produce six ENDOR lines from the eight tungsten neighbors if there is no lattice distortion. Instead, 16 ENDOR lines were observed that indicated lattice distortion and that were analyzed to obtain the isotropic coupling and dipolar tensors for two types of tungsten neighbors.

The theoretical spin-echo ENDOR response was first treated in terms of level population effects for an $I = \frac{1}{2}$ system.[65] It was pointed out that the ENDOR response would cause a decrease in the echo intensity and that the ENDOR response varied with τ, the time between the first and second microwave pulses. A more rigorous theoretical treatment valid for arbitrary nuclear spin based on the density matrix formalism was later given by Liao and Hartmann.[31] Some characteristics of the spin-echo ENDOR response are as follows. (1) The echo effect varies sinusoidally with τ with a period determined by the difference in the nuclear frequencies corresponding to the upper and lower electron spin states. Thus the period is proportional to the strength of the hyperfine interaction with a particular nucleus. (2) The echo ENDOR effect also depends on the nuclear rotation induced by the rf pulse. Theoretically the echo ENDOR effect disappears for a 360° rf pulse because the nucleus returns to its initial state and causes no change in the local field at the unpaired electron. In practice the ENDOR effect does not disappear completely because of inhomogeneities in the rf field, but the variation with rf field amplitude is easily observed. (3) The resolution of the echo ENDOR response is partly determined by the rf pulse duration through Heisenberg's uncertainty principle. Hence it is desirable to use long rf pulses so as not to broaden the line shapes.

The most detailed application of spin-echo ENDOR has been to a single crystal of Cr^{3+} doped Al_2O_3 (ruby) at 4.2 K.[31] By judicious use of the dependence of the echo ENDOR effect on τ and rf amplitude it was possible to assign the many ENDOR lines observed and to untangle overlapping resonances. For example, all the echo ENDOR lines belonging to a particular nuclear neighbor set have the same dependence on τ. Analysis of the data led to Cr-Al hyperfine and quadrupole tensors for 34 aluminum nuclear neighbors with a higher degree of accuracy than obtained by conventional continuous wave (cw) ENDOR. (See Fig. 22.)

One discrepancy with theory was found in the ruby study. For several equivalent nuclei the theoretical decrease in echo amplitude at nuclear resonance was typically several times greater than that actually observed. This may be due to an overlap of closely spaced nuclear transitions, as has been proposed for cw ENDOR.[66]

The echo ENDOR technique has some decided advantages over conventional ENDOR.[65] The echo ENDOR technique has good sensitivity, in many cases probably better than that in conventional ENDOR. In general, if a three-pulse spin echo can be observed τ can be adjusted to obtain an ENDOR response. Low radiofrequencies are readily applied in the echo ENDOR experiment, whereas this is difficult in conventional ENDOR. The echo detection method for ENDOR does not require the careful microwave bridge balance necessary for conventional ENDOR. However, the most important advantage of echo ENDOR is that it does not depend on a particular balance of relaxation rates, which is so critical for observing conventional ENDOR.[1] Relaxation rates are only important if they shorten the phase memory time enough to preclude echo detection.

Two-pulse spin-echo ENDOR methods have also been suggested although there have been no detailed applications. Brown, Sloop and Ames[67] studied phosphorous doped silicon by means of an alternating sequence of two microwave and two rf pulses and demonstrated an ENDOR response on the two pulse echo. Davies[68] described applying a saturating microwave pulse followed by an rf pulse followed by a two-pulse echo sequence. In $Ce^{3+}:CaF_2$ he demonstrates that this sequence gives two to five times greater ENDOR sensitivity for larger hyperfine interactions than the three-pulse echo ENDOR method. However for the smaller hyperfine interactions typically studied by spin echo methods, there is apparently little or no sensitivity advantage.

Although spin-echo ENDOR has not yet been much exploited, it appears to have considerable potential and deserves further study. The only applications to date have been to single-crystal systems. It would be interesting to see if this method can contribute new information to the analysis of paramagnetic species in disordered systems.

ACKNOWLEDGMENT

This work was supported by the U.S. Department of Energy under contract no. EY-76-02-2086.A002. I am grateful to Drs. P. A. Narayana, M. K. Bowman, T. Ichikawa, and R. N. Schwartz for helpful discussions and to W. B. Mims and M. K. Bowman for preprints in advance of publication.

REFERENCES

1. L. Kevan and L. D. Kispert, *Electron Spin Double Resonance Spectroscopy*, Wiley-Interscience, New York, 1976.
2. E. L. Hahn, *Phys. Rev.*, **80**, 580 (1950).

3. E. L. Hahn and D. E. Maxwell, *Phys. Rev.*, **88**, 1070 (1952).

4. L. G. Rowan, E. L. Hahn, and W. B. Mims, *Phys. Rev.*, **137**, A61 (1965).

5. F. C. Newman and L. G. Rowan, *Phys. Rev.* **B5**, 4231 (1972).

6. D. Grischkowsky and S. R. Hartmann, *Phys. Rev.* **B2**, 60 (1970).

7. M. Huisjen and J. S. Hyde, *Rev. Sci. Instrum.*, **45**, 669 (1974).

8. W. E. Blumberg, W. B. Mims, and D. Zuckerman, *Rev. Sci. Instrum.*, **44**, 546 (1973).

9. W. B. Mims, *Rev. Sci. Instrum.*, **36**, 1472 (1965).

10. G. M. Zhidomirov and K. M. Salikhov, *Theor. Exp. Chem.*, **4**, 332 (1968).

11. W. B. Mims, *Phys. Rev.*, **B5**, 2409 (1972).

12. W. B. Mims, *Phys. Rev.*, **B6**, 3543 (1972).

13. C. P. Slichter, *Principles of Magnetic Resonance, Harper & Row, New York*, 1963, pp. 127–142.

14. L. Kevan, M. K. Bowman, P. A. Narayana, R. K. Boeckman, V. F. Yudanov, and Yu. D. Tsvetkov, *J. Chem. Phys.*, **63**, 409 (1975).

15. W. B. Mims, J. Peisach, and J. L. Davis, *J. Chem. Phys.*, **66**, 5536 (1977).

16. M. K. Bowman, private communication.

17. W. B. Mims and J. L. Davis, *J. Chem. Phys.*, **64**, 4836 (1976).

18. P. A. Narayana and L. Kevan, *J. Mag. Resonance*, **26**, 437 (1977).

19. K. M. Salikhov and Yu. D. Tsvetkov, Chapter 7 of this book.

20. T. Ichikawa and L. Kevan, unpublished work; M. K. Bowman, private communication.

21. V. F. Yudanov, S. A. Dikanov, Yu. A. Grishin, and Yu. D. Tsvetkov, *J. Struct. Chem.*, **17**, 387 (1976).

22. P. A. Narayana, M. K. Bowman, L. Kevan, V. F. Yudanov, and Yu. D. Tsvetkov, *J. Chem. Phys.*, **63**, 3365 (1975).

23. V. F. Yudanov, S. A. Dikanov, and Yu. D. Tsvetkov, *J. Struct. Chem.*, **17**, 451 (1976).

24. P. A. Narayana and L. Kevan, *J. Mag. Resonance*, **23**, 385 (1976).

25. (a). W. B. Mims, K. Nassau, and J. D. McGee, *Phys. Rev.*, **23**, 2059 (1961).

25. (b). J. A. Cowen and D. E. Kaplan, *Phys. Rev.*, **124**, 1098 (1961).

26. R. Bottcher, W. Brunner, and W. Windsch, in I. Ursu, Ed., *Magnetic Resonance and Related Phenomena*, Proc. XVI Congress Ampere, Publ. House of Academy of Science of Romania, 1971, p. 714.

27. V. F. Yudanov, A. M. Raitsimring, and Yu. D. Tsvetkov, *Theoret. Exp. Chem.*, **4**, 335 (1968).

28. V. F. Yudanov, V. P. Soldatov, and Yu. D. Tsvetkov, *J. Struct. Chem.*, **15**, 510 (1974).

29. D. Grischkowsky and S. R. Hartmann, *Phys. Rev. Lett.*, **20**, 41 (1968).

30. P. F. Liao and S. R. Hartmann, *Solid State Commun.*, **10**, 1089 (1972).

31. P. F. Liao and S. R. Hartmann, *Phys. Rev.*, **B8**, 69 (1973).

32. V. F. Yudanov, K. M. Salikhov, G. M. Zhidormirov, and Yu. D. Tsvetkov, *Theoret. Exp. Chem.*, **5**, 451 (1969).

33. I. M. Brown and R. W. Kreilick, *J. Chem. Phys.*, **62**, 1190 (1975).

34. G. M. Zhidomirov, K. M. Salikhov, Yu. D. Tsvetkov, V. F. Yudanov and A. M. Raitsimring, *J. Struct. Chem.*, **9**, 704 (1968); see also Abstracts of the 8th International Symposium on Free Radicals, Novosibirsk, 1967.

35. V. F. Yudanov, K. M. Salikhov, G. M. Zhidomirov, and Yu. D. Tsvetkov, *J. Struct. Chem.*, **10,** 625 (1969).

36. K. M. Salikhov, V. F. Yudanov, A. M. Raitsimring, G. M. Zhidomirov, and Yu. D. Tsvetkov, *Proc. Colloque Ampere XV,* North Holland, Amsterdam, 1969, 278.

37. V. F. Yudanov, Yu. A. Grishin and Yu. D. Tsvetkov, *J. Struct. Chem.*, **17,** 55 (1976).

38. M. K. Bowman, L. Kevan, and I. M. Brown, *Chem. Phys. Lett.*, **22,** 16 (1973).

39. V. F. Yudanov, Yu. A. Grishin, and Yu. D. Tsvetkov, *J. Struct. Chem.*, **16,** 694 (1975).

40. D. G. Burkhard and D. M. Dennison, *Phys. Rev.*, **84,** 408 (1951).

41. R. Livingston and H. Zeldes, *J. Chem. Phys.*, **44,** 1245 (1966).

42. L. Kevan, *Adv. Radiat. Chem.*, **4,** 181 (1974).

43. P. A. Narayana and L. Kevan, *J. Chem. Phys.*, **65,** 3379 (1976).

44. S. Schlick, P. A. Narayana, and L. Kevan, *J. Chem. Phys.*, **64,** 3153 (1976).

45. B. L. Bales, M. K. Bowman, L. Kevan, and R. N. Schwartz, *J. Chem. Phys.*, **63,** 3008 (1975).

46. P. A. Narayana, D. Becker, and L. Kevan, *J. Chem. Phys.*, **68,** 652 (1978).

47. L. Kevan, P. A. Narayana, and T. Ichikawa unpublished results.

48. (a) L. Kevan, H. Hase, and K. Kawabata, *J. Chem. Phys.*, **66,** 3834 (1977); (b) L. Kevan, *J. Chem. Phys.*, **69,** 3444 (1978).

49. H. Zeldes and R. Livingston, *Phys. Rev.*, **96,** 1702 (1954).

50. J. Zimbrick and L. Kevan, *J. Chem. Phys.*, **47,** 5000 (1967).

51. M. Bowman, L. Kevan, and R. N. Schwartz, *Chem. Phys. Lett.*, **30,** 208 (1975).

52. S. A. Dikanov, V. F. Yudanov, and Yu. D. Tsvetkov, *J. Struct. Chem.*, **18,** 370 (1977).

53. V. F. Yudanov, A. M. Raitsimring, S. A. Dikanov, and Yu. D. Tsvetkov, *J. Struct. Chem.*, **17,** 139 (1976).

54. S. A. Dikanov, V. F. Yudanov, R. I. Samiolova, and Yu. D. Tsvetkov, *Chem. Phys. Lett.*, **52,** 520 (1977).

55. W. B. Mims and J. Peisach, *Biochemistry*, **15,** 3863 (1976).

56. B. Mondovi, M. T. Graziani, W. B. Mims, R. Oltzik and J. Peisach, *Biochemistry*, **16,** 4198 (1977).

57. D. J. Kosman, J. Peisach, and W. B. Mims, private communication.

58. G. H. Rist, J. S. Hyde, and T. Vanngard, *Proc. Nat. Acad. Sci. U.S.A.*, **67,** 79 (1970).

59. R. D. Bereman and D. J. Kosman, *J. Am. Chem. Soc.*, **99,** 7322 (1977).

60. W. B. Mims and J. Peisach, private communication.

61. M. J. Hunt, A. L. Mackay, and D. T. Edmonds, *Chem. Phys. Lett.*, **34,** 473 (1975).

62. M. K. Bowman, S. A. Dikanov, J. R. Norris, M. C. Thurnauer, J. Warden, and Yu. D. Tsvetkov, *Chem. Phys. Lett.*, **55,** 570 (1978).

63. T. Shimizu, W. B. Mims, J. Peisach, and J. L. Davis, private communication.

64. J. Peisach, N. R. Orme-Johnson, W. B. Mims and W. H. Orme-Johnson, *J. Biol. Chem.*, **252,** 5643 (1977).

65. W. B. Mims, *Proc. Roy. Soc.*, **283,** 452 (1965).

66. R. Allendoerfer and A. H. Maki, *J. Mag. Resonance*, **3,** 396 (1970).

67. I. M. Brown, D. J. Sloop and D. P. Ames, *Phys. Rev. Lett.*, **22,** 324 (1969).

68. E. R. Davies, *Phys. Lett.*, **47A,** 1 (1974).

9 TRANSIENT ELECTRON SPIN RESONANCE STUDIES OF MOLECULAR TRIPLET STATES IN ZERO FIELD

J. Schmidt and J. H. van der Waals

Center for the Study of the Excited States of Molecules, Huygens Laboratorium, University of Leiden, Leiden, The Netherlands

1   INTRODUCTION

The present chapter is devoted to the study of metastable triplet states by transient electron spin resonance (ESR) methods. Such triplet states are a common feature of aromatic molecules where, under suitable conditions, they may be populated via the absorption of ultraviolet (uv) light and decay with the emission of a long-lived (1 ms – 10 s) visible phosphorescence.[1] The first successful application of magnetic resonance to the study of these paramagnetic excited states was reported in 1958, when Hutchison and Mangum detected the photoexcited triplet state of naphthalene incorporated in a durene single-crystal host.[2] In the 20 years that have passed a great number of systems have been investigated in a similar way by ESR in an external field.

Although furnishing a wealth of structural information on photoexcited triplet states, these "slow-passage"[3] experiments by conventional ESR were not particularly suited for the study of dynamic processes. One of the exceptions is the work by Schwoerer and Sixl,[4] who studied the rates for populating and decay of the three sublevels of the triplet state of naphthalene and the relaxation between them from the transient response of the ESR signals to changes in the experimental conditions.

The situation changed with the introduction of resonance experiments between the triplet sublevels in zero field,[5] for which optical detection methods proved particularly well suited.[6] It appeared that because of the "quenching" of the electron spin angular momentum in zero field, the transitions are characterized by linewidths extremely small (≈ 1 MHz) in comparison with those of ESR signals of solids in an external field. It was realized[7] that, as a consequence, the situation in zero field is particularly promising to evoke phenomena that depend on the mutual phase coherence of the triplet spins in the ensemble, and successful adiabatic inversion[8] and phosphorescence modulation (transient nutation)[9] experiments were reported. Since then there has been a flurry of activity in Berkeley and Leiden to explore the different aspects of the interaction of systems of photoexcited triplet states with coherent, resonant microwave fields.

The aim of the present chapter is to provide a systematic introduction to transient ESR experiments on photoexcited triplet states in zero field, with a particular emphasis on those types of experiment in which the phase coherence between the individual spins manifests itself. The treatment is based on the geometrical representation of the Schrödinger equation, first explicitly introduced for maser problems by Feynman, Vernon, and Hellwarth.[10] This approach has the great merit that it brings out the relationship of the triplet state problem with the thoroughly known transient effects for $I = \frac{1}{2}$ systems in nuclear magnetic resonance (NMR),[3] and with coherent optical phenomena.

In the discussion we limit ourselves to the application to phosphorescent (aza)-aromatic molecules and triplet excitons in organic crystals. It is to be realized, however, that the present techniques have a wider range of applicability. They have, for instance, already been used in an extensive study of the triplet state of color centers in CaO[11] and might, equally well, be applied in the investigation of low-lying triplet states of small molecules such as SO_2.

In the past few years it has also been possible to carry out electron spin-echo experiments on photoexcited triplet states at a fixed microwave frequency in an external magnetic field.[12] These spin-echo experiments in a magnetic field hold considerable promise for further work, but since their principles are analogous to those of the $S = \frac{1}{2}$ systems discussed in other chapters of this treatise, they are not described here. By way of exception we briefly discuss in Section 4.5 the work of Botter et al.,[13] who studied the triplet state of the "miniexciton" of an excited naphthalene $C_{10}H_8$ pair in a $C_{10}D_8$ host with spin echoes in an external magnetic field.

2 THE PHOSPHORESCENT TRIPLET STATE

2.1 Magnetic Properties

On the magnetic properties of the lowest triplet state of organic molecules some extensive review papers[14,15] and a book[16] have been published. Here we only give a brief summary. One of the molecules of interest is quinoxaline and as an example we reproduce in Fig. 1 its energy level scheme. On the left are the singlet ground state S_0 and the lower excited singlet states. On the right are the lowest, metastable triplet state T_0 and the approximate position of a higher triplet. The labels at the left side of the levels give the orbital symmetries in C_{2v} (the symmetry group of quinoxaline).

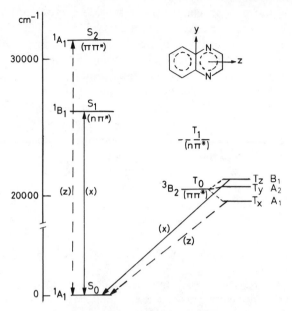

Figure 1 The lower electronic states of quinoxaline (1,4-diazanaphthalene). The labels at the left of the energy levels are orbital symmetries in C_{2v}; for the singlets these also represent the total symmetries. For the sub-levels of the lowest triplet the total symmetries are indicated by the labels at the right.
The polarization of the two symmetry-allowed components of phosphorescene are indicated together with two singlet-singlet transitions from which they derive intensity. The labels $n\pi^*$, $\pi\pi^*$ refer to the two singly occupied orbitals in a simple molecular orbital picture. (See reference 16.)

When irradiated by uv light at low temperature in a dilute crystalline matrix the molecule is excited from the ground state S_0 into the singlet system. In general the molecule may return from the S_1 singlet state to the ground state by emission of fluorescence or relax to the lowest triplet state T_0. From there it then decays to the ground state S_0, either by emission of phosphorescence or by nonradiative processes. In the case of quinoxaline there is hardly any fluorescence and the relaxation into the triplet state must be relatively fast as compared with the radiative decay of S_1 which is of order of $10^7 \, \text{s}^{-1}$; the decay rate of T_0 is about $3 \, \text{s}^{-1}$.

The usefulness of the scheme of Fig. 1 derives from the fact that for the organic molecules with which this chapter is concerned, spin-orbit coupling (SOC) is a weak interaction. Thus, although the orbital angular momentum **L** is "quenched," the total spin angular momentum **S** remains an acceptable quantum number.

The splitting in zero magnetic field of the metastable triplet state T_0, for which $S = 1$, is a manifestation of the lower symmetry of a molecule as compared with an atom. For the molecules with which we shall be concerned the extent of the splitting is mainly determined by the magnetic dipole-dipole interaction between the electron spins, with the SOC contributing to a minor extent. Exceptions may occur in molecules with (near) orbital degeneracy, or in those containing heavy atoms.[17]

The dipole-dipole interaction $\mathcal{H}_{ss}$ is given by

$$\mathcal{H}_{ss} = (\gamma\hbar)^2 \sum_{i<j} \left\{ \frac{(s_i \cdot s_j)}{r_{ij}^3} - \frac{3(s_i \cdot r_{ij})(s_j \cdot r_{ij})}{r_{ij}^5} \right\}, \tag{1}$$

where s_i and s_j are the spin angular momentum operators of electrons i and j and r_{ij} gives the position of electron i with respect to electron j. To calculate this interaction energy one must integrate over the spatial distribution of the electrons in the phosphorescent state T_0. It has been shown by van Vleck[18] that within the triplet multiplet the microscopic Hamiltonian (1) can be written as a phenomenological spin Hamiltonian $\mathcal{H}_0$ in which only operators for the total spin angular momentum occur. By introducing $S = \sum_i s_i$ we obtain

$$\mathcal{H}_0 = S \cdot T \cdot S, \tag{2}$$

where T is the zero-field splitting tensor; its elements involve integrals of the operator (1) over the electron wave functions and, possibly, additional contributions arising from SOC.

If an axis system x, y, z is chosen such that T is diagonal and terms involving $S_x S_y$, and so on, disappear, $\mathcal{H}_0$ reduces to

$$\mathcal{H}_0 = -XS_x^2 - YS_y^2 - ZS_z^2, \tag{3}$$

with $X + Y + Z = 0$.[15] The principal axes or "spin axes" of the tensor T coincide with the molecular axes x, y, z if the molecule in the excited state belongs to the symmetry group C_{2v} or to a symmetry group that contains C_{2v} as a subgroup. We assume in general that this occurs and for the particular case of quinoxaline choose the axes as in Fig. 1.

Whenever the spin axes are not predetermined by molecular symmetry, or if they are pulled away from the molecular symmetry axes (or planes) by interaction with the environment, we label them with primes.

As we shall presently see, the parameters X, Y, Z are equal to the zero-field energies of the three components and, therefore, the form (3) of the spin Hamiltonian is particularly well suited for the discussion of experiments in zero field. For the description of a conventional ESR experiment in a magnetic field it is customary to express $\mathcal{H}_0$ in the alternative form

$$\mathcal{H}_0 = D(S_z^2 - \tfrac{1}{3}S \cdot S) + E(S_x^2 - S_y^2), \tag{4}$$

where $D = -\tfrac{3}{2}Z$ and $2E = (Y - X)$.

The eigenfunctions of (3) are labeled T_x, T_y, T_z with the eigenfunction T_x corresponding to the eigenenergy X, and so on; for the sake of the present discussion we assume their order to be as in Fig. 1 for quinoxaline. To get some physical insight into the nature of these eigenfunctions it is helpful to make a sidestep and look at a system with $L = 1$, rather than $S = 1$.

If an orbitally degenerate (spinless) ion with $L = 1$ is placed in a crystal field of low symmetry, the degeneracy is lifted, and the three eigenfunctions then are usually designated as p_x, p_y, p_z. In the most general case the directions of the corresponding axes $\mathbf{x}$, $\mathbf{y}$, $\mathbf{z}$ have to be determined by diagonalizing an interaction Hamiltonian, but if the site has rhombic symmetry, for instance, these axes are fixed by symmetry. We also know that by choosing suitable phase factors, the functions p_x, p_y, p_z are related to the eigenfunctions p_1, p_0, p_{-1} of L_z according to[19]

$$p_x = 2^{-1/2}(p_- - p_+), \qquad p_y = 2^{-1/2}i(p_- + p_+), \qquad p_z = p_0, \qquad (5)$$

whereas their behavior under the angular momentum operators* is

$$L_x p_x = 0, \qquad L_x p_y = i p_z, \qquad L_x p_z = -i p_y, \qquad (6)$$

and so on, by cyclic permutation. The significance of the latter relations can be summarized in the rules

1. p_x is an eigenfunction of L_x with eigenvalue zero,
2. The angular momentum operators have no diagonal elements in the basis p_x, p_y, p_z,
3. For the off-diagonal elements of the angular momentum operators one has $\langle p_z | L_x | p_y \rangle = i$, and so on, by cyclic permutation of indices.

The preceding rules may seem trivial. But, since the rotation group in three dimensions has only one three-dimensional irreducible representation, we know that identical rules must hold for our triplet spin states of a polyatomic molecule in zero field: one merely has to substitute T_x for p_x, and S_x for L_x, and so on! In fact, the same reasoning can be applied to the three eigenstates of a nucleus with $I = 1$ where the quadrupole moment interacts with an (anisotropic) molecular or crystalline electric field.

Apparently, we have three triplet eigenfunctions in zero field which obey the relations

$$S_u T_u = 0 \qquad (u = x, y, z). \qquad (7)$$

* We express angular momenta in $\hbar$ as unit.

Or, in words, in the eigenstate T_u the spin angular momentum lies in the plane $u = 0$. Taking T_x as an example, we have

$$S_x T_x = 0, \qquad S_x T_y = i T_z, \qquad S_x T_z = -i T_y, \tag{8}$$

and so on, by cyclic permutation of indices. These relations express the fact that in the present kind of system the spin angular momentum is "quenched" in zero field. On the other hand, the components of **S** have off-diagonal matrix elements of the form

$$\langle T_z | S_x | T_y \rangle = i. \tag{9}$$

For a simple (orbitally nondegenerate) two-electron system the preceding rules can be verified easily. One may then write

$$T_u = \psi(\mathbf{r}_1, \mathbf{r}_2) \tau_u, \tag{10}$$

where ψ is an orbital function that is antisymmetric with respect to the interchange of the positions $\mathbf{r}_1$, $\mathbf{r}_2$ of the two electrons. The τ_u are eigenfunctions of $\mathbf{S}^2$ with eigenvalue 2 that by analogy with (5) are given by

$$\tau_x = 2^{-1/2}\{\beta(1)\beta(2) - \alpha(1)\alpha(2)\},$$
$$\tau_y = 2^{-1/2}i\{\beta(1)\beta(2) + \alpha(1)\alpha(2)\}, \tag{11}$$
$$\tau_z = 2^{-1/2}\{\alpha(1)\beta(2) + \beta(1)\alpha(2)\}.$$

By operating with $\mathcal{H}_0$ on any one of the T_u and making repeated use of the relations (8), one may further verify that the parameters X, Y, Z in Eq. (3) really represent the eigenenergies. Their magnitudes are determined by the energy of interaction of electron spin magnetic moments at distances comparable with the molecular dimensions. This gives rise to splittings between the levels corresponding to frequencies in the range of 0.1–10 GHz and, thus, the transitions between the zero-field components fall in the microwave range.[14] In this chapter we are concerned with the study of such transitions between pairs of zero-field levels.

Let us take the transition $T_y \leftrightarrow T_z$ by way of example. It has an angular frequency $\omega_0 = |Z - Y| \hbar^{-1}$, and as implied by Eq. (9), it may occur as a magnetic dipole transition in a microwave magnetic field with an amplitude H_1 along the x-direction

$$H_x(t) = -H_1 \cos \omega t, \tag{12}$$

where $\omega \approx \omega_0{}^*$. The interaction of our triplet systems with such a microwave field gives rise to a time-dependent term in the Hamiltonian of the form

$$V(t) = \gamma \hbar H_1 S_x \cos \omega t. \tag{13}$$

Here γ is the magnetogyric ratio of an electron; for the systems of present interest γ is very close to the free electron value $\gamma = -2.0023$.

In this review two distinct classes of experiments are considered: the incoherent experiments of Section 3, and the coherent experiments of Section 4. For the first type of experiments the conditions are such that they may be discussed by means of time-dependent perturbation theory in its usual form.[20] One then supposes that the perturbation $V(t)$ gives rise to a *time-independent transition probability* that is proportional to the absolute square of the transition moment

$$|\langle T_z | \gamma \hbar H_1 S_x | T_y \rangle|^2 = (\gamma \hbar H_1)^2. \tag{14}$$

In the second type of experiment one is dealing with the characteristic transient effects that arise when an ensemble of spins is driven coherently by a resonant microwave field. The interaction $V(t)$ then no longer may be accounted for by means of perturbation theory and one has to integrate the time-dependent Schrödinger equation for the ensemble. By using the "geometrical representation of the Schrödinger equation" suggested by Feynman, Vernon, and Hellwarth,[10] the problem of this integration is transformed into that of finding the solutions of the Bloch equations. This trick has the great merit that one obtains a pictorial representation of the evolution in time of an ensemble of triplet spins that is identical with the rotating-frame description familiar in the discussion of transient experiments in NMR.

In the present context there is no need to discuss the problem of the line shape of the zero-field transitions in any detail. Here we merely indicate the principal features; for a more extensive analysis see reference[21].

In the molecules under consideration we expect a magnetic dipolar coupling term in the Hamiltonian of the form

$$\mathcal{H}_{HF} = \sum_K \mathbf{S} \cdot \mathbf{A}_K \cdot \mathbf{I}_K, \tag{15}$$

where $\mathbf{A}_K$ is the hyperfine coupling tensor for nucleus K and the summation is over all nuclei with a nonvanishing angular momentum $\mathbf{I}_K$. However, the zero-field lines cannot show the familiar hyperfine structure

---

* The arbitrary phase factor in the driving field has been chosen equal to -1 for later convenience so as to get a positive amplitude Ω_2 in Eq. (42) and Fig. 7.

of ordinary ESR spectra: because of the relations (8), the operator $\mathcal{H}_{HF}$ can only give matrix elements between different zero-field states and no first-order splittings will occur in the electron spin levels. If a single nuclear spin $I = \frac{1}{2}$ is present one even expects no splitting at all because a state having a half integral angular momentum is at least twofold degenerate in zero-magnetic field (Kramers' theorem).

Though a single proton does not lead to any splitting, groups of protons may still affect the line shape. This arises because their combined states may have a resultant integral spin angular momentum (e.g., $I = 0$ or 1 for a pair of equivalent protons) that couples to the electron spin. Because of this effect the three zero-field transitions of naphthalene $C_{10}H_8$ are inhomogeneously broadened with linewidths of 1–2 MHz;[22] their shapes have been quantitatively interpreted by Hutchison et al.[23] on the basis of (15). For perdeutero naphthalene $C_{10}D_8$ these widths are reduced to somewhat less than 1 MHz.

Although the magnetic dipolar coupling $\mathcal{H}_{HF}$ by itself (with very few exceptions—e.g., glyoxal[24]) does not cause a resolved hyperfine splitting of

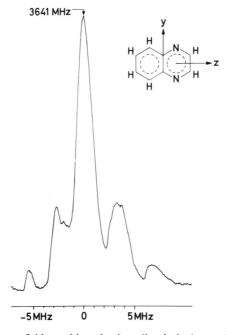

Figure 2 The 3641 zero-field transition of quinoxaline-h_6 in durene. The central line is the "allowed" transition broadened by second-order hyperfine interaction with the nitrogen nuclei. The four satellites are "forbidden" transitions: the inner two correspond with the simultaneous flip of one nitrogen nucleus, the outer two with the flip of two nitrogen nucleus. From Schmidt and van der Waals. (Reference 21.)

the zero-field transitions, it may indirectly lead to a very characteristic power-dependent structure of the zero-field spectrum if the molecule contains a nucleus or nuclei with $I_K \geq 1$ having a sizable quadrupole moment. By virtue of the off-diagonal matrix elements of $\mathcal{H}_{HF}$ between different electron spin states, satellites may then occur in which nuclear spins "flip" simultaneously with the electron spin.[21,25] An example of this effect is shown in Fig. 2 where in the $T_z - T_x$ transition of quinoxaline in durene at 3641 MHz four satellites occur that correspond with simultaneous flips of the nitrogen nuclei with $I = 1$.

For our purpose the detailed structure of the zero-field transitions is not of prime interest. It is important, however, that because of the quenching of the magnetic moment of the triplet electron spin in zero field, all magnetic dipolar couplings become second- (or higher-) order perturbations in the zero-field basis T_x, T_y, T_z. Hence the zero-field transitions have very small inhomogeneous linewidths as compared to the widths of conventional ESR transitions in solids. As a result, it becomes far easier in zero field to create experimental situations in which the coupling to the radiation field through the term $V(t)$ is comparable in magnitude to the inhomogeneous width of a transition.

2.2 Radiative Properties of the Zero-Field Components

Optical detection has proved particularly fruitful for electron resonance experiments on phosphorescent triplets in zero field.[6] Let us, therefore, briefly examine what determines the radiative properties of the three spin components. When, for the time being, talking in terms of a two-electron system, it is clear from Eq. (8) and (10) that the triplet spin functions τ_u (11) transform like the elementary rotations R_u about the coordinate axes $u = x, y, z$.[15] Hence in the point group C_{2v} applicable to quinoxaline τ_z, τ_y, τ_x belong to the irreducible representations A_2, B_1, B_2, respectively. In this molecule, where from symmetry one expects the spin axes to coincide with the molecular axes, the irreducible representation to which each of the triplet components belongs then is obtained as the direct product of orbital and spin symmetry. This is indicated in Fig. 1 at the right of the triplet spin levels; since these assignments are imposed by the symmetry of the point group concerned they hold in a more sophisticated many-electron description.

The radiative properties of the individual spin components of the phosphorescent triplet state are determined by the spin-orbit coupling in the molecule. This interaction mixes singlets with triplets and allows electric dipole transitions to occur between the lowest triplet state and the singlet ground state. Hence the phosphorescence acquires its intensity from allowed singlet-singlet and triplet-triplet transitions. The mixing is

Table 1 Character Table of the Group C_{2v}

C_{2v}			E	C_2	σ_y	σ_x
	z	A_1	1	1	1	1
T_z		A_2	1	1	-1	-1
T_y	x	B_1	1	-1	1	-1
T_x	y	B_2	1	-1	-1	1

very selective because spin-orbit interaction only gives matrix elements between singlet and triplet states of the same total symmetry. As a result, the radiative transition probabilities $k_u^r(u = x, y, z)$ of the three spin components of a phosphorescent triplet state in general are quite different. For quinoxaline, for instance, one concludes from group theory that, in principle, the top level (B_1) and bottom level (A_1) may decay to the ground state via x and z polarized phosphorescence, respectively, while decay from the middle level by electric dipole radiation is forbidden.[15] In reality, the decay mode originating from T_z is highly favored, and the radiative decay rates for the three levels are in the ratio $k_z^r : k_y^r : k_x^r = 1.00 : 0.02 : 0.015$.[26]

In the optical detection technique one makes use of these differing radiative decay rates. When magnetic dipole transitions are induced between two components of the triplet state, the populations of the levels are affected and a change in the phosphorescence intensity occurs. This effect will be more pronounced the larger the differences in radiative decay rates of the individual levels. Thus in the study of the zero-field transitions one may often circumvent the complications inherent in microwave detection techniques at variable frequency by detecting the much more energetic photons.

Physically, the inequivalence in the radiative decay rates of the individual components of the triplet state is related to the anisotropy of spin-orbit coupling in polyatomic molecules: The degree to which a given component T_u is contaminated with singlet character is quite sensitive to the orientation of the plane $u = 0$ relative to the molecular frame. Mathematically it can be expressed as follows. When allowing for spin-orbit coupling the wavefunctions of the zero-field components take the general form

$$\left.\begin{aligned}
T_x &= T_x^o + i \sum_j \lambda_j S_j^o, & S_j^o &\in \Gamma_{T_x} \\[6pt]
T_y &= T_y^o + i \sum_k \mu_k S_k^o, & S_k^o &\in \Gamma_{T_y} \\[6pt]
T_z &= T_z^o + i \sum_l \nu_l S_l^o, & S_l^o &\in \Gamma_{T_z}
\end{aligned}\right\} \qquad (16)$$

Zero superscripts here indicate "pure spin" singlet or triplet states and the real constants λ_j, μ_k, ν_l depend on the matrix elements of spin-orbit coupling. The summation in the expression of a given T_u is over those singlet states that belong to the same irreducible representation as $T_u^o(\Gamma_{T_u})$.

When applying these ideas to quinoxaline, it is clear that the component T_z^o is allowed to interact by spin-orbit coupling with the first excited singlet state S_1 and a whole series of further singlet states of species B_1. Similarly, the components T_y^o, T_x^o may interact with singlet states of species A_2 and A_1, respectively. However, when investigating the size of the coupling constants in (16) it turns out that the absolute value of the coefficient ν_1 (which couples the T_z component of the lowest $\pi\pi^*$ triplet state to the $n\pi^*$ state S_1) exceeds that of all the other coefficients by at least one order of magnitude.[15,16] Hence, for quinoxaline one effectively has

$$\left.\begin{array}{l} T_x \approx T_x^o \\ T_y \approx T_y^o \\ T_z \approx T_z^o + i\nu_1 S_1^o \end{array}\right\}, \qquad (17)$$

with $|\nu_1| \approx 2 \times 10^{-3}$.[27] It is because of the smallness of the coupling coefficients such as ν_1 that one may ignore the distinction between the T_u and T_u^o in the discussion of the magnetic properties of Section 2.1.

For the transition moment of the dominant, x-polarized component of the phosphorescence one, accordingly, has

$$\langle T_z |ex| S_0 \rangle \approx i\nu_1 \langle S_1 |ex| S_0 \rangle. \qquad (18)$$

Or, in familiar language, the phosphorescence intensity is "stolen" from the $S_1 \to S_0$ transition. (To keep the discussion simple we have here neglected the breaking of electronic selection rules due to vibronic coupling. Although nontotally symmetric vibrations often are very important in the structure of the $T_0 \to S_0$ optical spectrum,[28–30] they usually contribute only a small fraction to the total intensity of the phosphorescence.)

The dominance of the spin-orbit coupling route between T_z^o and S_1^o in quinoxaline and similar compounds is related to the electron configuration around the nitrogen nuclei in azaaromatic molecules. Whereas in (planar) aromatic hydrocarbons spin-orbit coupling between the $\pi\pi^*$ states is exceedingly weak because all one and two-center terms vanish on expanding the spin-orbit matrix elements in sums of integrals over atomic integrals,[31,32] an effective route is opened with the insertion of a nitrogen atom carrying an electron "lone pair."[33,34] The coefficient $i\nu_1$ in (17) by perturbation theory is equal to

$$iv_1 = \frac{\langle S_1^o(n\pi^*) |\mathcal{H}_{SO}| T_z^o(\pi\pi^*)\rangle}{(E_{T_z^o} - E_{S_1^o})}. \tag{19}$$

When the numerator in (19) is reduced to a sum of integrals over atomic orbitals, one-center one-electron terms on nitrogen remain; for quinoxaline these arise between the $2p_x$ π-type AO's and the trigonal lone-pair hybrids $t_{\pm y} = 3^{-1/2}(2s) \pm (\frac{2}{3})^{1/2}(2p_y)$ involved in the η-type MO. Because of this "local symmetry"[32] one expects the spin-orbit coupling to be a maximum for a triplet state T_η for which the plane $\eta = 0$ coincides with the "$n - \pi$ plane" determined by the out-of-plane axis and the direction of the lone-pair orbital.[35] In quinoxaline T_η corresponds with the stationary zero-field component T_z; in molecules of lower symmetry it may correspond to a linear combination of two (or, conceivably, three) zero-field components. The latter situation has recently become of interest since one then has two (or three) phosphorescence decay channels with the same polarization, and quantum interference effects between them may arise (see reference 36 and the end of Section 5).

3 INCOHERENT EXPERIMENTS

The optical detection of magnetic dipole transitions between pairs of zero-field levels of a phosphorescent triplet state T_0 was first realized in 1968.[6] At that time it was conceived as a steady-state method for the observation of the zero-field resonances within T_0 and as such it yields detailed information about the energy level structure. It soon appeared, however, that with experiments in which transient microwave fields were applied one can measure the dynamic properties of the individual spin levels: the decay rates, the relative radiative decay rates, the relative populating rates, and the spin-lattice relaxation (SLR) rates. This is of considerable interest because a knowledge of the first three parameters allows one to study the radiative and nonradiative processes that arise in the optical pumping cycle $S_0 \rightarrow S_1 \cdots T_0 \rightleftarrows S_0$ (here the dotted and full arrows indicate radiative and nonradiative processes, respectively). Further, a knowledge of the SLR rates gives insight into the coupling of the triplet spins with the phonons of the host crystal.

The general nature of the problem may be illustrated with the aid of Fig. 3, in which the optical pumping cycle is represented in a schematic manner taking the example of quinoline (1-azanaphthalene) in a durene host. To keep things simple the figure has been drawn as a five-level system: the ground state S_0, the first excited singlet state S_1, and the three

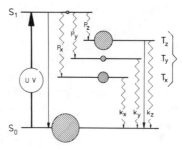

Figure 3 A schematic drawing of the optical pumping cycle in quinoline (1-azanaphthalene in durene. The spin level T_z carries almost the entire radiative activity for decay to the ground state S_0.

sublevels T_u of the metastable triplet state ($u = x, y, z$). In contrast to the triplet levels, the state S_1 is very short-lived (less than 10^{-7} s) and it will carry a negligible fraction of the total population; the same holds *a fortiori* for other, more highly excited (singlet and triplet) states that have been omitted from the figure. Thus, for the experiments presently to be discussed with a time resolution of the order of 10^{-3} s, our system behaves like a pseudo-four-level system, and the optical pumping process may be described in terms of a set of linear differential equations in which the following rate parameters appear ($u, v = x, y, z$):

1. The populating rate P_u of each of the components T_u on excitation "via" S_1,
2. The rates k_u for the decay $T_u \rightarrow S_o$,
3. The rates W_{uv} of spin reorientation via the process $T_u \rightarrow T_v$ (not drawn).

Whereas, in the absence of quantum yield experiments, the populating rates can only be determined up to a common unknown constant that depends on the condition of excitation, the k_u and W_{uv} represent absolute rates; the problem in its most general form thus contains eight unknown parameters.

Each of the decay rates k_u is the sum of a radiative and a radiationless part,

$$k_u = k_u^r + k_u^d, \tag{20}$$

and the phosphorescence intensity I is given by

$$I(t) = c \sum_u k_u^r N_u(t), \tag{21}$$

where c is an instrumental constant and N_u denotes the population of T_u at a particular time t. In practice, the success of the dynamic experiments derives from the circumstance that the k_u^r tend to be greatly different for the three sublevels and, consequently, when by the application of resonant microwaves molecules are transferred from one triplet sublevel to another, appreciable changes of the phosphorescence intensity may occur. For quinoline, for instance, the spin-orbit coupling between the T_z component of the lowest $\pi\pi^*$ triplet state to the $n\pi^*$ state S_1 is by far the most important, exactly as for quinoxaline [see Eq. (17)] and, as a result, T_z carries almost the entire activity for decay to the (vibrationless) ground state.

The great variety of experiments designed to unravel the dynamics of the optical pumping cycle all have in common that one analyzes the change in phosphorescence intensity that occurs in response to a "tickling" of the system of excited molecules by resonant microwaves. But, although the mathematics of the problem are well understood, the accuracy with which the rate parameters may be determined for a given system from its dynamical response to a number of imposed perturbations (switching the excited source on or off, and microwave contacts) will depend greatly on the characteristics of the system. In particular one should realize that the response of the system is monitored via a single property only: the phosphorescence intensity. In practice this means for quinoline, for instance, that one probes the population of the radiative level, T_z.

From what has been said, two things should be clear. First the design of the experiment must be tailored to the properties of the system under examination and no universal "best" method can prescribed. Second, it will be an exception if one succeeds in determining all eight independent rate parameters with reasonable accuracy without making some *a priori* assumptions. Here we do not consider the many varieties of dynamic experiments that have been proposed, but treat a single one by way of example: the microwave-induced delayed phosphorescence (MIDP) method.

The MIDP techniques in its simple form[26] takes advantage of the circumstance that for many systems, by merely lowering the temperature to below a few Kelvin, one can reach a situation where relaxation practically has ceased. That is, one has "isolation" of the three spin levels whenever the condition

$$W_{uv} \ll k_x, k_y, k_z \qquad \text{(for all } u, v = x, y, z) \qquad (22)$$

is met. In this manner the size of the problem is reduced and it then proves feasible, in general, to measure the absolute decay rates k_u, the relative radiative decay rates k_u^r and the relative populating rates P_u with adequate precision. It would go too far to discuss here all details of this method but in order to get a feeling for the essence of things we treat the determination of the k_u of the spin levels of quinoline-h_7 in durene. For a complete description of the method the reader is referred to reference 26.

The sample is first irradiated by uv light at 1.1 K until a stationary distribution of the excited molecules over the three spin levels of the lowest triplet state is established. At time $t = 0$ the illumination is terminated and the subsequent phosphorescence decay displayed on an oscilloscope. (See Fig. 4.) The signal one initially observes is almost entirely due to the rapid decay of the radiative level T_z with a rate k_z. At a time t_1 chosen long compared with the lifetime k_z^{-1} of T_z, one suddenly sweeps through the T_z-T_x transition, for instance. This causes a repopulation of the, almost empty, radiative level T_z from the slowly decaying "dark" level T_x, and an increase in phosphorescence intensity results.

The height of the leading edge of this MIDP signal is given by

$$\Delta I(t_1) = cf\{N_x(t_1) - N_z(t_1)\}(k_z^r - k_x^r). \tag{23}$$

Here c is an instrumental constant and f the microwave transfer factor, which is the fraction of the population difference between T_x and T_z transferred by the microwaves. The value of f depends on the microwave power and the sweep rate. At high power levels an adiabatic inversion of the population difference can even be achieved[8] (see also Section 4), but usually the power level is adjusted so that saturation occurs, that is, $f \approx 0.5$.

For sufficiently long delay times, $t_1 \gg k_z^{-1}$, the term $N_z(t_1)$ in (23) becomes negligible and

$$\Delta I(t_1) = cf N_x(0) \exp\{-k_x t_1\}(k_z^r - k_x^r). \tag{24}$$

In Fig. 5 $\log \Delta I(t_1)$ has been plotted against t_1. For times $t_1 \geq 2$ s the

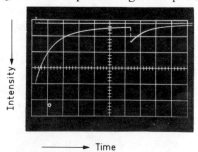

Figure 4 MIDP signal for quinoline-h_7 in durene on sweeping through the T_z-T_x resonance at 3585 MHz, 1.28 s after shutting off the exciting light. $T = 1.25$ K. Horizontal 0.2 s/div. (From Schmidt *et al.*, reference 26.)

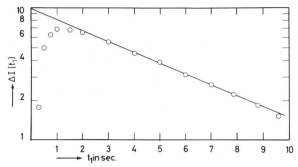

Figure 5 A logarithmic plot of $\Delta I(t_1)$ for a series of experiments on quinoline-h_7 in durene in which the delay time t_1 was varied: $T = 1.1$ K. For $t_1 = 0$ the signal becomes negative indicating that in equilibrium T_x is underpopulated with respect to T_z.[26] (From Schmidt *et al.*, reference 26.)

condition $N_x \gg N_z$ applies and a straight line results. If relaxation is negligible the slope κ_x of this line must be equal to the rate constant k_x. It will be clear that one can obtain a similar rate κ_y for the other dark level T_y by sweeping through the T_z–T_y transition at 1000.5 MHz.

If the relaxation rates W_{uv} are not completely negligible, then the observed slopes κ_x and κ_y will not be equal to the absolute decay rates k_x

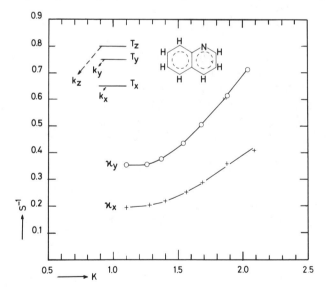

Figure 6 The decay rates κ_y and κ_z of quinoline-h_7 in durene, derived from the slopes of the decay curves of the MIDP signals, as a function of the temperature. (From Schmidt *et al.*, reference 26.)

and k_y. To verify that isolation is achieved one repeats the measurements at higher temperatures. For quinoline in durene these results are shown in Fig. 6. Here it is seen that κ_y remains constant from 1.1 to 1.25 K but that in the same temperature region κ_x already slightly increases. At temperatures higher than 1.25 K the observed values increase rapidly with temperature, because spin relaxation then opens an extra decay route for the two dark levels via the fast-decaying level T_z. If we assume that W_{uv} still decreases on cooling from 1.25 to 1.1 K where the observed decay rates remain constant, it may be concluded that at 1.1 K relaxation has a small effect. Hence we assume that the rates observed at 1.1 K are the absolute decay rates k_y and k_x of the levels T_y and T_x.

As we have mentioned already, we shall not go into the details of the methods for determining the relative radiative rates k_u^r and the relative populating rates P_u. They also are based on the idea that the application of resonant microwave fields in the isolation regime produces changes in the phosphorescence intensity that are related in a relatively simple way to the parameters sought. In Table 2 we present the results of a series of these experiments for quinoline-h_7 in durene that are illustrative of the accuracy that may be obtained in this case. In particular it is seen that in quinoline the spin level T_z dominates in the populating as well as in the decay process because of its effective SOC to the excited singlet state S_1.

From the example given earlier, it appears that the MIDP method allows a simple interpretation of the signals, because one level dominates in the radiation and the total decay rates are very different. Generally speaking, the method is applicable if the triplet state has at least one slowly decaying spin component. In systems with smaller differences in absolute and radiative decay rates the situation is more complex. It may then be more advantageous to follow the technique developed by Schweitzer et al.,[37] who applied transient, resonant microwave fields under continuous illumination and analyzed the resulting time-dependent changes in the phosphorescence.

Table 2 The decay rates k_u in s^{-1}, the relative radiative rates k_u^r and the relative populating rates P_u for the spin components of the triplet state of quinoline-h_7 in durene

	$k_u(s^{-1})$	k_u^r	P_u
T_z	3.1	1	1
T_y	0.32	0.036	0.020
T_x	0.19	0.025	0.034

Finally we want to mention a very interesting method for measuring the dynamic properties of nonphosphorescing triplet states via microwave-induced changes in the *fluorescence*. The first experiment of this kind has been performed by van Dorp et al.[38] on the triplet state of free base porphin. The idea of the experiment can be understood by considering the optical pumping cycle in Fig. 3. Here, under continuous illumination, molecules are pumped to the triplet state T_0 and a steady-state population distribution results, as indicated in Fig. 3. When saturating one of the zero-field transitions one affects the decay from T_0 and thus the steady-state population of S_0. This change in population of S_0 is transferred immediately into a change of the fluorescence from S_1 by the presence of the pumping light source. By applying transient microwave fields van Dorp et al.[38] obtained the dynamic properties of the triplet state of free base porphin by analyzing the time-dependent changes in the fluorescence. It was soon realized that the fluorescence detection technique might even make it feasible to determine the kinetic properties of triplet states in *in vivo* photosynthetic systems. The first experiments were reported by Clarke et al.,[39] who succeeded in observing zero-field transitions in photosynthetic bacteria. Soon further results followed for a variety of systems, including plant chloroplasts and algae.[40]

4 COHERENT EXPERIMENTS

4.1 The FVH Model Applied to a Transition between Two Zero-Field Components of a Molecular Triplet State

In transient experiments on a micro-second time scale, that is, short relative to the lifetime and spin-lattice relaxation times of the metastable triplet state, the effect of the interaction with the radiation field usually cannot be accounted for by time-dependent perturbation theory. To see what happens one has to integrate the Schrödinger equation.

Let us take the T_z-T_y transition with angular frequency $\omega_0 = |Z - Y|/\hbar$ by way of example. If the microwave field has a frequency ω close to ω_0, while it is far off resonance with the two other transitions (T_z-T_y and T_y-T_x) one may neglect the presence of the component T_x.[3] We then have to solve the two-level problem familiar from the theory of lasers and masers for our particular experiment with the simplification that the size of the sample may be considered small relative to the wavelength $\lambda = 2\pi c/\omega_0$. To do this we first think in terms of a single molecule, an arbitrary state of which is described by the wavefunction

$$\psi(t) = a(t)T_z + b(t)T_y. \tag{25}$$

As we are interested in the behavior of the triplet system in time intervals short compared with the lifetimes of the spin levels, we assume ψ to be normalized; $aa^* + bb^* = 1$.

The evolution of the system in time is determined by the Hamiltonian

$$\mathcal{H}(t) = \mathcal{H}_0 + V(t) = \mathcal{H}_0 + \gamma \hbar H_1 S_x \cos \omega t, \tag{26}$$

and, by introducing the one-particle density matrix ρ, the time-dependent Schrödinger equation can be written in the form

$$\frac{d\rho}{dt} = i\hbar^{-1}[\rho, \mathcal{H}]. \tag{27}$$

In general, the integration of this equation poses an awkward problem, especially when it is remembered that in echo experiments the amplitude H_1 of the microwave magnetic field is a step function of the time. Now Feynman, Vernon, and Hellwarth's (FVH) geometrical model[10] represents a "trick" by which Eq. (27) is transformed into an alternative form amenable to well-established methods of solution.

The FVH model introduces a three-dimensional abstract vector space spanned by the orthonormal vectors e_1, e_2, e_3. In this space the state of the system is described by a vector r and the Hamiltonian by a vector Ω. These vectors are defined in terms of their components as

$$\left. \begin{array}{l} r_1 = ab^* + a^*b \\ r_2 = i(ab^* - a^*b) \\ r_3 = aa^* - bb^* \end{array} \right\}, \tag{28}$$

$$\left. \begin{array}{l} \Omega_1 = \dfrac{(V_{zy} + V_{yz})}{\hbar} \\[2mm] \Omega_2 = \dfrac{i(V_{zy} - V_{yz})}{\hbar} \\[2mm] \Omega_3 = \dfrac{(Z - Y)}{\hbar} = \omega_0 \end{array} \right\}. \tag{29}$$

With the aid of this description the equation of motion (27) is transformed into

$$\frac{d\mathbf{r}}{dt} = \Omega \times \mathbf{r}. \tag{30}$$

Or, in words: the vector $\mathbf{r}$, which according to the definition (28) has a constant length $r = aa^* + bb^* = 1$, precesses about the vector Ω.

The proof of the equivalence of the two forms of the equation of motion, (27) and (30), may be indicated briefly. Let $\mathbf{E}$ represent the 2×2

unit matrix and $\frac{1}{2}(Z + Y)$ be the zero of energy. Then r_i and Ω_i as defined in (28) and (29) are proportional to the coefficients that appear when writing the traceless matrices $\boldsymbol{\rho} - \frac{1}{2}\mathbf{E}$ and $\mathcal{H}$ as linear combinations of the Pauli matrices $\boldsymbol{\sigma}_x$, $\boldsymbol{\sigma}_y$, $\boldsymbol{\sigma}_z$:

$$\boldsymbol{\rho} - \tfrac{1}{2}\mathbf{E} = \tfrac{1}{2}(r_1\boldsymbol{\sigma}_x + r_2\boldsymbol{\sigma}_y + r_3\boldsymbol{\sigma}_z), \tag{31}$$

$$\mathcal{H} = \tfrac{1}{2}\hbar(\Omega_1\boldsymbol{\sigma}_x + \Omega_2\boldsymbol{\sigma}_y + \Omega_3\boldsymbol{\sigma}_z). \tag{32}$$

By substituting (31) and (32) into (27) and making use of the commutation rules of the Pauli matrices,

$$(\tfrac{1}{2}\boldsymbol{\sigma}_x\tfrac{1}{2}\boldsymbol{\sigma}_y - \tfrac{1}{2}\boldsymbol{\sigma}_y\tfrac{1}{2}\boldsymbol{\sigma}_x) = \tfrac{1}{2}i\boldsymbol{\sigma}_z, \tag{33}$$

and so on, the equivalence of (27) and (30) is established.

In passing we may draw attention to a further feature of the FVH model. It has been shown[41,42] that the operators for infinitesimal rotations about the $\mathbf{e}_1$, $\mathbf{e}_2$, $\mathbf{e}_3$ axes in the FVH vector space are represented by the expressions $\frac{1}{2}(S_yS_z + S_zS_y)$, $\frac{1}{2}S_x$, and $\frac{1}{2}(S_y^2 - S_z^2)$, which obey commutation rules identical to those of the Pauli matrices,

$$\left.\begin{aligned}
[\tfrac{1}{2}(S_yS_z + S_zS_y), \tfrac{1}{2}S_x] &= \tfrac{1}{2}i(S_y^2 - S_z^2) \\
[\tfrac{1}{2}(S_y^2 - S_z^2), \tfrac{1}{2}(S_yS_z + S_zS_y)] &= \tfrac{1}{2}iS_x \\
[\tfrac{1}{2}S_x, \tfrac{1}{2}(S_y^2 - S_z^2)] &= \tfrac{1}{2}i(S_yS_z + S_zS_y)
\end{aligned}\right\}. \tag{34}$$

With the aid of the operators for the infinitesimal rotations in FVH space one may construct unitary transformations, which correspond to rotations of the FVH frame about one of its three axes. For example, the exponential operator

$$\exp\{-i\omega\tfrac{1}{2}(S_y^2 - S_z^2)t\} \tag{35}$$

induces a rotation with frequency ω about the $\mathbf{e}_3$-axis. This rotation is equivalent to a transformation to the interaction representation via the exponential operator

$$\exp(-i\mathcal{H}_0 t). \tag{36}$$

This can be seen by writing $\mathcal{H}_0$ in a more appropriate form

$$\mathcal{H}_0 = -\tfrac{3}{2}X\{S_x^2 - \tfrac{1}{3}S(S+1)\} + \tfrac{1}{2}(Z - Y)(S_y^2 - S_z^2). \tag{37}$$

Here the first term at the right-hand side commutes with S_x, $S_yS_z + S_zS_y$, and $S_y^2 - S_z^2$. Hence any Hamiltonian containing these three operators is transformed by (36) in a way similar to a rotation about the $\mathbf{e}_3$-axis of the FVH frame related to the $T_z \leftrightarrow T_y$ transition.

So far we have only considered the set of commuting spin operators related to the $T_z \leftrightarrow T_y$ transition. It will be clear that the appropriate sets of operators for the other two transitions follow directly from (34) by cyclic permutation of the indices.

The attractive feature of this description is that one can find the form of the effective Hamiltonian in the rotating frame. This may be of importance when studying the effect of hyperfine interactions on the dephasing of the triplet spins.[43]

From (30) we see that the evolution in time of our triplet system as described by the vector $\mathbf{r}$ is identical in form to the equation of motion of a magnetic moment $\mathbf{M}$ resulting from a collection of spin-$\frac{1}{2}$ particles in a magnetic field $\mathbf{H}$,

$$\frac{d\mathbf{M}}{dt} = \mathbf{M} \times \gamma \mathbf{H}. \tag{38}$$

By our definition $\boldsymbol{\rho}$ and $\mathbf{r}$ referred to a single system but the same equations hold when in (28) and (29) one takes the averages over an ensemble. Accordingly, we shall talk further in terms of a whole ensemble of triplet spins. The tendency of the ensemble toward thermodynamic equilibrium may, tentatively, be accounted for by adding phenomenological relaxation terms to (30), in exactly the same manner as the reversible mechanical equation (38) is extended to the Bloch equations.[3]

By their trick, Feynman, Vernon, and Hellwarth have made the two-level problem in a radiation field amenable to the methods developed for analyzing time-dependent phenomena in NMR,[3] in particular to the transformation to a rotating frame in which $\boldsymbol{\Omega}$ is stationary. But it is to be realized that whereas (38) describes the behavior of the magnetization in the laboratory, (30) describes the fate of the ensemble in an abstract vector space spanned by $\mathbf{e}_1, \mathbf{e}_2, \mathbf{e}_3$. To make the description useful we first have to relate the components of $\mathbf{r}$ to physical observables of our system of triplet spins.

The component r_3, which is the analogue of the longitudinal magnetization in an NMR experiment, is proportional to the population difference of the two triplet levels T_z and T_y

$$r_3 = \overline{aa^* - bb^*} = -\langle \Psi | S_z^2 - S_y^2 | \Psi \rangle \propto N_z - N_y. \tag{39}$$

The component r_2 represents the expectation value of the operator S_x connecting the levels T_z and T_y and thus measures the "transverse" magnetization M_x

$$r_2 = \overline{i(ab^* - a^*b)} = -\overline{\langle \Psi | S_x | \Psi \rangle} \propto \frac{\langle M_x \rangle}{|\gamma| \hbar}. \tag{40}$$

Finally, the component r_1 is the expectation value of a quadrupole moment operator connecting the levels T_z and T_y

$$r_1 = \overline{ab^* + a^*b} = -\overline{\langle \Psi | S_y S_z + S_z S_y | \Psi \rangle}. \tag{41}$$

These equalities may be verified by substituting (25) into the matrix elements of the spin operators in Eq. (39)–(41) and making repeated use of relations such as (8).

As we shall see later the physical observable $r_2 \propto M_x/|\gamma| \hbar$ plays an important part in a Hahn echo experiment, and the observation of a rotary echo is related to $r_3 \propto N_z - N_y$. In our application of the FVH model we shall assume throughout that Ψ_a refers to the upper level. This implies that for an ensemble at thermodynamical equilibrium r_3 has a negative value determined by the Boltzmann factor, whereas $r_1 = r_2 = 0$.

As regards the components of $\boldsymbol{\Omega}$ we have for the system here considered

$$\left.\begin{array}{l} \Omega_1 = 0 \\ \Omega_2 = 2\,|\gamma|\,H_1 \cos \omega t \\ \Omega_3 = \omega_0 \end{array}\right\}, \tag{42}$$

where $|\gamma|\,H_1 = \omega_R$ represents the "Rabi frequency."

The pictorial representation of the precession equation (30) is shown in Fig. 7. In the absence of a microwave field ($V(t) = 0$) we have $\Omega_1 = \Omega_2 = 0$ and the vector $\mathbf{r}$ then precesses about the $\mathbf{e}_3$ axis with Larmor frequency $\Omega_3 = \omega_0$. This is illustrated in Fig. 7a; clearly the zero-field splitting here plays the role of the external magnetic field applied in NMR. To see what happens when the perturbation $V(t)$ is switched on, it is advantageous to

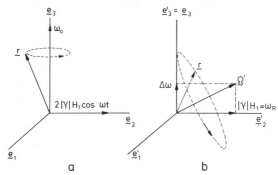

Figure 7 (a) The FVH vector space with r referred to the fixed axes $\mathbf{e}_1$, $\mathbf{e}_2$, $\mathbf{e}_3$. (b) The frame rotating at the microwave angular frequency ω about the $\mathbf{e}_3$ axis. The driving term $|\gamma|H_1 = \omega_R$ is stationary along $\mathbf{e}_2$. The vector $\mathbf{r}$ precesses about the effective field $\boldsymbol{\Omega}'$ with angular frequency ω_{eff}

follow the practice customary in magnetic resonance of transforming to a primed frame rotating at the microwave frequency ω about $\mathbf{e}_3$.

$$(\mathbf{e}'_1, \mathbf{e}'_2) = (\mathbf{e}_1, \mathbf{e}_2)\begin{pmatrix} \cos \omega t & -\sin \omega t \\ \sin \omega t & \cos \omega t \end{pmatrix}; \qquad \mathbf{e}'_3 = \mathbf{e}_3. \tag{43}$$

In the rotating frame the equation of motion becomes

$$\frac{\delta \mathbf{r}}{\delta t} = (\boldsymbol{\Omega} - \omega \mathbf{e}'_3) \times \mathbf{r}. \tag{44}$$

It thus turns out that by the transformation the component of $\boldsymbol{\Omega}'$ along the $\mathbf{e}'_3 = \mathbf{e}_3$-axis has been reduced to $\Omega'_3 = \omega_0 - \omega = \Delta\omega$. The linearly polarized field $-H_1 \cos \omega t$ along the x-axis in the laboratory can be decomposed into a left and a right rotating component. Only one of these has the proper sense of rotation (i.e., the same sense as the rotating frame) and, as usual in resonance phenomena, we neglect the other.[3] In the rotating frame the microwave perturbation now appears as a stationary vector $\omega_R = |\gamma| H_1$ along the $\mathbf{e}'_2$-axis and the equation of motion (44) reduces to

$$\frac{\delta \mathbf{r}}{\delta t} = (\omega_R \mathbf{e}'_2 + \Delta\omega \mathbf{e}'_3) \times \mathbf{r} = \boldsymbol{\Omega}' \times \mathbf{r}. \tag{45}$$

Hence the vector $\mathbf{r}$ precesses with angular frequency

$$\omega_{\text{eff}} = |\boldsymbol{\Omega}'| = [\omega_R^2 + (\Delta\omega)^2]^{1/2} \tag{46}$$

about the appropriate direction of the effective field $\boldsymbol{\Omega}'$ in the rotating frame, as illustrated in Fig. 7b. At exact resonance $\omega = \omega_0$ and the vector $\mathbf{r}$ nutates with angular velocity $\omega_{\text{eff}} = \omega_R$ about the $\mathbf{e}'_2$-axis.

Thus far we have assumed that all triplet spins of our ensemble have exactly the same resonance frequency ω_0. However, the zero-field resonance lines are broadened by second-order hyperfine interaction, by nuclear quadrupole interaction, and also by slightly different environments of the molecules in the host crystal. These effects give rise to a finite linewidth $\Delta\omega_{1/2}$ (half-width at half-height). We assume that this broadening is inhomogeneous, that is, that the line consists of a distribution of spectral components, or "spin packets," all independent of each other. In the absence of a microwave field the spin packets precess with distinct Larmor frequencies about the $\mathbf{e}_3$-axis in the laboratory frame, and in the equilibrium situation the components r_1 and r_2 of the spin packets cancel out.

When a microwave field is applied the different resonance frequencies of the spin packets lead to a spread in the effective field $\boldsymbol{\Omega}'$ in the precession equation (45). Hence each spin packet has its own dynamical history and the collective behavior of the whole ensemble of spins is complicated. This is discussed in more detail in Section 4.2.3. The essential features of the coherence phenomena, however, can be described conveniently by considering the special case in which the amplitude of the microwave field considerably exceeds the linewidth of the zero-field transition ($\omega_R \gg \Delta\omega_{1/2}$). Then all spin packets approximately feel the same effective field and the equation of motion (45) may be applied to the overall ensemble of spin packets.

The influence of relaxation processes on the behavior of the triplet spins has not been considered so far. Although, in general, these processes are of a complicated nature and may have a variety of origins, one expects that their effect on the component r_3 is different from the effect on r_1 and r_2. Whereas a change of r_3 corresponds with a change of the energy of the ensemble, the transverse components r_1 and r_2 measure the mutual phase relation of the triplet spins in the ensemble.

One may try to account for the relaxation toward thermodynamical equilibrium by extending the equation of motion (30) with two phenomenological terms as in the Bloch equations

$$\frac{d\mathbf{r}}{dt} = \mathbf{\Omega} \times \mathbf{r} - \frac{r_1}{T_2^*}\mathbf{e}_1 - \frac{r_2}{T_2^*}\mathbf{e}_2 - \frac{(r_3 - r_3^{eq})}{T_1}\mathbf{e}_3. \tag{47}$$

The last term defines the "spin-lattice" relaxation time T_1, which determines the rate at which the population difference tends to equilibrium. The relaxation time T_2^* accounts for the disappearance of any nonzero value of r_1 and r_2.

For a collection of triplet spins in a crystalline sample the relaxation time T_2^*, which refers to the whole ensemble, just will be determined by inhomogeneous line broadening, as in other magnetic resonance experiments in solids. Hence the rate at which the phase coherence is lost is determined by the spread in resonance frequencies of the individual spins in the ensemble, for example, that arising from hyperfine coupling. As for NMR, one is primarily interested in the spin memory time T_2 (often indicated by T_M) with which a nonzero value of r_1 and r_2 for the individual *homogeneous* "spin packets" tends to zero as a result of irreversible phase-destructive events.

According to equation (47) one would expect a purely exponential decay of r_1 and r_2, which is not always in accordance with the experimental observations. The reason is that the Bloch formalism is only valid for spins in fast relative motion with respect to each other (e.g., paramagnetic ions in liquid solution). As we shall see, the dephasing in a solid is a complicated matter and does not always give an exponential decay of the coherence. Nevertheless we often find a behavior that can be described in terms of a phase memory time that is characteristic for the system and the conditions under which the experiment has been carried out.

We shall discuss the coherence experiments on photoexcited triplet states in the following sequence. In Sections 4.2.1 and 4.2.2 we begin by treating the two basic types of echo experiments, (1) phosphorescence modulation and its refocusing as a rotary echo; and (2) free-induction decay and its refocusing as a two-pulse Hahn echo. Then, in Section 4.2.3., we investigate how inhomogeneous line broadening manifests itself in these experiments and in Section 4.2.4 we discuss the instrumentation required.

When examining the rate of dephasing determined via two-pulse Hahn echo and rotary echo experiments it appears that in many systems of interest the dephasing is dominated by nuclear spin diffusion. Such diffusion then tends to mask other dephasing processes that are more characteristic for ensembles of photoexcited triplet states, like excitation transfer and vibronic relaxation. In Section 4.3 we therefore consider the nuclear spin diffusion process and then introduce two further types of coherence experiments designed to (partly) suppress the effect of this diffusion, (3) Carr–Purcell multiple echo trains; and (4) spin-locking.

4.2 Rotary Echoes and Two-Pulse Hahn Echoes

4.2.1 THE TRANSIENT NUTATION AND ROTARY ECHO AS MODULATIONS OF THE PHOSPHORESCENCE INTENSITY

We consider the experiment where a continuous microwave field, resonant with the $T_z \leftrightarrow T_y$ transition of a phosphorescent molecule, is applied from $t = 0$ onward. For the time being we assume that the condition $\omega_R \gg \Delta\omega_{1/2}$ is fulfilled so that the vector $\mathbf{r}$, initially aligned along the $\mathbf{e}_3$-axis with a length r_3^0, starts a precession about the $\mathbf{e}_2'$-axis with angular frequency $\omega_R = |\gamma| H_1$ (Fig. 7b) and hence

$$r_3(t) = r_3^0 \cos \omega_R t. \tag{48}$$

This "transient nutation" is a matter of common knowledge in nuclear magnetic resonance. In our triplet state it means, according to (39), that the population difference of the two spin levels becomes time dependent: dependent:

$$N_z(t) - N_y(t) = (N_z^0 - N_y^0) \cos \omega_R t. \tag{49}$$

Whenever the radiative decay rates k_y^r and k_z^r differ significantly, the oscillation of the population difference manifests itself as a modulation of the phosphorescence intensity.

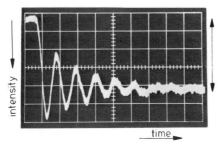

Figure 8 Modulation of the phosphorescence ("transient nutation") of quinoline-d_7 in a durene host produced by a microwave field with the $T_z \leftrightarrow T_y$ transition at 1001.0 MHz. Horizontal 1 μs per division. $T = 1.2$ (From van't Hof *et al.*, reference 53)

An example of this experiment on the T_z-T_y transition at 1001.0 MHz of quinoline-d_7 in a durene host at $T = 1.2$ K is shown in Fig. 8. Here the microwave field is applied 5 sec after shutting off the exciting light. The fast-decaying level T_z then is empty ($N_z^0 \approx 0$) while the long-lived state T_y still carries a population. (See Section 3.) The signal in Fig. 8, in fact, is the leading edge of the microwave-induced delayed phosphorescence signal observed with a short time constant and it can be described by

$$I(t) = c\tfrac{1}{2}k_z^r N_y^0 (1 - \cos \omega_R t), \tag{50}$$

where c again is an instrumental constant.*

In the experiment of Fig. 8 one observes the coherent nutation of the triplet population via the "window" of the radiative level T_z. The first extremum reflects the inversion of the population at $t = \pi/\omega_R$. The modulation frequency ω_R, attained with the maximum microwave power here available, appears to be 0.9 MHz. Since the ESR linewidth is about 0.35 MHz, the condition $\omega_R \gg \Delta\omega_{1/2}$ actually is not quite fulfilled.

It is seen that the modulation of the phosphorescence decays within a few periods. This damping is caused by a spread in the effective field ω_{eff} over the triplet spins. According to (46) this spread may be due to the finite linewidth $\Delta\omega_{1/2}$ and (or) to an inhomogeneity of the microwave amplitude H_1. In this particular experiment, however, the inhomogeneity of H_1 is responsible for the damping and the spread in resonance frequencies only has a negligible effect. (We shall come back to the problem of the finite linewidth in Section 4.2.3.) In Fig. 9a we have indicated that because of the inhomogeneity of H_1 the spin packets composing the vector **r** will precess with slightly different angular frequencies ω_R. In the course of time the triplet spins completely get out of

---

* Here we have taken $t = 0$ as the moment that the microwave field is switched on, in contrast to the MIDP experiments where $t = 0$ corresponds with the closure of the shutter.

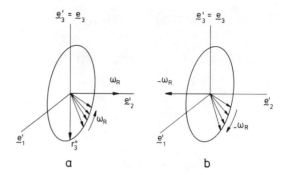

a   b

Figure 9 Description of the transient untation and the rotary echo in the FVH rotating frame. The resonant microwave field is stationary along the $\mathbf{e}_2'$ axis. (a) The dephasing of the vector $\mathbf{r}$ in the time interval $0 < t < \tau$. (b) The refocusing of $\mathbf{r}$ in the time interval $\tau < t < 2\tau$ as a result of a π phase shift in the microwave field at $t = \tau$. [From van't Hof, reference 42.]

phase and, consequently, the coherence reflected by the modulation of the phosphorescence disappears. The phosphorescence intensity tends to a value of about one-half the first maximum. In this state of "saturation" the levels T_z and T_y are equally populated.

The rotary echo in photoexcited triplet states is the analogue of an experiment first carried out in NMR by Solomon,[44] and it represents a trick to restore the modulation of the phosphorescence.[45] This is illustrated by Fig. 10, where we have shown the result of a rotary echo experiment on the $T_z \leftrightarrow T_y$ transition of quinoline-d_7 in durene. For the explanation we return to Fig. 9a where we have seen that the different spin packets fan out owing to the spread in ω_R. If one now performs a phase shift of π on the microwave field at $t = \tau$ the component of the effective field along the $\mathbf{e}_2'$-axis in the rotating frame suddenly is reversed

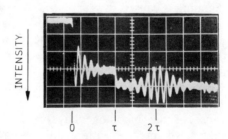

Figure 10 An optically detected rotary echo of quinoline-d_7 in a durene host. The microwave field is resonant with the $T_z \leftrightarrow T_y$ transition at 1001.0 MHz. Horizontal 2 μs per division. $T = 1.2$ K. Microwave irradiation starts at $t = 0$ and a π phase shift occurs at the time $\tau = 5$ μs.

from $\Omega_2' = |\gamma| H_1 = \omega_R$ to $\Omega_2' = -|\gamma| H_1 = -\omega_R$ (Fig. 9b). From then on each spin packet is precessing at the same rate as before but in the opposite direction, so that the system behaves as if it were going back in time. Therefore the coherence and thus the modulation of the phosphorescence gradually increase, reaching a maximum at $t = 2\tau$ when all spin packets are again in phase along the $\mathbf{e}_3$ axis.

We yet have to consider the effect of relaxation processes on the phase relation of the triplet spins. In a real experiment this relaxation will lead to a decay of the rotary echo intensity as a function of the delay time 2τ, with a characteristic time that is usually designated by $T_{2\rho}$. In Section 4.3 and 4.4 we go into the details of the mechanisms causing this relaxation. Here it is important to realize that the rotary echo allows us to measure the dephasing rate $T_{2\rho}^{-1}$ of the triplet spins in the vertical $\mathbf{e}_1' - \mathbf{e}_3'$ plane of the rotating FVH frame.

4.2.2 THE FREE PRECESSION AND HAHN ECHO

The next coherence experiment to be described is the analogue of the nuclear spin echo discovered by Hahn.[46] Let us again take a microwave field resonant with the $T_z \leftrightarrow T_y$ transition of a photoexcited triplet state for which the condition $\omega_R \gg \Delta\omega_{1/2}$ holds. Contrary to the situation in a rotary echo experiment where from $t = 0$ onward the microwave field is applied continuously, the echo in the present experiment is induced by the application of two short pulses. We suppose that the pulses last for times t_{p1} and t_{p2} such that $\omega_R t_{p1} = \pi/2$ and $\omega_R t_{p2} = \pi$. The sequence of events is illustrated in Fig. 11. At the time $t = 0$ the vector $\mathbf{r}$ merely has a component r_3^0 along the (negative) $\mathbf{e}_3$-axis, representing the population difference between the levels T_z and T_y. We assume that this vector $\mathbf{r}$ can be decomposed into the contributions of a number of spin packets with slightly different resonance frequencies. Along the $\mathbf{e}_3$-axis of the rotating frame in Fig. 11a we have indicated schematically the spectral distribution of these spin packets.

At time $t = 0$ the $\pi/2$ pulse is switched on. Since we have assumed that $\omega_R \gg \Delta\omega_{1/2}$ all spin packets start a precession about the $\mathbf{e}_2'$-axis with approximately the same Rabi frequency and at the end of the pulse they are in alignment with the (negative) $\mathbf{e}_1'$-axis of the rotating frame (Fig. 11b).

Hence, as a result of the pulse, the vector $\mathbf{r}$ now lies in the $\mathbf{e}_1 - \mathbf{e}_2$ plane and in the fixed frame it rotates with an angular frequency ω_0. This gives rise to an oscillating component r_2, which, according to (40) and (43), may be observed in the laboratory as an *oscillating magnetic moment*,

$$\langle \overline{M_x} \rangle \propto -\gamma \hbar r_2(t) = \gamma \hbar r_3^0 \sin \omega_0 t. \tag{51}$$

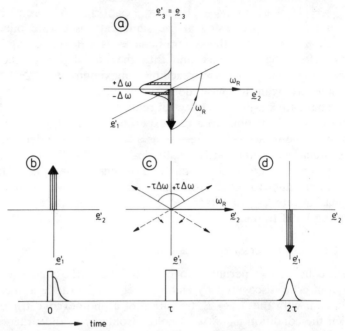

Figure 11 Description of the formation of the two-pulse electron spin echo in the FVH rotating frame. (From van't Hof, reference 42.) (*a*) The spectral distribution of the spins along the $\mathbf{e}_3$-axis and the action of the $\pi/2$ pulse. (*b*) The situation in the $\mathbf{e}_1'$-$\mathbf{e}_2'$ plane immediately after the $\pi/2$ pulse when the spins are aligned along the $\mathbf{e}_1'$-axis. In the time interval $0 < t < \tau$ the spins fan out because the spread in resonance frequency. (*c*) The angular positions of the spin packets at $\pm \Delta \omega$ from exact resonance before and after the π pulse applied at $t = \tau$. Refocusing of the spins in the time interval $\tau < t < 2\tau$. (*d*) The echo along the positive $\mathbf{e}_1'$ axis at $t = 2\tau$. At the bottom of this figure we have indicated the driving pulses, the free precession after the $\pi/2$ pulse and the echo.

This "free precession," which manifests itself by an emission of microwaves by the sample, decays very quickly, because, as soon as the driving microwave field has ceased, the approximate uniformity of the effective fields vanishes and the spin packets then start to precess with different angular velocities $\Delta \omega$ about the $\mathbf{e}_3$-axis in the rotating frame. They will attain an isotropic distribution in a characteristic time $T_2^* = 2/\Delta \omega_{1/2}$.

As first shown by Hahn in the original NMR experiment, the phase coherence between the individual spin packets can be restored by the application of the π pulse at time $t = \tau$. By way of illustration we follow the behavior of the pair of isochromatic spin packets, precessing at frequency $\Delta \omega$ in opposite directions (Fig. 11*c*). Through the application of the π pulse, the whole pancake of spins is turned about the $\mathbf{e}_2'$-axis by an angle of π and the spins that had arrived at $t = \tau$ in the angular positions $\phi = \pm \tau \, \Delta\omega$ now suddenly find themselves in the positions $\phi = \pi \mp \tau \, \Delta\omega$. However, they maintain their original directions of rotation and at $t = 2\tau$ all moments are aligned along the positive e_1'-axis, provided the individual spins have maintained their characteristic values of $\Delta\omega$ throughout the experiment (Fig. 11d). This phenomenon is what we call the *Hahn echo*.

At this point we like to remark that any two-pulse sequence gives rise to an echo. The $\pi/2$, π combination is a special case that gives the largest echo amplitude and is the easiest to describe. However, in experimental practice pulses of equal length often are more convenient and it can be shown that then the condition $\omega_R t_p = 2\pi/3$ yields the largest echo signal, which is 65% of that obtainable by means of a $\pi/2$, π sequence.[47]

Immediately after the $\pi/2$ pulse, and again in the echo, the triplet system spontaneously emits microwaves that can be detected by a microwave receiver. In the lower trace of Fig. 12 we present the first result[48] of such an electron spin-echo experiment for a phosphorescent triplet state, carried out on the $T_z \leftrightarrow T_y$ transition of quinoline-d_7 in a single crystal of durene-d_{14} at $T = 1.2$ K. The upper trace shows the same experiment with the exciting light shut off. It is seen that during the microwave pulses, which are of equal duration with $\omega_R t_p = 2\pi/3$, the receiver is completely overloaded. The proper pulse length of 0.4 μs had first been determined from the modulation period $2\pi/\omega_R$ in a transient nutation experiment (Fig. 8).

The shape of the echo generally depends on the ratio $\omega_R/\Delta\omega_{1/2}$. However, if the condition $\omega_R/\Delta\omega_{1/2} \gg 1$ is fulfilled the echo simply is the Fourier transform of the frequency distribution. In Section 4.2.3 we go into the influence of the linewidth on the echo shape.

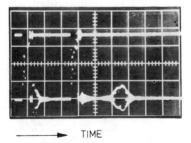

TIME

Figure 12 A two-pulse electron spin echo of quinoline-d_7 in durene-d_{14} detected with the aid of a microwave receiver. The pulses are resonant with the $T_z \leftrightarrow T_y$ transition at 1001.0 MHz. $T = 1.2$ K. Horizontal 2 μs per division. Pulse width 0.4 μs. The upper trace represents the result of an experiment with the exciting light shut off. (From Schmidt, reference 48.)

As an alternative to microwave detection, which requires a fairly sophisticated system, the Hahn echo can also be detected optically, as first shown by Breiland, Harris, and Pines[49] for 2,3-dichloroquinoxaline. The method is illustrated in Fig. 13. In the lower trace we show a sequence of three microwave pulses resonant with the $T_z \leftrightarrow T_x$ transition at 3598 MHz of quinoline-d_7 in durene and in the upper trace the effect of these pulses on the phosphorescence intensity at $T = 1.2$ K. Prior to the application of the pulses, the system is prepared in a state where the radiative level T_z is empty and hence no phosphorescence is emitted. This can be achieved in a similar way as in the MIDP experiments described in Section 3. In the FVH rotating frame the vector $\mathbf{r}$ along the negative $\mathbf{e}_3$-axis thus represents the population of the "dark" level T_x. The first $\pi/2$ pulse, applied at $t = 0$, brings $\mathbf{r}$ in alignment with the negative $\mathbf{e}_1'$-axis. As a result, the phosphorescence intensity increases to one-half its maximum value. [See Eq. (50).] The second pulse, applied at $t = \tau$, is a π refocusing pulse that leads to the formation of an echo at $t = 2\tau$. The echo manifests itself as a vector $\mathbf{r}$ along the positive $\mathbf{e}_1'$-axis, which, for its optical detection, is brought back along the negative $\mathbf{e}_3$-axis by a recording $\pi/2$ pulse applied at $t = 2\tau$. By using this trick the echo in the transverse plane is transformed to a new population difference where T_x is more populated than T_z and hence we observe a decrease of the phosphorescence emission. The echo waveform can be recorded by scanning the third pulse through a small time interval around $t = 2\tau$ while measuring the change of the phosphorescence intensity.

The effect of relaxation processes on the behavior of the two-pulse Hahn echoes has not yet been taken into account. As a result of this relaxation the echo intensity will fall off according to some decay function $E(2\tau)$. This decay of the phase relation of the triplet spins in the horizontal $\mathbf{e}_1'$–$\mathbf{e}_2'$ plane of the FVH frame usually can be characterized by the dephasing time T_2, often called phase memory time T_M.

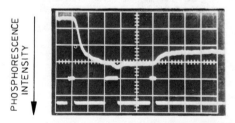

Figure 13 An optically detected Hahn echo in the $T_z \leftrightarrow T_x$ transition (3598 MHz) of quinoline-d_7 in durene-h_{14} at $T = 1.2$ K. Lower trace: a sequence of a $\pi/2$, π and $\pi/2$ pulse. Upper trace: the effect of the pulses on the phosphorescence intensity. Horizontal: 0.5 μs per division. (From van't Hof, reference 42.)

With the two types of coherence experiments discussed so far we have two techniques at our disposal for studying phase-destroying processes. The rotary echoes measure the phase coherence in the vertical e'_1–e'_3 plane of the rotating FVH frame and the two-pulse Hahn echoes in the horizontal e'_1–e'_2 plane. The difference in dephasing rate in the two types of experiment is mainly due to hyperfine interaction.[43] This problem is discussed in Section 4.3.

4.2.3   THE INFLUENCE OF AN INHOMOGENEOUSLY BROADENED LINE

To keep the description of the coherence phenomena simple we have assumed thus far that $\omega_R \gg \Delta\omega_{1/2}$. In actual practice, however, one often meets situations where ω_R is small compared to the linewidth. The dynamical behavior of a single spin packet then depends on its position in the resonance line, and the overall evolution in time cannot be visualized so easily in the rotating frame. Here we discuss the influence of the (inhomogeneous) linewidth on the shape of the rotary echo and the two-pulse Hahn echo.

Rotary echo. In Fig. 14 we present simulations of the time development of the population difference $r_3(t)$ in the case of the rotary echo for two representative cases: the *broad line* ($\omega_R/\Delta\omega_{1/2}=0.2$) and an intermediate case ($\omega_R/\Delta\omega_{1/2}=1$).[42] The curves have been calculated by starting from the assumption that the spin packets compose a resonance line having a Gaussian profile, with the microwave frequency ω_0 coinciding

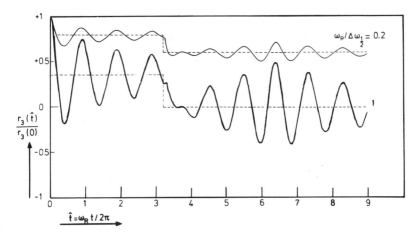

Figure 14   The calculated shape of the rotary echo after a phase shift π in the microwave field applied at $\hat{t}=3.2$. The dimensionless parameter $\omega_R/\Delta\omega_{1/2}$ has been taken equal to 0.2 and 1 for the two curves. (From van't Hof, reference 42.)

with the center of the line. First the precession Eq. (45) has been solved for an arbitrary spin packet in the rotating frame with the appropriate initial conditions. Subsequently, the total population difference has been calculated by integrating the contribution of the individual spin packets over the line shape. To get a conveniently arranged set of graphs the time t has been replaced by the dimensionless parameter $\hat{t} = \omega_R t/2\pi$. The phase shift of π in the microwave field is applied at the reduced time $\hat{t} = t\omega_R/2\pi = 3.2$.

The first thing to note from Fig. 14 is that the depth of the modulation, and its damping, strongly depends on the ratio $\omega_R/\Delta\omega_{1/2}$. This is a consequence of destructive interference of the various spin packets with their different effective fields $\mathbf{\Omega}'$; the depth of the modulation becoming smaller and the damping faster, the broader the inhomogeneous line. The second thing to note is that at time $\hat{t} = 6.4$ an echo is indeed induced, but there is no question of a complete restoration of the original modulation. This can be understood with the help of the rotating frame in Fig. 15. For the spins, outside the direct environment of the line center, the application of the phase shift does not lead to a complete inversion of the effective field $\mathbf{\Omega}'$. Hence the contribution of these spins to the echo is smaller than to the modulation in the beginning of the curve.

Further, it is seen in Fig. 14 that immediately after the phase shift a jump is predicted in the average level of the signal, which decreases in relative size when the number of spins that effectively cooperate in the modulation increases. Such a jump is a common experimental phenomenon, and it can be seen, for instance, in the rotary echo experiment of Fig. 10. A simplified explanation of this feature can be given by merely considering the spins for which $|\Delta\omega| \approx \omega_R$. In Fig. 15 we consider a packet

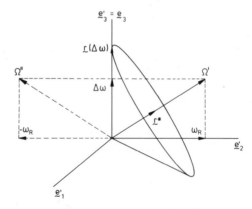

Figure 15 The behavior of an off-resonance spin packet $\mathbf{r}(\Delta\omega)$ in the FVH rotating frame during a rotary echo experiment.

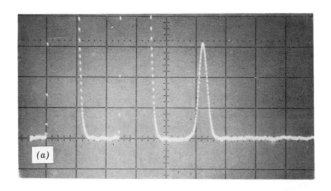

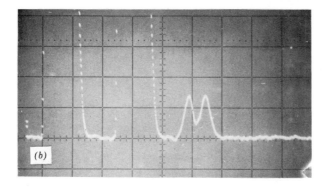

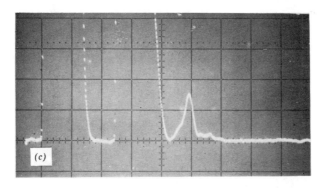

Figure 16 Hahn echoes of phenanthrene in a biphenyl host excited by two pulses of equal duration t_p. The microwave frequency is tuned to the $T_z \leftrightarrow T_y$ transition at 1613 MHz. $T = 1.2$ K. Horizontal: 1 μs per division. The ratio $\omega_R/\Delta\omega_{1/2} \approx 0.06$. (From van't Hof, reference 42.) (a) $t_p = 0.31$ μs ($2\pi/3$ pulses), (b) $t_p = 0.46$ μs (π pulses), (c) $t_p = 0.62$ μs ($4\pi/3$ pulses).

of such spins $r(\Delta\omega)$ that starts at $t = 0$ a precession on a cone about the effective field Ω'. After some time there is a random distribution of the spins in the packet over the conical surface and only along the direction of Ω' a nonzero component, denoted by r^*, remains. It is clear that for this spin packet the populations of the two triplet levels involved in the resonance have not been equalized, since the vector r^* still has a component along the positive e_3-axis. When the phase shift is applied at $t = \tau$ the vector r^* starts a precession about the new effective field Ω''. After some time only along Ω'' a nonzero component of r^* results, which corresponds with a new population difference between the levels. This explanation is supported by the observation of a modulation after the jump in Fig. 10 that is a manifestation of the precession of r^* about the new field Ω''. Moreover, the jump is found to become larger when the microwave power is reduced. In the case of a broad resonance line it is advantageous to use the maximum available microwave power because the number of spins that contribute to the formation of the echo is proportional to H_1,

Hahn echo. For an extensive discussion of the influence of the line-width on the Hahn echo we refer to the work of Mims,[47] who has calculated the shape and the intensity of the Hahn echo as a function of the pulse duration t_p in the case of a broad line as well as in the intermediate case. It turns out that for a broad line the echo intensity is proportional to H_1 and thus, here also, a strong microwave field can be used to advantage. It appears, however, that the shape of the echo depends strongly on the pulse width. To illustrate what kind of echo shapes may appear we show in Fig. 16 the results of an experiment on phenanthrene in a biphenyl host. In this experiment the microwave field was tuned to the $T_z \leftrightarrow T_y$ transition at 1613 MHz which is characterized by a very large width $(\Delta\omega_{1/2}/2\pi = 15\ \text{MHz})$. The observed echo shapes correspond very well with those predicted by Mims for a sequence of two $2\pi/3$, π, or $4\pi/3$ pulses, respectively.

4.2.4   EXPERIMENTAL TECHNIQUES AND SENSITIVITY CONSIDERATIONS

Here we describe some experimental techniques used for the generation and detection of the two-pulse Hahn echoes and the rotary echoes, and compare the sensitivity of optical versus microwave detection. In Fig. 17 we show a schematic drawing of the arrangement used by van't Hof.[42] In the Hahn echo experiment the microwave power from a cw signal oscillator (SO) is fed directly to a traveling wave tube amplifier (TWT), which amplifies the microwave power to about 1 W. The continuous wave is converted by a PIN diode modulator into short pulses, which are

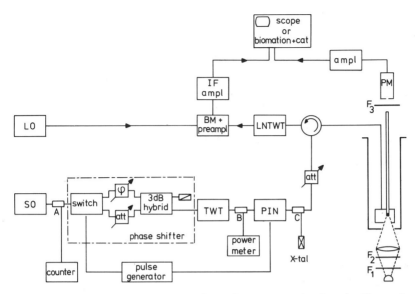

Figure 17 Schematic diagram of the experimental arrangement for detecting Hann echoes and rotary echoes as used by van't Hof. (See reference 42.)

guided via a circulator to the cavity. In the rotary echo experiments a phase shifter assembly is incorporated between the SO and the TWT.

The microwave detection of the echoes proceeds in the following way. The echoes generated by the sample in the cavity, together with the reflected driving pulses, pass via the circulator through a low-noise traveling wave tube amplifier (LNTWT) that serves as an amplifier for the echoes and as a limiter for the driving pulses. The microwave signals are transferred to an intermediate frequency (IF) of 30 MHz in a balanced mixer (BM) with the help of a local oscillator tuned 30 MHz away from the signal frequency. After amplification in the IF amplifier the echoes can be observed on an oscilloscope at the IF frequency or, further rectified, as a "video" signal. The system has a threshold sensitivity of 10^{-12} W with a bandwidth of 8 MHz.

To increase the signal-to-noise ratio of the microwave signals emitted by the sample it is essential to use a cavity. In the arrangement of Fig. 17 a so-called reentrant cavity described by Erickson[50] has been used. In Fig. 18 such a cavity is shown together with the approximate electric and magnetic field configurations. By changing the position of the plunger the resonance frequency can be varied over one octave. To obtain a short ringing time commensurate with the microwave pulse duration of the order of 0.5 μs the quality factor Q of the cavity usually is reduced artificially to a value between 250 and 500, depending on the resonance

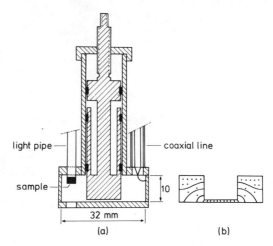

Figure 18 (*a*) A tunable microwave cavity. (From van't Hof, reference 42.) (*b*) A schematic drawing of the electric and magnetic field configuration. The electric fields are indicated by the lines and the circumferential magnetic fields by dots. (See reference 50.)

frequency. With a microwave power of 1 W incident on this cavity the maximum amplitude of the microwave magnetic field H_1 is about 0.4 G. The sample, fixed against a light pipe, is positioned in the region of the highest magnetic field and is irradiated through a hole in the bottom of the cavity.

The rotary echoes may be observed with the same setup. The phosphorescence then is monitored by a photomultiplier via the light pipe to which the sample is attached. For this experiment, however, the reentrant cavity is not essential and, as shown by Harris et al.,[45] the rotary echoes can also be observed with a helix. In a helix the microwave field is rather inhomogeneous and, as a result, a transient nutation signal generated in a helix damps out much faster than in a cavity.[42]

The choice of optical versus conventional (microwave) detection is determined by the properties of the system to be studied. In principle both methods can be applied because one may always transform a transverse magnetization into a population difference, and vice versa, by a $\pi/2$ pulse of the correct phase. From the point of view of sensitivity the conventional microwave technique is more successful in long-lived, weakly phosphorescent triplet states, whereas in strongly emitting triplets the optical detection is preferable. This question of sensitivity has already been brought up by Harris et al.,[45] who claim that the optical detection is by far superior because as few as 10^4 spins can be detected. One should realize that this is only possible if the echo signal can be converted in a pulse of light against a zero background. In practice, there is usually "noise" arising from background radiation, for example, from impurity emission or the unmodulated part of the phosphorescence, even if a monochromator is used in the detection system. For these reasons it is often more advantageous in weakly phosphorescing systems to use a microwave detector where the signal is discriminated against the noise level of the microwave receiver. For instance, in naphthalene in durene it has thus far not been possible to detect the coherence optically, whereas strong spin-echo signals can be observed when using microwave detection.

4.3   Spectral Diffusion through Hyperfine Interactions and Methods to Avoid Its Influence on Dephasing

4.3.1   THE EVIDENCE FOR SPECTRAL DIFFUSION

The first spin-echo experiments[48] already proved that the dephasing time T_2 often is determined by hyperfine interactions with surrounding nuclear spins. This is shown very clearly in Fig. 19, where we present results of a study by van't Hof[42,43] on quinoline present as a dilute deep trap in a single crystal of durene for various isotopic combinations of guest and host. The figure shows the decay curves for two-pulse Hahn echo measurements and it is seen that by deuterating the guest as well as the host

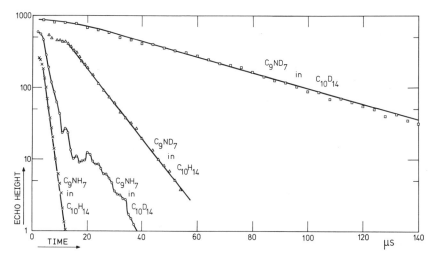

Figure 19   Decay curves derived from two-pulse Hahn echo experiments in the $T_z \leftrightarrow T_x$ transition of quinoline C_9HN_7 or C_9ND_7 in a single crystal of durene $C_{10}H_{14}$ or $C_{10}D_{14}$ at $T = 1.2$ K. The echo height is plotted as a function of the time interval 2τ between the first of the two driving pulses and the formation of the echo. (From van't Hof, reference 42.)

molecules, T_2 increases from 1.5 to 39 μs! A similar lengthening has been observed for the decay time $T_{2\rho}$ measured by means of rotary echoes. A second important thing to note is that the curves in Fig. 19 do not follow an exponential behavior and that one can only speak about a phase memory time T_2 in a somewhat loose manner.

The destruction of the phase coherence of the triplet spins by the nuclear magnetic moments is thought to arise in the following way. As a result of mutual dipole-dipole interactions the nuclei undergo flip-flop motions and thereby modulate the hyperfine fields at the site of the triplet spins. Hence the resonance frequencies of the triplet spins are no longer constant but fluctuate slightly in the course of time. By analogy with the effect of spatial diffusion of molecules in liquid solution in an inhomogeneous magnetic field, this effect is called spectral diffusion.[47]

The nuclear spin-induced dephasing of the triplet spins often forms a threshold for the observation of other dephasing mechanisms like vibronic relaxation or detrapping. Hence in the course of the last few years one has looked for methods by which one can suppress the effect of the nuclei without affecting other relaxation processes that one wants to study. Two of these techniques—Carr–Purcell multiple echoes and spin-locking—are treated here. For a detailed discussion of the effect of spectral diffusion on the phase coherence as measured in the various coherence experiments the reader is referred to the paper cited in reference 43.

4.3.2 Carr–Purcell Multiple Electron Spin Echoes

In 1954 Carr and Purcell[51] have shown that in a liquid the effect of spatial diffusion of nuclear spins in an inhomogeneous magnetic field can be (partly) eliminated by using a multiple spin-echo train. In these experiments the sample is subjected to a resonant $\pi/2$ pulse at $t = 0$ and then to a series of π pulses at times τ, 3τ, 5τ, and so on, which give rise to the formation of echoes at times 2τ, 4τ, 6τ, and so on. In designing such a multiple-pulse experiment a complicating feature has to be taken into consideration. Because of the inhomogeneity of H_1 over the sample, the nominal "π pulses" in the train will not be felt as such by all molecules, but there will be a spread around this mean value. Now it is known from the NMR experiments that the pulses in the Carr–Purcell sequence must be of a duration exactly corresponding to π for all spins, since a systematic deviation has a cumulative effect and spoils the refocusing of the echoes in the train. This complication has been investigated by Meiboom and Gill,[52] and they have shown that a small deviation of the pulse duration from the nominal value π is acceptable in a multiple-echo

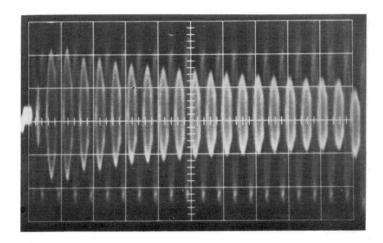

Figure 20 Multiple spin-echo signal of quinoline C_9ND_7 in durene $C_{10}H_{14}$. The microwave pulses are resonant with the $T_z \leftrightarrow T_x$ transition at 3598 MHz. Horizontal 4 μs per division. $T = 4.2$ K. (From van't Hof *et al.*, reference 53.)

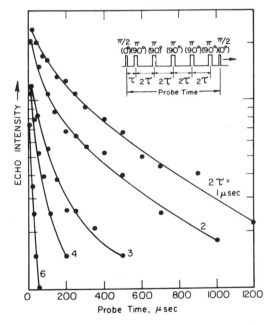

Figure 21 The decay of the echo maxima in the optically detected Carr-Purcell Meiboom-Gill spin echo train in the T_y–T_z transition at 1744.6 MHz for 1,2,4,5-tetrachlorobenzene in durene.

383

experiment, provided the pulses are coherent and a $\pi/2$ phase shift is introduced between the $\pi/2$ pulse and the first of the π pulses.

It has been possible to perform similar multiple spin-echo experiments on photoexcited triplet systems[49,53] and by analogy with the results of the NMR experiments one observes that the effect of spectral diffusion on the dephasing of the triplet spins can be greatly reduced. In Fig. 20 an example is shown of a Carr–Purcell multiple spin-echo train with $\tau = 1$ μs on the T_z-T_x transition of quinoline-d_7 in durene; the high, narrow spikes are the driving pulses, and the signals in between are the echoes. (To illustrate the power of the Meiboom–Gill "trick," which was incorporated in the experiment of Fig. 20, it is worth mentioning that without a $\pi/2$ phase shift only a few echoes could be observed.) Whereas from a single-echo experiment at 4.2 K one derives $T_2 = 9.2$ μs, the multiple-echo experiment yields echoes during more than 500 μs!

The decay of the echoes in a similar experiment performed by Breiland et al.[49] on one of the zero-field transitions of 1,2,4,5-tetrachlorobenzene in durene is presented in Fig. 21 for values of τ ranging from 1 to 6 μs. The remarkable lengthening of the phase coherence up to hundreds of microseconds, apparent from Fig. 21 for tetrachlorobenzene and displayed in Fig. 4 of reference 53 for quinoline-d_7, indicates that the spectral diffusion within the zero-field ESR line may be partly circumvented by a multiple-pulse sequence. As in Carr and Purcell's original NMR experiments, the loss of phase coherence of the triplet spins due to "diffusion" effects is gradually eliminated by making τ shorter. But contrary to their results, the decay of our echoes is not exponential.

We shall give a qualitative explanation of the lengthening of the phase coherence with the aid of Fig. 22. In this picture the evolution of the phase of one triplet spin, off resonance by an amount $\Delta\omega$, is followed during both the single-echo and the multiple-echo experiment. The phase is defined as the angular position $\phi = \int \Delta\omega \, dt$ in the horizontal plane of the rotating frame. First we consider a single-echo experiment without fluctuations in $\Delta\omega$. After the $\pi/2$ pulse the phase increases proportional with time: $\phi = t \, \Delta\omega$. The effect of a π pulse at $t = \tau$ can be considered as an inversion of the phase: $\tau \, \Delta\omega \rightarrow -\tau \, \Delta\omega$. After the π pulse the phase again increases with time and becomes zero at $t = 2\tau$. Then the spins are all again in phase and we observe the echo. However, if we allow for some frequency jumps (i.e., spectral diffusion) during the experiment, the phase at $t = 2\tau$ will differ substantially from zero. Since the contribution of each spin to the formation of the echo is proportional to $\cos[\phi(2\tau)]$, the echo will be attenuated. From Fig. 22 it is clear that this attenuation will depend on the number and size of the jumps in frequency.

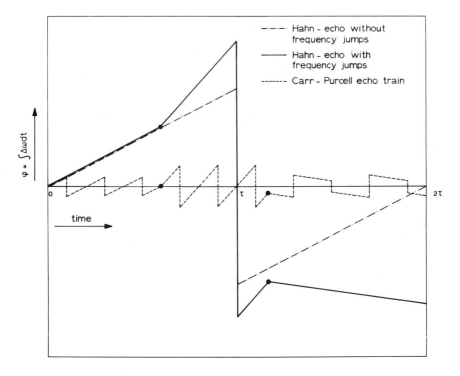

Figure 22 Comparison between the phase behavior of a single triplet spin in a two-pulse Hahn echo and a multiple spin-echo experiment. The phase is defined as the angular position $\phi = \int \Delta\omega \, dt$ in the the horizontal plane of the FVH rotating frame after the $\pi/2$ pulse and is inverted ($\phi \rightarrow -\phi$) by the π pulses. Two random phase jumps are assumed to occur (at $t \approx 0.60\tau$ and $t \approx 1.17\tau$, respectively). (From van't Hof, reference 42.)

Now we consider the Carr–Purcell cycle for frequency jumps identical to those just considered. It has the property of dividing the original time interval 2τ into many small parts, in which the phase is alternated owing to the action of the π pulses. Figure 22 illustrates the effect: The phase is forced to stay closer to $\phi = 0$, the smaller the time elapsed between the pulses. As a result the attenuation of the echo due to frequency jumps can be reduced considerably. In the limit $\tau \rightarrow 0$ this effect will even disappear completely.

The effect of spectral diffusion on the dephasing during rotary echo trains has also been studied by van't Hof.[43] One again observes a lengthening of the phase memory time $T_{2\rho}$ on reducing the pulse interval τ. The effect can be explained in a similar way as the lengthening of T_2 in a multiple Hahn echo experiment.

4.3.3 SPIN LOCKING

The Carr–Purcell–Meiboom–Gill method in the limiting case $\tau \rightarrow 0$ is equivalent to spin-locking.[54] In the photoexcited triplet state this experiment was first performed by Harris, Schlupp, and Schuch[45] in the following way. A $\pi/2$ pulse, applied at the resonance frequency of one of the zero-field transitions along the $\mathbf{e}_2'$-axis of the FVH rotating frame, turns the pseudomagnetization $\mathbf{r}$ from the $\mathbf{e}_3$-axis along $\mathbf{e}_1'$. (See Fig. 23.) Then, after, the completion of this pulse, the microwave irradiation is not terminated, but a phase change of $\pi/2$ is applied. By this phase change the H_1 field in the rotating frame is shifted over an angle $\pi/2$, locking the spins along $\mathbf{e}_1'$ and preventing the free-induction decay. With the magnetization parallel to the H_1 field, the spin alignment in the rotating frame decays to an equilibrium value with a characteristic time $T_{1\rho}$.

In an actual experiment the decay time $T_{1\rho}$ is measured from the height of the free-induction signal immediately after switching off the locking microwave field, as a function of the locking time. In phosphorescent triplet states, however, $T_{1\rho}$ can also be detected optically just as for the two-pulse Hahn echo. After the locking field has been switched off, a second $\pi/2$ pulse then is applied at the same phase as the first one. Hereby the remainder of $\mathbf{r}$ is placed along the positive $\mathbf{e}_3$-axis, resulting in a new value of the triplet sublevel populations and hence in a change of the phosphorescence intensity. The measurement of this change as a function of the spin-locking time τ again yields $T_{1\rho}$. For tetrachlorobenzene-d_2 in a durene-d_{14} host $1/T_{1\rho}$ is found to be only slightly higher than the mean rate of decay to the ground state of the two triplet sublevels involved.[45,55] Hence it appears that in this case the effect of spectral diffusion is completely eliminated by the spin-locking.

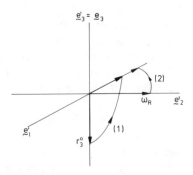

Figure 23 Schematic representation of the spin-locking technique: (1) $\pi/2$ rotation of the vector $\mathbf{r}_3^0$; (2) $\pi/2$ phase shift of the microwave field.

signal can be understood by realizing that the fluctuations of the precessional frequency $\Delta\omega$, which are the result of the spectral diffusion, are perpendicular to the lock field $-\omega_R \mathbf{e}_1'$ in Fig. 23. As a result, one expects that effective spin locking occurs as soon as ω_R exceeds the amplitude of the time-dependent part of $\Delta\omega_{1/2}$, a condition that is fairly easy to fulfill. For other reasons, related to the kinetics of populating and depopulating the triplet state, it nevertheless is advantageous to have $\omega_R \gg \Delta\omega_{1/2}$. For a more detailed discussion of this problem the reader is referred to reference 43.

The application of the multiple-echo or spin-lock techniques is only of some use if the effect of other dephasing mechanisms that are more characteristic for ensembles of photoexcited triplet state molecules, like vibronic relaxation and excitation transfer, in which one is usually interested, are not impaired. Fortunately, the properties of these two dephasing mechanisms are so different from the effect of spectral diffusion that this is indeed the case. The effect of spectral diffusion is characterized by discontinuities in the resonance frequencies, which result in stochastic changes in the rate at which phase differences are accumulated in the course of time. (See Fig. 22.) The loss of phase coherence by vibronic relaxation or energy transfer in many instances is caused by transitions to other triplet states where the triplet spins during a very short time experience a quite different resonance frequency. The phase accumulated in this other state then manifests itself as a random discontinuity in the phase itself. The effect of such phase jumps is not reduced by applying the multiple-echo or spin-lock techniques.

4.4 Applications

From the preceding sections it should be clear that the spin coherence experiments offer a unique way for measuring phase-destructive processes. Here we shall indicate a number of applications showing that the coherence experiments may supply important information about thermally activated processes that transfer triplet excitation to higher-lying vibronic levels or to neighboring molecules.

4.4.1 THE EFFECT OF DETRAPPING AND VIBRONIC RELAXATION ON THE SPIN COHERENCE

The first applications of coherence experiments have been in the study of dynamic processes between different vibronic states and detrapping in mixed crystals. An example of the investigation of detrapping by means of a spin-locking experiment has been reported by Harris and co-workers.[56-58] The attractive feature of the spin-locking technique is that it allows one to suppress almost completely the effect of spectral diffusion caused by the nuclear spins. (See Section 4.3.) It is further claimed that the decay rate of the spin-locked components of the vector **r** is only affected by the detrapping process; any additional molecules entering the trap state after the phase shift has occurred do not contribute to the spin-locked signal. The authors studied the spin-locked signals of the tetrachlorobenzene-h_2 (TCB-h_2) traps in TCB-d_2 and obtained quantitative information about the detrapping rates. It appears that in isotopically mixed crystals with a low concentration of TCB-h_2, trap-to-trap communication takes place via the exciton band, whereas in highly concentrated crystals a direct energy transfer occurs between TCB-h_2 and TCB-hd traps.

An example of vibronic relaxation has been studied by van't Hof and Schmidt[59] in the system parabenzoquinone present as a deep trap in a single crystal of paradibromobenzene. In this system a vibronic level is present 16 cm^{-1} above the lowest triplet state T_0, as was shown by optical studies. It appears that the spin coherence in T_0 is strongly influenced by the presence of this vibronic level T_e. The phase memory decay rate $T_{2\rho}^{-1}$ as measured by rotary echo experiments proves to be temperature dependent and when plotted as a function of the reciprocal temperature it points to a relaxation mechanism with an activation energy of 16 cm^{-1}.

In the theoretical model for explaining the observations one assumes[59] that the triplet excitation jumps between the two states T_0 and T_e, which are separated by an energy interval δE of 16 cm^{-1}. When solving the equations of motion one finds the following expression for the temperature dependence of $T_{2\rho}^{-1}$

$$T_{2\rho}^{-1} = \frac{1}{\tau}\frac{\Delta^2\tau^2}{1+\Delta^2\tau^2}\exp\left(-\frac{\delta E}{kT}\right). \tag{52}$$

Here τ is the lifetime of the upper state T_e and Δ the difference in resonance frequency between two corresponding zero-field transitions in T_0 and T_e. Further a shift ε in the resonance frequency is predicted, according to

$$\varepsilon = \frac{\Delta}{1+\Delta^2\tau^2}\exp\left(-\frac{\delta E}{kT}\right), \tag{53}$$

which indeed is observable. By combining the two results, on the dephasing rate $T_{2\rho}^{-1}$ and the frequency shift ε, one derives the lifetime as well as the zero-field splitting of the vibronic state T_e. For the specific case of parabenzoquinone in paradibromobenzene one finds for the lifetime τ of the vibronic level T_e, $\tau = (3.2 \pm 0.5) \times 10^{-12}$ s. The zero-field splitting parameters in T_e are $D_e = (-9100 \pm 700)$ MHz and $E_e = (78.0 \pm 0.2)$ MHz as compared to $D_0 = (-5299.3 \pm 0.3)$ MHz and $E_0 = (78.0 \pm 0.2)$ MHz in T_0.

The experiment nicely illustrates a general feature of the coherence experiments: the spin memory time $T_{2\rho}$ (or T_2) that one measures is the result of a dynamical process for the description of which a sophisticated model is required; it usually cannot be directly related to the "lifetime" of some physical state of the system.

4.4.2 Spin Coherence in Triplet Exciton States

The electron spin-echo technique has contributed to our understanding of energy transfer processes via triplet excitons in molecular crystals as shown by two experiments on dimer states in naphthalene isotopically mixed crystals and on linear chain excitons in tetrachlorobenzene. In the system naphthalene-h_8 in naphthalene-d_8 Botter et al.[60] succeeded in observing two-pulse electron spin-echo signals in the lowest (antisymmetric) state of naphthalene-h_8 pairs.* The phase memory decay rate T_2^{-1} appears to be temperature dependent and, by analogy with the experiments on parabenzoquinone, a shift in the zero-field resonance frequencies is observed when varying the temperature. By using the model developed by van't Hof the authors derived the lifetime and zero-field frequencies of the upper (symmetric) dimer state and concluded that in the case of this naphthalene pair it is correct to speak of a "miniexciton."

The first successful spin-echo experiment in which the signals arising from a triplet exciton were observed directly has been reported by Botter et al.[61] In the system tetrachlorobenzene, which is an example of a linear chain exciton, they observed a temperature dependence of the phase memory time T_2 in the liquid helium temperature range. Further, they observed at higher temperatures that the linewidth of the zero-field transition becomes temperature dependent. The authors conclude that the two effects are caused by scattering of the exciton k-states by phonons. At low temperature the phase memory time T_2 is directly related to the lifetime of a k-state. When increasing the temperature the scattering rate becomes faster and one reaches the situation of motional narrowing (narrowing of the zero-field line). Finally at temperatures higher than 8 K the scattering is so fast that the exciton motion becomes diffusive.

---

* In this particular experiment the spin-echo signals have been obtained in an X-band spectrometer operating with an external magnetic field.

5   QUANTUM BEATS IN THE PHOSPHORESCENCE OF PHOTOEXCITED TRIPLET STATES

The phenomenon of quantum beats is well known in atomic spectroscopy and arises when two excited states, which are brought into a coherent superposition, can decay to the ground state via the emission of photons of the same polarization. This effect has also recently been demonstrated in the phosphorescence of tetramethylpyrazine (TMP) present as a guest in a single crystal of durene by Schadee et al.[36] They reported the observation of the quantum beats at a frequency equal to the splitting of two zero-field levels after preparing the triplet state with a resonant microwave field. Here we shall show that these quantum beats are intimately related to the coherence effects described in the preceding sections and that they also may conveniently be described in the geometrical model of Feynman, Vernon, and Hellwarth.

A prerequisite for the observation of quantum beats in phosphorescent triplet states (at least without using a polarizer) is the presence of two triplet spin levels that are connected to the ground state S_0 by the same component of the electric dipole moment operator.[62,63] As we saw at the end of Section 2.2 this, for instance, is expected to occur in azaaromatic molecules if the principal axes of the zero-field splitting tensor, the "spin axes," do not coincide with the local symmetry axes of the dominant spin-orbit coupling (SOC) on the nitrogen nucleus. In such a situation Eq. (16) no longer reduces to the simple form (17), but instead one has to a good approximation.

$$\left.\begin{aligned} T_x &= T_x^0 \\ T_y &= T_y^0 + i\mu_1 S_1 \\ T_z &= T_z^0 + i\nu_1 S_1 \end{aligned}\right\}, \tag{54}$$

where, again, we have taken $\mathbf{x}$ as the out-of-plane axis. Further, we have neglected any additional SOC to higher singlet states, since its inclusion would not change things materially.

A quantum beat experiment is based on the possibility to prepare the system via the application of a $\pi/2$ microwave pulse resonant with the T_z–T_y transition in a nonstationary superposition of these two spin states. Let us suppose, for instance, that this superposition at $t=0$ is given by

$$\psi(0) = a_o T_z + b_o T_y, \tag{55}$$

then in the course of time this state evolves as

$$\psi(t) = a_o \exp\left(-i\tfrac{1}{2}\omega_0 t\right)T_z + b_o \exp\left(i\tfrac{1}{2}\omega_0 t\right)T_y. \tag{56}$$

The intensity of the phosphorescence that one observes is proportional to the square of the matrix element

$$I \propto |\langle S_0| \, ex \, |\psi(t)\rangle|^2. \tag{57}$$

Since both T_y and T_z are contaminated with singlet character, the nonstationary nature of (56) leads to an oscillation in the intensity of the emission. This follows on substitution of (54) into (56) and of the result into (57)

$$I \propto |M|^2 [\nu_1^2 a_o a_o^* + \mu_1^2 b_o b_o^* + \nu_1 \mu_1 (a_o b_o^* + a_o^* b_o) \cos \omega_o t] \tag{58}$$

in which M denotes the transition moment $\langle S_0| \, ex \, |S_1\rangle$. Thus, as is well known, the amplitude of the quantum beats, observed as the Fourier component of the phosphorescence at ω_0, is determined by the cross product of the coefficients μ_1 and ν_1. The amplitude becomes a maximum whenever $\mu_1 = \nu_1$, which, as we shall see, happens to obtain for tetramethylpyrazine in durene. For the sake of simplicity, the present discussion is first held in terms of the wavefunction of a single molecule. Later one has to consider the average over an ensemble in which an initial phase coherence between the systems at $t = 0$ will gradually be lost in time.

Let us now illustrate the interrelationship between the quantum beat phenomenon and the coherence experiments previously discussed, by taking the system TMP in durene as an example. For this system it is known from the ESR studies by de Groot et al.[64] that the in-plane spin axes y, z of the phosphorescent state are rotated by about 45° relative to the molecular symmetry axes ζ, η. (See Fig. 24.) This "skewness" is reflected in the radiative decay rates of the T_z and T_y spin levels, which prove to be equal, $k_z^r = k_y^r = k$. From this it is inferred that in the system TMP in durene the unknown coefficients on the right-hand side of (54) likewise are equal, $\mu_1 = \nu_1 = \nu$.

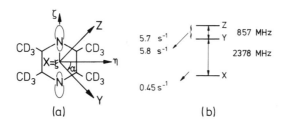

(a)   (b)

Figure 24 (a) The position of the spin axes x, y and z and the molecular symmetry axes ξ, η and ζ of the lowest triplet state T_0 of TMP-d_{12} in durene. The angle α is approximately 45°. (b) The zero-field splitting and the total decay rates of the same system. (From Schadee et al., reference 36a.)

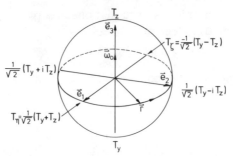

Figure 25 The represenation used in the FVH picture, where an arbitrary state $\psi = aT_z + bT_y$ describing a two-level system is related uniquely with a unit vector **r**. As an example, the states corresponding to each of the three unit vectors, $\mathbf{e}_1$, $\mathbf{e}_2$ and $\mathbf{e}_3$ are inserted in the picture. (From Schadee *et al.*, reference 36b.)

In the Feynman, Vernon, and Hellwarth (FVH) model every state of the TMP molecule of the form (55) is represented by a point with the Cartesian coordinates r_1, r_2, r_3 on a unit sphere. (See Fig. 25.) The north and south poles of this sphere correspond with the stationary states T_z, T_y; the points along the $\pm\mathbf{e}_1$-axis with

$$T_\eta = 2^{-1/2}[T_z + T_y], \qquad T_\zeta = 2^{-1/2}[T_z - T_y],$$

and those along the $\pm\mathbf{e}_2$-axis with

$$T_{\mp 1} = 2^{-1/2}[T_y \mp iT_z].$$

The examples illustrate that diametrical points on the sphere correspond with pairs of orthogonal states, since we assume the coefficients μ_1, ν_1 in (54) to be very small.

Let us now look at what happens when an ensemble of TMP molecules that initially has an excess population in the state T_y is subjected to a $\pi/2$ pulse. In the terminology of Section 4.2 the vector $\mathbf{r}_0 = (0, 0, r^0)$, which points toward the south pole of our sphere, then is tipped into the equatorial plane where it is brought along the $\mathbf{e}_1'$-axis in the rotating frame, as in Fig. 11 before. In a fixed system of axes the vector, subsequently, precesses in this plane with angular velocity ω_0,

$$r_1(t) = r^0 \cos \omega_0 t; \qquad r_2(t) = r^0 \sin \omega_0 t. \tag{59}$$

[See Eq. (43).] It is clear from Fig. 25 that, in its precession, the tip of the vector **r** moves along the equator and thereby passes periodically through the sequence of states $\cdots T_\eta \rightarrow T_{-1} \rightarrow T_\zeta \rightarrow T_{+1} \rightarrow T_\eta \cdots$. First of all this will give rise to a magnetic free-induction decay, since the state T_{-1} with a magnetic moment parallel to **x** is periodically converted to the state T_{+1} with a magnetic moment antiparallel to **x**, as discussed at length in Section 4.2.

Second, the precession along the equator will lead to the appearance of "beats" in the phosphorescence intensity. In a molecular picture this becomes obvious when writing out the wavefunction of the nonstationary states T_ζ, T_η in terms of "pure spin functions." According to Eq. (54) one then has for TMP where $\mu_1 = \nu_1 = \nu$,

$$T_\eta = 2^{-1/2}[T_z^0 + T_y^0 + 2i\nu S_1^0], \tag{60}$$
$$T_\zeta = 2^{-1/2}[T_z^0 - T_y^0].$$

For the relative radiative decay rates of these two states one subsequently finds by substitution of Eq. (60) into Eq. (57)

$$k_\eta^r = |\langle S_0| \, \text{ex} \, |T_\eta\rangle|^2 = 2\nu^2 \, |M|^2 = 2k, \tag{61}$$
$$k_\zeta^r = |\langle S_0| \, \text{ex} \, |T_\zeta\rangle|^2 = 0.$$

The result (61) finds a simple, intuitive explanation. When the triplet state molecule in its precession along the equator passes through the state T_η, its spin lies in the $n\pi$ plane $\eta = 0$ and the coupling to the radiative decay channel $S_1 \to S_0$ is maximal. While the molecule is in the state T_ζ, however, its spin lies in the plane $\zeta = 0$, SOC becomes ineffective and the radiative decay channel is shut. The probability for radiative decay thus oscillates between $2k$ and zero with angular frequency ω_0.

In the language of the FVH model our result can be quantitatively formulated by writing Eq. (57) in terms of the **r** vector. For the simple case of TMP one then obtains [for a more general expression see Eq. (17) and (18) of reference 36b],

$$I = I_{av}(1 + r_1), \tag{62}$$

in which I_{av} is the phosphorescence intensity averaged over one precession period. For the experiment we have just considered (62) is reduced by substitution of (59) to

$$I = I_{av}(1 + r^0 \cos \omega_0 t), \tag{63}$$

which is the replica of Eq. (58).

In the first quantum beat experiment on TMP-d_{12} in durene, reproduced in Fig. 26, the preceding has been demonstrated in a convincing manner. Before considering the design of this experiment, it may be helpful to note that the vertical axis in the photograph represents the strength of a dc signal. This signal is obtained at the detector of a narrow-banded receiver connected to a fast photomultiplier and tuned to 857 MHz. The signal as displayed thus measures the amplitude of the

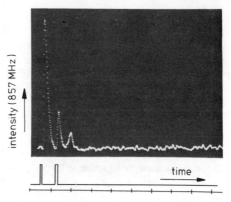

Figure 26   The upper trace represents the d.c. output of the 857 MHz detection system monitoring the quantum beats in the phosphorescence of TMP in durene during and after the application of two microwave pulses resonant with the T_z-T_y transition. The figure represents the average of 32 experiments. Timescale 1 μs/div. As in many Hahn-type spin-echo experiment the pulse lengths were not equal to $\pi/2$ and π, respectively, but were both equal to $2\pi/3$ because of instrumental reasons. This weakens the echo signal somewhat but does not change the essence of the effect. The time intervals at which the microwaves are present are indicated in the lower trace. The decrease in height of the echo as a function of the delay between the two pulses yields a dephasing time $T_z = 1.2$ μs. (From Schadee et al., reference 36a.)

Fourier component of this frequency in the emission; each of the peaks in Fig. 26 envelops some 200 beats.

Prior to the actual quantum beat experiment the system first had to be prepared in a state where the T_y level carries a substantial excess population relative to T_z. This has been achieved in the following way. To begin with, a steady-state population is established by irradiating the crystal with ultraviolet light at a temperature of 1.3 K, where spin-lattice relaxation is negligible; since the populating and depopulating rates of the two levels are about equal the difference in the steady-state populations of T_z and T_y is very small. Then the exciting light is shut off and 1.0 s later the two radiative, fast-decaying levels T_z and T_y are empty. At this moment T_y is selectively repopulated from the slowly decaying level T_x by a microwave field that is swept suddenly through the T_y-T_x resonance frequency of 2378 MHz. In this way a large population difference is built up between T_z and T_y.

Immediately after the completion of the previous step, a microwave pulse is applied at the frequency $\omega_0/2\pi = 857$ MHz of the T_z-T_y transition. By this pulse the state vector of the ensemble is brought close to the equatorial plane of the FVH sphere. As soon as the ensemble is in this nonstationary state beats appear in the emission decay in accordance with Eq. (63) and this gives rise to the first peak in Fig. 26.

Just as in a free-induction decay experiment, the beats gradually disappear because the phase coherence between the spins is lost. But when, as in a Hahn-type spin-echo experiment, this phase coherence is restored by a second refocusing microwave pulse, the beats reappear as an echo signal (the peak in Fig. 26). The spin-echo signal generated in the experiment of Fig. 26 serves to remove any lingering doubts that might exist as to the cause of the effects here described; whereas the first two peaks in Fig. 26 accompany a strong burst of microwave power, the third peak is observed when all is silent on the microwave front.

In passing it is worth mentioning that an alternative way of preparing the molecules in the "superemitting" T_η might be feasible, namely, via optical excitation with a laser flash short compared to the beat period ω_0. Such a picosecond laser flash experiment might provide an answer to the question of whether spin coherence is created (and conserved) during the competing intersystem crossing process.*

In conclusion the following may be said. The quantum beat experiments and the coherence effects of Section 4.2.2 have in common that one studies the evolution in time of a nonstationary state of the ensemble of triplet spins of the general form (25), and characterized by a nonvanishing horizontal component of the FVH vector $\mathbf{r}$. In the electron resonance experiments of Section 4.2.4 this evolution is followed by interaction with the *magnetic dipole transition between the T_y and T_z spin levels* with the aid of a microwave detection system tuned to $\omega_0 = |Z - Y| \hbar^{-1}$; as seen from Eq. (40) the microwave system here measures the component r_2. Alternatively, in this type of experiment one may follow Breiland et al.[49] and try to monitor the horizontal component of $\mathbf{r}$ optically with the aid of a probe pulse. As we have seen at the end of Section 4.2.2 such a probe pulse transfers the horizontal component of $\mathbf{r}$ into a population difference that in turn may be monitored by measuring the phosphorescence intensity with a conventional photomultiplier (time constant $< \omega_0^{-1}$).

The quantum beat experiments, on the contrary, are characterized by the fact that one observes the nonstationary nature of the emitting state in the *electric dipole radiation to the ground state* by looking at the phosphorescence with a detection system capable of following the recurrence frequency ω_0. As shown by Eq. (62), the amplitude of the beats is a measure of the component r_1. Of course, the quantum beat technique is less general, since it requires both levels to be emissive. But, if applicable,

* Note added in proof.

This experiment has recently been performed. Following flash excitation of TMP in durene with a 25 ps laser pulse at 266 nm in the singlet manifold a free induction signal has been observed at the resonance frequency of the $T_z - T_y$ zero-field transition.[65]

it offers the nice feature that one may study resonant phenomena with a detection system that is not directly sensitive to the microwave radiation field. This is illustrated by the system TMP-d_{12} in durene, where the quantum beats are observed with a satisfactory signal-to-noise ratio, whereas it proved impossible to detect the Hahn echo directly with the instrumentation presently available.

ACKNOWLEDGMENT

The present work reflects a considerable part of the sustained efforts of C. A. van't Hof and R. A. Schadee in the field of spin coherence in photoexcited triplet states and is part of the research program of *de Stichting voor Fundamenteel Onderzoek der Materie (FOM)*, financially supported by *de Nederlandse Organisatie voor Zuiver–Wetenschappelijk Onderzoek (ZWO)*.

The authors gratefully acknowledge the permission granted by Messrs. Taylor and Francis, North-Holland Publishing Co., and The American Institute of Physics to reproduce a number of figures. Figures 4, 5, 6, 24, 25, 26 have been taken from papers published in *Molecular Physics*; Figures 8, 12, 20, from *Chemical Physics Letters*; and Figure 21, from the *Journal of Chemical Physics*.

REFERENCES

1. G. N. Lewis and M. Kasha, *J. Am. Chem. Soc.*, **66,** 2100 (1944); **67,** 994 (1945).

2. C. A. Hutchison and B. W. Mangum, *J. Chem. Phys.*, **29,** 952 (1958); **32,** 1261 (1960).

3. A. Abragam, *The Principles of Nuclear Magnetism*, Clarendon Press, Oxford, 1961.

4. M. Schwoerer and H. Sixl, *Chem. Phys. Lett.*, **2,** 14 (1968); *Z. Naturforsch.* **A24,** 952 (1969); H. Sixl, Thesis, University of Stuttgart, 1971.

5. R. W. Brandon, R. E. Gerkin, and C. A. Hutchison, *J. Chem. Phys.*, **41,** 3717 (1964).

6. J. Schmidt and J. H. van der Waals, *Chem. Phys. Lett.* **2,** 640 (1968); J. Schmidt, Thesis, University of Leiden, 1971.

7. C. B. Harris, *J. Chem. Phys.*, **54,** 972 (1971).

8. C. B. Harris and R. J. Hoover, *J. Chem. Phys.*, **56,** 2199 (1972).

9. J. Schmidt, W. G. van Dorp and J. H. van der Waals, *Chem. Phys. Lett.*, **8,** 345 (1971).

10. R. P. Feynman, F. L. Vernon, and R. W. Hellwarth, *J. Appl. Phys.* **28,** 49 (1957).

11. D. J. Gravesteijn, J. H. Scheijde, and M. Glasbeek, *Phys. Rev. Lett.* **39,** 105 (1977).

12. B. J. Botter, D. C. Doetschman, J. Schmidt, and J. H. van der Waals, *Mol. Phys.*, **30,** 609 (1975).

13. B. J. Botter, C. J. Nonhof, J. Schmidt, and J. H. van der Waals, *Chem. Phys. Lett.*, **43,** 210 (1976).

14. C. A. Hutchison, in A. B. Zahlan, Ed., *The Triplet State*, Cambridge University Press, New York, 1967.

15. J. H. van der Waals and M. S. de Groot, in A. B. Zahlan, Ed., *The Triplet State*, Cambridge University Press, New York, 1967.

16. S. P. McGlynn, T. Azumi, and M. Kinoshita, *Molecular Spectroscopy of the Triplet State*, Prentice-Hall, Englewood Cliffs, N. J., 1969.

17. J. A. Kooter, G. W. Canters, and J. H. van der Waals, *Mol. Phys.*, **33**, 1545 (1971).

18. J. H. van Vleck, *Rev. Mod. Phys.*, **23**, 213 (1951).

19. M. Tinkham, *Group Theory and Quantum Mechanics*, McGraw-Hill, New York, 1964, p. 113.

20. L. I. Schiff, Quantum Mechanics, McGraw-Hill, New York, 1955, Chap. 8.

21. J. Schmidt and J. H. van der Waals, *Chem. Phys. Lett.*, **3**, 546 (1969).

22. C. A. Hutchison and G. W. Scott, *J. Chem. Phys.* **61**, 2240 (1974).

23. C. A. Hutchison, J. V. Nicholas, and G. W. Scott, *J. Chem. Phys.* **53**, 1906 (1970).

24. I. Y. Chan and K. R. Walton, *Mol. Phys.*, **34**, 65 (1977).

25. M. J. Buckley and C. B. Harris, *J. Chem. Phys.*, **56**, 137 (1972).

26. J. Schmidt, D. A. Antheunis, and J. H. van der Waals, *Mol. Phys.*, **22**, 1 (1971).

27. W. S. Veeman and J. H. van der Waals, *Mol. Phys.*, **18**, 63 (1970).

28. G. Herzberg and E. Teller, *Z. Phys. Chem.*, **B21**, 410 (1933); A. C. Albrecht, *J. Chem. Phys.*, **33**, 156 (1960).

29. G. Herzberg, *Molecular Spectra and Molecular Structure*, Vol. 3, Van Nostrand, New York, 1966.

30. D. S. Tinti and M. A. El-Sayed, *J. Chem. Phys.*, **54**, 2529 (1971).

31. D. S. McClure, *J. Chem. Phys.*, **20**, 682 (1952).

32. H. F. Hameka, in A. B. Zahlan, Ed., *The Triplet State*, Cambridge University Press, New York, 1967.

33. (a) J. W. Sidman, *J. Mol. Spect.* **2**, 333 (1958); (b) E. Clementi and M. Kasha, *J. Mol. Spect.* **2**, 297 (1958); (c) L. Goodman and V. G. Krishna, *Rev. Mod. Phys.*, **35**, 541 (1963).

34. H. F. Hameka and L. J. Oosterhoff, *Mol. Phys.*, **1**, 358 (1958).

35. R. A. Schadee, J. Schmidt, and J. H. van der Waals, *Chem. Phys. Lett.* **41**, 435 (1976).

36. (a) R. A. Schadee, C. J. Nonhof, J. Schmidt, and J. H. van der Waals, *Mol. Phys.*, **34**, 171, 1497 (1977); (b) R. A. Schadee, J. Schmidt, and J. H. van der Waals, *Mol. Phys.*, **36**, 177 (1978).

37. D. Schweitzer, J. Zuclich, and A. H. Maki, *Mol. Phys.*, **25**, 193 (1973).

38. W. G. van Dorp, T. J. Schaafsma, M. Soma, and J. H. van der Waals, *Chem. Phys. Lett.*, **21**, 47 (1973); W. G. van Dorp, W. H. Schoemaker, M. Soma, and J. H. van der Waals, Mol. Phys. **30**, 1701 (1975).

39. R. H. Clarke and R. E. Connors, *Chem. Phys. Lett.*, **42**, 69 (1976).

40. A. J. Hoff and J. H. van der Waals, *Biochim. Biophys. Acta*, **412**, 615 (1976); S. J. v. d. Bent, T. J. Schaafsma, and J. C. Goedheer, *Biochim. Biophys. Res. Comm.*, **71**, 1147 (1976).

41. S. Vega and A. Pines, *J. Chem. Phys.*, **66**, 5624 (1977).

42. C. A. van't Hof, Thesis, University of Leiden, 1976.

43. C. A. van't Hof and J. Schmidt, *Mol. Phys.* in press (1979).

44. I. Solomon, *Phys. Rev. Lett.*, **2**, 301 (1959).

45. C. B. Harris, R. L. Schlupp, and H. Schuch, *Phys. Rev. Lett.*, **30**, 1019 (1973).

46. E. L. Hahn, *Phys. Rev.*, **80**, 580 (1950).

47. W. B. Mims, *Rev. Sci. Instrum.* **36**, 1472 (1965); S. Geschwind, Ed., *Electron Paramagnetic Resonance*, Plenum, New York, 1972.

48. J. Schmidt, *Chem. Phys. Lett.*, **14**, 411 (1972).

49. (a) W. G. Breiland, C. B. Harris, and A. Pines, *Phys. Rev. Lett.*, **30**, 158 (1973); (b) W. G. Breiland, H. C. Brenner, and C. B. Harris, *J. Chem. Phys.*, **62**, 3458 (1975).

50. L. E. Erickson, *Phys. Rev.*, **143**, 295 (1966). See further T. Moreno, *Microwave Transmission Design Data*, McGraw-Hill, New York, 1948.

51. H. Y. Carr and E. M. Purcell, *Phys. Rev.*, **94**, 630 (1954).

52. S. Meiboom and D. Gill, *Rev. Sci. Instrum.* **29**, 688 (1958).

53. C. A. van't Hof, J. Schmidt, P. J. F. Verbeek, and J. H. van der Waals, *Chem. Phys. Lett.*, **21**, 437 (1973).

54. A. G. Redfield, *Phys. Rev.*, **98**, 1787 (1955).

55. H. Schuch and C. B. Harris, *Z. Naturforsch.*, **30a**, 361 (1975).

56. M. D. Fayer and C. B. Harris, *Phys. Rev.*, **B9**, 748 (1974).

57. H. C. Brenner, J. C. Brock, M. D. Fayer, and C. B. Harris, *Chem. Phys. Lett.*, **33**, 471 (1975).

58. M. T. Lewellyn, A. H. Zewail, and C. B. Harris, *J. Chem. Phys.*, **63**, 3687 (1976).

59. C. A. van't Hof and J. Schmidt, *Chem. Phys. Lett.*, **42**, 73 (1976); **36**, 460 (1975).

60. B. J. Botter, C. J. Nonhof, J. Schmidt, and J. H. van der Waals, *Chem. Phys. Lett.*, **43**, 210 (1976).

61. B. J. Botter, A. I. M. Dicker, and J. Schmidt, *Mol. Phys.*, **36**, 129 (1978).

62. M. I. Podgoretskii and O. A. Khrustalev, *Sov. Phys. Usp.*, **6**, 682 (1964).

63. M. Bixon, J. Jortner, and Y. Dothan, *Mol. Phys.*, **17**, 109 (1969).

64. M. S. de Groot, I. A. M. Hesselmann, F. J. Reinders, and J. H. van der Waals *Mol. Phys.*, **29**, 37 (1975).

65. C. J. Nonhof, F. L. Plantenga, J. Schmidt, C. A. G. O. Varma, and J. H. van der Waals, *Chem. Phys. Lett.*, **60**, 353 (1979).

AUTHOR INDEX

Numbers in parentheses following a page number are reference numbers citing the particular author.

SUBJECT INDEX